TIME SERIES ANALYSIS AND FORECASTING BY EXAMPLE

TIME SERIES ANALYSIS AND FORECASTING BY EXAMPLE

Søren Bisgaard

Murat Kulahci
Technical University of Denmark

A JOHN WILEY & SONS, INC., PUBLICATION

Published by John Wiley & Sons, Inc., Hoboken, New Jersey.
Published simultaneously in Canada

For general information on our other products and services or for technical support, please contact our
Customer Care Department within the United States at (800) 762-2974, outside the United States at
(317) 572-3993 or fax (317) 572-4002.

Wiley also publishes its books in a variety of electronic formats. Some content that appears in print
may not be available in electronic formats. For more information about Wiley products, visit our web
site at www.wiley.com.

Library of Congress Cataloging-in-Publication Data:
Bisgaard, Søren, 1938-
  Time series analysis and forecasting by example / Søren Bisgaard, Murat Kulahci.
      p. cm. -- (Wiley series in probability and statistics)
  Includes bibliographical references and index.
  ISBN 978-0-470-54064-0 (cloth)
1.   Time-series analysis. 2.   Forecasting.   I. Kulahci, Murat. II. Title.
  QA280.B575 2011
  519.5′5--dc22
                                                        2010048281

oBook ISBN: 978-1-118-05694-3
ePub ISBN: 978-1-118-05695-0

10 9 8 7 6 5 4 3 2 1

To the memory of
Søren Bisgaard

CONTENTS

PREFACE

Data collected in time often shows serial dependence. This, however, violates one of the most fundamental assumptions in our elementary statistics courses where data is usually assumed to be independent. Instead, such data should be treated as a time series and analyzed accordingly. It has, unfortunately, been our experience that many practitioners found time series analysis techniques and their applications complicated and subsequently were left frustrated. Recent advances in computer technology offer some help. Nowadays, most statistical software packages can be used to apply many techniques we cover in this book. These often user-friendly software packages help the spreading of the use of time series analysis and forecasting tools. Although we wholeheartedly welcome this progress, we also believe that statistics welcomes and even requires the input from the analyst who possesses the knowledge of the system being analyzed as well as the shortfalls of the statistical techniques being used in this analysis. This input can only enhance the learning experience and improve the final analysis.

Another important characteristic of time series analysis is that it is best learned by applications (as George Box used to say for statistical methods in general) akin to learning how to swim. One can read all the theoretical background on the mechanics of swimming, yet the real learning and joy can only begin when one is in the water struggling to stay afloat and move forward. The real joy of statistics comes out with the discovery of the hidden information in the data during the application. Time series analysis is no different.

It is with all these ideas/concerns in mind that Søren and I wrote our first *Quality Quandaries* in Quality Engineering in 2005. It was about how the stability of processes can be checked using the variogram. This led to a series of *Quality Quandaries* on various topics in time series analysis. The main focus has always been to explain a seemingly complicated issue in time series analysis by providing the simple intuition behind it with the help of a numerical example. These articles were quite well received and we decided to write a book. The challenge was to make a stand-alone book with just enough theory to make the reader grasp the explanations provided with the example from the Quality Quandaries. Therefore, we added the necessary amount of theory to the book as the foundation while focusing on explaining the topics through examples. In that sense, some readers may find the general presentation approach of this book somewhat unorthodox. We believe, however, that this informal and intuition-based approach will help the readers see the time series analysis for what it really is—a fantastic tool of discovery and learning for real-life applications.

As mentioned earlier, throughout this book, we try to keep the theory to an absolute minimum and whenever more theory is needed, we refer to the seminal books by Box et al. (2008) and Brockwell and Davis (2002). We start with an introductory chapter where we discuss why we observe autocorrelation when data is collected in time with the help of the simple pendulum example by Yule (1927). In the same chapter we also discuss why we should prefer parsimonious models and always seek the simpler model when all else is the same. Chapter 2 is somewhat unique for a time series analysis book. In this chapter, we discuss the fundamentals of graphical tools. We are strong believers of these tools and always recommend using them before attempting to do any rigorous statistical analysis. This chapter is inspired by the works of Tufte and particularly Cleveland with particular focus on the use of graphical tools in time series analysis. In Chapter 3, we discuss fundamental concepts such as stationarity, autocorrelation, and partial autocorrelation functions to lay down the foundation for the rest of the book. With the help of an example, we discuss the autoregressive moving average (ARMA) model building procedure. Also, in this chapter we introduce the variogram, an important tool that provides insight about certain characteristics of the process. In real life, we cannot expect systems to remain around a constant mean and variance as implied by stationarity. For that, we discuss autoregressive integrated moving average (ARIMA) models in Chapter 4. With the help of two examples, we go through the modeling procedure. In this chapter, we also introduce the basic principles of forecasting using ARIMA models. At the end of the chapter, we discuss the close connection between EWMA, a popular smoothing and forecasting technique, and ARIMA models. Some time series, such as weather patterns, sales and inventory data, and so on, exhibit cyclic behavior that can be analyzed using seasonal ARIMA models. We discuss these models in Chapter 5 with the help of two classic examples from the literature. In our modeling efforts, we always keep in mind the famous quote by George Box "All models are wrong, some are useful." In time series analysis, sometimes more than one model can fit the data equally well. Under those circumstances, system knowledge can help to choose the more relevant model. We can also make use of some numerical criteria such as AIC and BIC, which are introduced in Chapter 6 where we discuss the model identification issues in ARIMA models. Chapter 7 consists of many sections on additional issues in ARIMA models such as constant term and cancellation of terms in ARIMA models, overdifferencing and underdifferencing, and missing values in the data. In Chapter 8, we introduce an input variable and discuss ways to improve our forecasts with the help of this input variable through the transfer function–noise models. We use two examples to illustrate in detail the steps of the procedure for developing transfer function–noise models. In this chapter, we also discuss the intervention models with the help of two examples. In the last chapter, we discuss additional topics such as spurious relationships, autocorrelation in regression, multiple time series, and structural analysis of multiple time series using principal component analysis and canonical analysis.

This book would not have been possible without the help of many friends and colleagues. I would particularly like to thank John Sølve Tyssedal and Erik Vanhatalo who provided a comprehensive review of an earlier draft. I would also like to thank Johannes Ledolter for providing a detailed review of Chapters 3 and 7. I have tried to incorporate their comments and suggestions into the final version of the manuscript.

Data sets and additional material related to this book can be found at ftp://ftp.wiley.com/public/sci_tech_med/times_series_example.

I would also like to extend special thanks to my wife, Stina, and our children Minna and Emil for their continuing love, support, and patience throughout this project.

I am indebted to the editors of *Quality Engineering* as well as Taylor and Francis and the American Society for Quality (ASQ), copublishers of *Quality Engineering* for allowing us to use the *Quality Quandaries* that Søren and I wrote over the last few years as the basis of this book.

In the examples presented in this book, the analyses are performed using R, SCA, SAS JMP version 7 and Minitab version 16. SCA software is a registered trademark of Scientific Computing Associates Corp. SAS JMP is a registered trademark of SAS Institute Inc., Cary, NC, USA. Portions of the output contained in this book are printed with permission of Minitab Inc. All material remains the exclusive property and copyright of Minitab Inc. All rights reserved.

While we were writing this book, Søren got seriously ill. However, he somehow managed to keep on working on the book up until his untimely passing last year. While finishing the manuscript, I tried to stay as close as I possibly can to our original vision of writing an easy-to-understand-and-use book on time series analysis and forecasting. Along the way, I have definitely missed his invaluable input and remarkable ability to explain in simple terms even the most complicated topics. But more than that, I have missed our lively discussions on the topic and on statistics in general. This book is dedicated to the memory of my mentor and dear friend Søren Bisgaard.

Murat Kulahci
Lyngby, Denmark

CHAPTER 1

TIME SERIES DATA: EXAMPLES AND BASIC CONCEPTS

1.1 INTRODUCTION

In many fields of study, data is collected from a system (or as we would also like to call it a *process*) over time. This sequence of observations generates a time series such as the closing prices of the stock market, a country's unemployment rate, temperature readings of an industrial furnace, sea level changes in coastal regions, number of flu cases in a region, inventory levels at a production site, and so on. These are only a few examples of a myriad of cases where time series data is used to better understand the dynamics of a system and to make sensible forecasts about its future behavior.

Most physical processes exhibit inertia and do not change that quickly. This, combined with the sampling frequency, often makes consecutive observations correlated. Such correlation between consecutive observations is called *autocorrelation*. When the data is autocorrelated, most of the standard modeling methods based on the assumption of independent observations may become misleading or sometimes even useless. We therefore need to consider alternative methods that take into account the serial dependence in the data. This can be fairly easily achieved by employing time series models such as autoregressive integrated moving average (ARIMA) models. However, such models are usually difficult to understand from a practical point of view. What exactly do they mean? What are the practical implications of a given model and a specific set of parameters? In this book, our goal is to provide intuitive understanding of seemingly complicated time series models and their implications. We employ only the necessary amount of theory and attempt to present major concepts in time series analysis via numerous examples, some of which are quite well known in the literature.

1.2 EXAMPLES OF TIME SERIES DATA

Examples of time series can be found in many different fields such as finance, economics, engineering, healthcare, and operations management, to name a few.

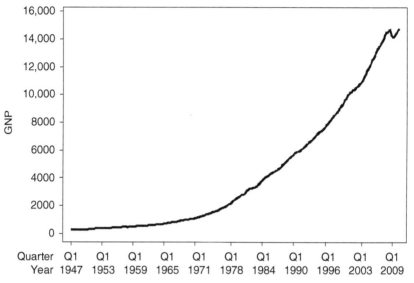

Figure 1.1 GNP (nominal) of the United States from 1947 to 2010 (in billion dollars).
Source: US Department of Commerce, http://research.stlouisfed.org/fred2/data/GNP.txt.

Consider, for example, the gross national product (GNP) of the United States from 1947 to 2010 in Figure 1.1 where GNP shows a steady exponential increase over the years. However, there seems to be a "hiccup" toward the end of the period starting with the third quarter of 2008, which corresponds to the financial crisis that originated from the problems in the real estate market. Studying such macroeconomic indices, which are presented as time series, is crucial in identifying, for example, general trends in the national economy, impact of public policies, or influence of global economy.

Speaking of problems with the real estate market, Figure 1.2 shows the median sales prices of houses in the United States from 1988 to the second quarter of 2010. One can argue that the signs of the upcoming crisis could be noticed as early as in 2007. However, the more crucial issue now is to find out what is going to happen next. Homeowners would like to know whether the value of their properties will fall further and similarly the buyers would like to know whether the market has hit the bottom yet. These forecasts may be possible with the use of appropriate models for this and many other macroeconomic time series data.

Businesses are also interested in time series as in inventory and sales data. Figure 1.3 shows the well-known number of airline passengers data from 1949 to 1960, which will be discussed in greater detail in Chapter 5. On the basis of the cyclical travel patterns, we can see that the data exhibits a seasonal behavior. But we can also see an upward trend, suggesting that air travel is becoming more and more popular. Resource allocation and investment efforts in a company can greatly benefit from proper analysis of such data.

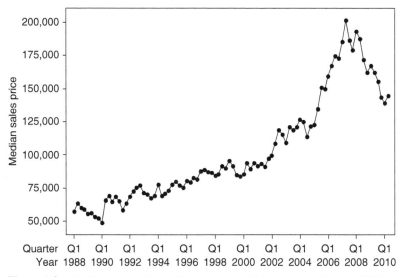

Figure 1.2   Median sales prices of houses in the United States. *Source:* US Bureau of the Census, http://www.census.gov/hhes/www/housing/hvs/historic/index.html.

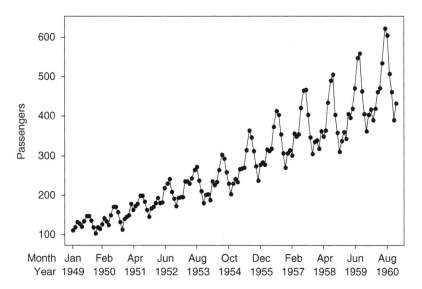

Figure 1.3   The number of airline passengers from 1949 to 1960.

In Figure 1.4, the quarterly dollar sales (in $1000) data of Marshall Field & Company for the period 1960 through 1975 also shows a seasonal pattern. The obvious increase in sales in the fourth quarter can certainly be attributed to Christmas shopping sprees. For inventory problems, for example, this type of data contains invaluable information. The data is taken from George Foster's

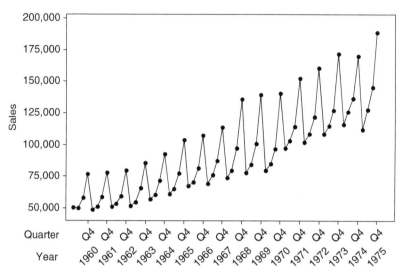

Figure 1.4 Quarterly dollar sales (in $1000) of Marshall Field & Company for the period 1960 through 1975.

Financial Statement Analysis (1978), where Foster uses this dataset in Chapter 4 to illustrate a number of statistical tools that are useful in accounting.

In some cases, it may also be possible to identify certain leading indicators for the variables of interest. For example, building permit applications is a leading indicator for many sectors of the economy that are influenced by construction activities. In Figure 1.5, the leading indicator is shown in the top panel whereas

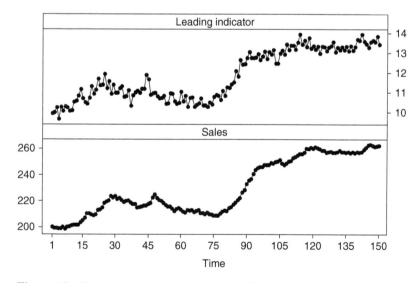

Figure 1.5 Time series plots of sales and a leading indicator.

the sales data is given at the bottom. They exhibit similar behavior; however, the important task is to find out whether there exists a lagged relationship between these two time series. If such a relationship exists, then from the current and past behavior of the leading indicator, it may be possible to determine how the sales will behave in the near future. This example will be studied in greater detail in Chapter 8.

Sometimes, the natural course of time series is interrupted because of some known causes such as public policy changes, strikes, new advertisement campaigns, and so on. In Chapter 8, the classic example of the market share fight between Colgate–Palmolive's "Colgate Dental Cream" and Proctor and Gamble's "Crest Toothpaste" will be discussed. Before the introduction of Crest by Proctor and Gamble into the US market, Colgate enjoyed a market leadership with a close to 50% market share. However, in 1960, the Council on Dental Therapeutics of the American Dental Association (ADA) endorsed Crest as an "important aid in any program of dental hygiene." Figure 1.6 shows the market shares of the two brands during the period before and after the endorsement. Now is it possible to deduce from this data that ADA's endorsement had any impact on the market shares? If so, was the effect permanent or temporary? In our analysis of these series in Chapter 8, some answers to these questions have been provided through an "intervention analysis."

This book also covers many engineering examples, most of which come from Box *et al.* (2008) (BJR hereafter). The time series plot of hourly temperature readings from a ceramic furnace is given in Figure 1.7. Even though the time interval considered consists of only 80 observations, the series looks stationary in the sense that both the mean and the variance do not seem to vary over time. The analysis of this series has been performed in Chapter 4.

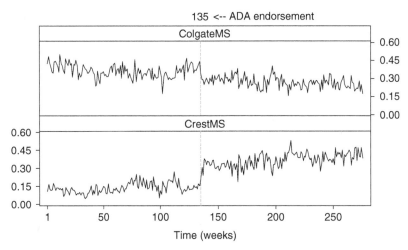

Figure 1.6 Time series plot of the weekly Colgate market share (ColgateMS) and Crest market share (CrestMS).

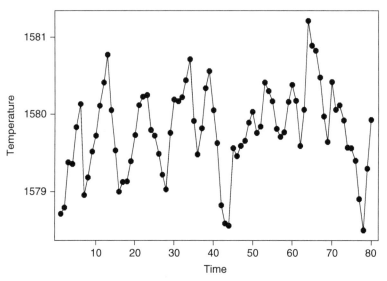

Figure 1.7 A time series plot of 80 consecutive hourly temperature observations from a ceramic furnace.

Figures 1.8 and 1.9 show the concentration and temperature readings, respectively, of a chemical process. The data come from series A and C of BJR. Both series exhibit nonstationary behavior in the sense that the mean seems to vary over time. This is to be expected from many engineering processes

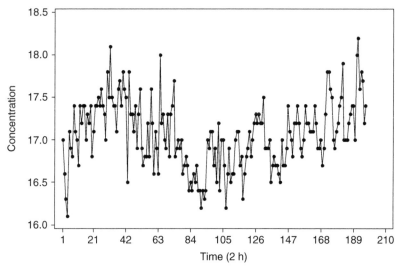

Figure 1.8 Time series plot of chemical process concentration readings sampled every 2 h (BJR series A).

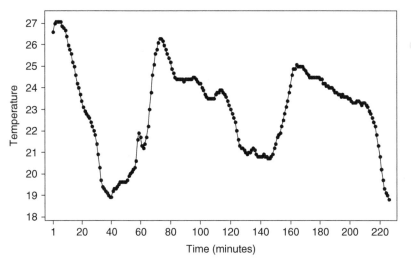

Figure 1.9   Time series plot of the temperature from a pilot plant observed every minute (BJR series C).

that are not tightly controlled. The analysis of both series has been provided in Chapter 4.

Data for another engineering example is given in the series J of BJR where the dynamic relationship between the input variable, methane gas rate, and the output, CO_2 concentration, in a pilot plant is discussed. In Figure 1.10, we can observe an apparent relationship between these two variables but a rigorous analysis is needed to fit a so-called transfer function–noise model to quantify this

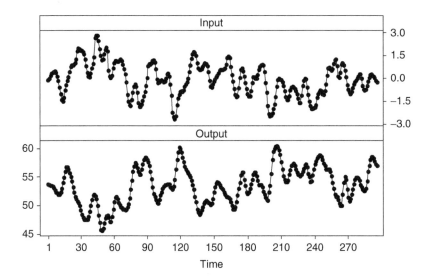

Figure 1.10   Time series plots of gas furnace data (BJR series J).

relationship. This is one of the examples used in Chapter 8 to illustrate some of the finer points in transfer function–noise models.

Time series data is of course not limited to economics, finance, business, and engineering. There are several other fields where the data is collected as a sequence in time and shows serial dependence. Consider the number of internet users over a 100-min period given in Figure 1.11. The data clearly does not follow a local mean but wanders around showing signs of "nonstationarity." This data is used in Chapter 6 to discuss how seemingly different models can fit a dataset equally well.

Figure 1.12 shows the annual sea level data for Copenhagen, Denmark, from 1889 to 2006. The data seems to have a stationary behavior with a subtle increase during the last couple of decades. What can city officials expect in the near future when it comes to sea levels rising? Can we make any generalizations regarding the sea levels all around the world based on this data? The data is available at www.psmsl.org. It is interesting to observe that the behavior we see in Figure 1.12 is only one of many different behaviors exhibited by similar datasets collected at various locations around the world. Note that in Figure 1.12, we observe missing data points, which is a surprisingly common problem with this type of data, and hence provides an excellent example to discuss the missing observations issue in Chapter 7.

There are also many examples in healthcare where time series data is collected and analyzed. In the fall of 2009, H1N1 flu pandemic generated a lot of fear throughout the world. The plot of the weekly number of reported cases in the United States is given in Figure 1.13. On the basis of this data, can we predict the number of flu cases in the autumn of 2010 and winter of 2011? What could

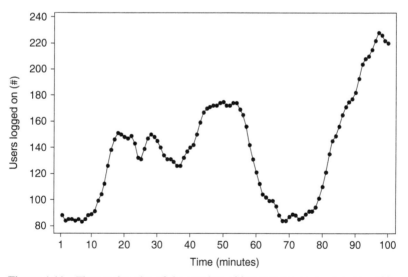

Figure 1.11 Time series plot of the number of internet server users over a 100-min period.

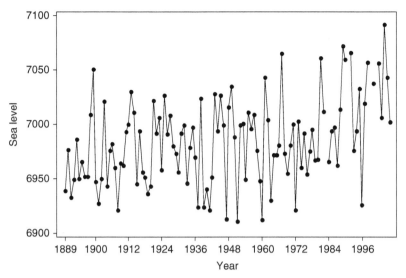

Figure 1.12 The annual sea levels in millimeters for Copenhagen, Denmark. *Source:* www.psmsl.org.

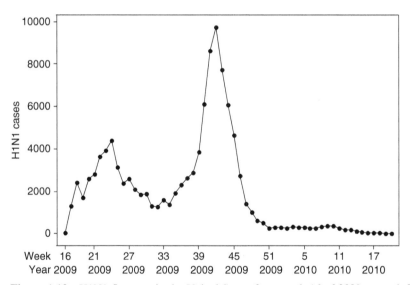

Figure 1.13 H1N1 flu cases in the United States from week 16 of 2009 to week 20 of 2010. *Source:* US Center for Disease Control CDC.

be the reason for a considerable decline in the number of cases at the end of 2009—a successful vaccination campaign, a successful "wash your hands" campaign, or people's improved immune system? Needless to say, an appropriate analysis of this data can greatly help to better prepare for the new flu season.

These examples can be extended to many other fields. The common thread is the data that is collected in time exhibiting a certain behavior, implying serial dependence. The tools and methodologies presented in this book will, in many cases, be proved very useful in identifying underlying patterns and dynamics in a process and allow the analyst to make sensible forecasts about its future behavior.

1.3   UNDERSTANDING AUTOCORRELATION

Modern time series modeling dates back to 1927 when the statistician G. U. Yule published an article where he used the dynamic movement of a pendulum as the inspiration to formulate an *autoregressive* model for the time dependency in an observed time series. We now demonstrate how Yule's pendulum analogue is an excellent vehicle for gaining intuition more generally about the dynamic behavior of time series models.

First, let us review the basic physics of the pendulum shown in Figure 1.14. If a pendulum in equilibrium with mass *m* under the influence of gravity is suddenly hit by a single impulse force, it will begin to swing back and forth. Yule describes this as a simple pendulum that is in equilibrium in the middle of the room, being pelted by peas thrown by some naughty boys in the room. This of course causes the harmonic motion that the pendulum displays subsequently. The frequency of this harmonic motion depends on the length of the pendulum,

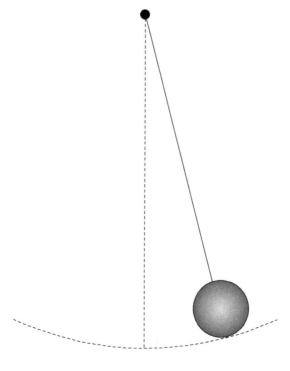

Figure 1.14   A simple pendulum.

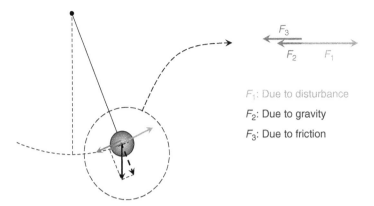

Figure 1.15 A simple pendulum in motion.

the amplitude of the mass of the bob, the impulse force, and the dissipative forces of friction and viscosity of the surrounding medium. The forces affecting a pendulum in motion are given in Figure 1.15.

After the initial impulse, the pendulum will gradually be slowed down by the dissipative forces until it eventually reaches the equilibrium again. How this happens provides an insight into the dynamic behavior of the pendulum—is it a short or long pendulum, is the bob light or heavy, is the friction small or large, and is the pendulum swinging in air or in a more viscous medium such as water?

An example for the displacement of the pendulum referenced to the equilibrium position at 0 is given in Figure 1.16. The harmonic movement $z(t)$ of a pendulum as a function of time t can be at least approximately described by a

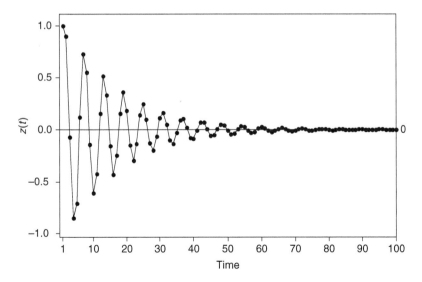

Figure 1.16 The displacement of a simple pendulum in motion.

second order linear differential equation with constant coefficients

$$m\frac{d^2z}{dt^2} + \gamma\frac{dz}{dt} + kz = a\delta(t) \tag{1.1}$$

where $\delta(t)$ is an impulse (delta) function that, like a pea shot, at time $t = 0$ forces the pendulum away from its equilibrium and a is the size of the impact by the pea. It is easy to imagine that the curve traced by this second order differential equation is a damped sinusoidal function of time although, if the friction or viscosity is sufficiently large, the (overdamped) pendulum may gradually come to rest following an exponential curve without ever crossing the centerline.

Differential equations are used to describe the dynamic process behavior in continuous time. But time series data is typically sampled (observed) at discrete times—for example, every hour or every minute. Yule therefore showed that if we replace the first- and second order differentials with discrete first- and second order differences, $\nabla z_t = z_t - z_{t-1}$ and $\nabla^2 z_t = \nabla(\nabla z_t) = z_t - 2z_{t-1} + z_{t-2}$, we can rewrite Equation (1.1) as a second order difference equation $\beta_2 \nabla^2 \tilde{z}_t + \beta_1 \nabla \tilde{z}_t + \beta_0 \tilde{z}_t = a_t$ where a_t mimics a random pea shot at time t and $\tilde{z}_t = z_t - \mu$ is the deviation from the pendulum's equilibrium position. After simple substitutions and rearrangements, this can be written as

$$\tilde{z}_t = \phi_1 \tilde{z}_{t-1} + \phi_2 \tilde{z}_{t-2} + a_t \tag{1.2}$$

which is called a *second order autoregressive time series model* where the current observation $\tilde{z}_t$ is regressed on the two previous observations $\tilde{z}_{t-1}$ and $\tilde{z}_{t-2}$ and the error term is a_t. Therefore, if observed in discrete time, the oscillatory behavior of a pendulum can be described by Equation (1.2).

The model in Equation (1.2) is called an *autoregressive model* as the position of the pendulum at any given time t can be modeled using the position of the same pendulum at times $t - 1$ and $t - 2$. Borrowing the standard linear regression terminology, this model corresponds to the one where the position of the pendulum at any given time is (auto)-regressed onto itself at previous times. The reason that the model uses only two positions that are immediately preceding the current time is that the governing physics of the behavior of a simple pendulum dictates that it should follow second order dynamics. We should not expect all systems to follow the same second order dynamics. Nor do we expect to have a prior knowledge or even a guess of such dynamics for any given system. Therefore, empirical models where the current value is modeled using the previous values of appropriate lags are deemed appropriate for modeling time series data. The determination of the "appropriate" lags will be explored in the following chapters.

1.4   THE WOLD DECOMPOSITION

It is possible to provide another intuitive interpretation of Equation (1.2). The current position is given as a function of not only two previous positions but also of the current disturbance, a_t. This is, however, an incomplete account of what is going on here. If Equation (1.2) is valid for $\tilde{z}_t$, it should also be valid for $\tilde{z}_{t-1}$ and

$\tilde{z}_{t-2}$ for which a_{t-1} and a_{t-2} will be used respectively as the disturbance term in Equation (1.2). Therefore, the equation for $\tilde{z}_t$ does not only have a_t on the right-hand side but also a_{t-1} and a_{t-2} through the inclusion of the autoregressive terms $\tilde{z}_{t-1}$ and $\tilde{z}_{t-2}$. Using the same argument, we can further show that the equation for $\tilde{z}_t$ contains all previous disturbances. In fact, the powers of the coefficients in front of the autoregressive terms in Equation (1.2), namely ϕ_1 and ϕ_2, serve as the "weights" of these past disturbances. Therefore, certain coefficients can lead to an unstable infinite sum as these weights can increase exponentially as we move back in the past. For example, consider $+2$ and $+3$ for ϕ_1 and ϕ_2, respectively. This combination will give exponentially increasing weights for the past disturbances. Hence, only certain combinations of the coefficients will provide stable behavior in the weights and lead to a *stationary* time series. Indeed, stationary time series provide the foundation for discussing more general time series that exhibit trend and seasonality later in this book.

Stationary time series are characterized by having a distribution that is independent of time shifts. Most often, we will only require that the mean and variance of these processes are constant and that autocorrelation is only lag dependent. This is also called *weak stationarity*.

Now that we have introduced stationarity, we can also discuss one of the most fundamental results of modern time series analysis, the Wold decomposition theorem (see BJR). It essentially shows that any stationary time series process can be written as an infinite sum of weighted random shocks

$$\tilde{z}_t = a_t + \psi_1 a_{t-1} + \psi_2 a_{t-2} + \ldots$$

$$= a_t + \sum_{j=1}^{\infty} \psi_j a_{t-j} \tag{1.3}$$

where $\tilde{z}_t = z_t - \mu$ is the deviation from the mean, a_t's are uncorrelated random shocks with zero mean and constant variance, and $\{\psi_i\}$ satisfies $\sum_{i=0}^{+\infty} \psi_i^2 < \infty$. For most practical purposes, the Wold decomposition involving an infinite sum and an infinite number of parameters ψ_j is mostly of theoretical interest but not very useful in practice. However, we can often generate the ψ_j's from a few parameters. For example, if we let $\psi_j = \phi_1^j$, we can generate the entire infinite sequence of ψ_j's as the powers of a single parameter ϕ_1. It should be noted that although this imposes a strong restriction on the otherwise unrelated ψ_j's, it also allows us to represent infinitely many parameters with only one. Moreover, for most processes encountered in practice, most of the ψ_j weights will be small and without much consequence except for a relatively small number related to the most recent a_t's. Indeed, one of the essential ideas of the groundbreaking Box–Jenkins approach to time series analysis (see BJR) was their recognition that it was possible to approximate a wide variety of ψ weight patterns occurring in practice using models with only a few parameters. It is this idea of "parsimonious" models that led them to introduce the autoregressive moving average (ARMA) models that will be discussed in great detail in Chapter 3.

It should also be noted that while the models for stationary time series, such as the ARMA models, constitute the foundation of many methodologies we present in this book, the assumption that a time series is stationary is quite unrealistic in real life. For a system to exhibit a stationary behavior, it has to be tightly controlled and maintained in time. Otherwise, systems will tend to drift away from a stationary behavior following the second law of thermodynamics, which, as George E. P. Box, one of the pioneers in time series analysis, would playfully state, dictates that everything goes to hell in a hand basket. What is much more realistic is to claim that the changes to a process, or the first difference, form a stationary process. And if that is not realistic, we may try to see if the changes of the changes, the second difference, form a stationary process. This observation is the basis for the very versatile use of time series models. Thus, as we will see in later chapters, simple manipulations such as taking the first difference, $\nabla z_t = z_t - z_{t-1}$ or $\nabla^2 z_t = \nabla(z_t - z_{t-1}) = z_t - 2z_{t-1} + z_{t-2}$, can make those first- or second order differences exhibit stationary behavior even if z_t did not. This will be discussed in greater detail in Chapter 4.

1.5   THE IMPULSE RESPONSE FUNCTION

We have now seen that a stationary time series process can be represented as the dynamic response of a linear filter to a series of random shocks as illustrated in Figure 1.17. But what is the significance of the ψ_j's? The reason we are interested in the ψ_j weights is that they tell us something interesting about the dynamic behavior of a system. To illustrate this, let us return to the pendulum example. Suppose we, for a period of time, had observed a pendulum swinging back and forth, and found "coincidentally" that the parameters were $\hat{\phi}_1 = 0.9824$ and $\hat{\phi}_2 = -0.3722$. (Note that these estimates are from the example that will be discussed in Chapter 3.) Now, suppose the pendulum is brought to rest, but then at time $t = 0$ it is suddenly hit by a single small pea shot and then again left alone. The pendulum, of course, will start to swing but after some time it will eventually return to rest. But how much will it swing and for how long? If we knew that, we would have a feel for the type and size of pendulum we are dealing

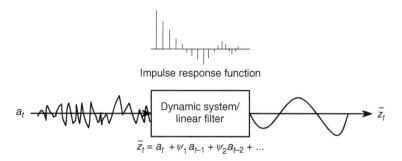

Figure 1.17   A time series model as a linear filter of random shock inputs.

with. In other words, we would be able to appreciate the dynamic behavior of the system under study, whether it is a pendulum, a ceramic furnace, the US economy, or something else. Fortunately, this question can directly be answered by studying the ψ_j's, also known as the *impulse response function*. Furthermore, the impulse response function can be computed easily with a spreadsheet program directly from the autoregressive model, $\tilde{z}_t = \phi_1 \tilde{z}_{t-1} + \phi_2 \tilde{z}_{t-2} + a_t$.

Specifically, suppose we want to simulate that our pendulum is hit from the left with a single pea shot at time $t = 0$. Therefore, we let $a_0 = 1$ and $a_t = 0$ for $t > 0$. To get the computations started, suppose we start a few time units earlier, say $t = -2$. Since the pendulum is at rest, we set $z_{-1} = 0$ and $z_{-2} = 0$ and then recursively compute the responses as

$$\tilde{z}_{-2} = 0$$
$$\tilde{z}_{-1} = 0$$
$$\tilde{z}_0 = 0.9824\tilde{z}_{-1} - 0.3722\tilde{z}_{-2} + a_0 = 0.9824 \times 0 - 0.3722 \times 0 + 1 = 1$$
$$\tilde{z}_1 = 0.9824\tilde{z}_0 - 0.3722\tilde{z}_{-1} + a_1 = 0.9824 \times 1 - 0.3722 \times 0 + 0 = 0.9824$$
$$\tilde{z}_2 = 0.9824\tilde{z}_1 - 0.3722\tilde{z}_0 + a_2$$
$$= 0.9824 \times 0.9824 - 0.3722 \times 1.0 + 0 = 0.59291$$

and so on   (1.4)

This type of recursive computation is easily set up in a spreadsheet. The response, $\tilde{z}_t, t = 1, 2, \ldots$, to the single impulse $a_0 = 1$ at $t = 0$ as it propagates through the system provides us with the ψ weights. The impulse response function is shown in Table 1.1 and plotted in Figure 1.18 where we see that the single pea shot causes the pendulum instantly to move to the right, then slowly returns back toward the centerline, crosses it at about $t = 4$, overshoots it a bit, again crosses the centerline about $t = 9$, and eventually comes to rest at about $t = 14$. In other words, our pendulum is relatively dampened as if it were moving in water or as if it were very long and had a heavy mass relative to the force of the small pea shot.

Now suppose we repeated the experiment with a much lighter and less damped pendulum with parameters $\phi_1 = 0.2$ and $\phi_2 = -0.8$.

The impulse response for this pendulum is shown in Figure 1.19. We see that it has a much more temperamental and oscillatory reaction to the pea shot and that the dynamic reaction stays much longer in the system.

1.6 SUPERPOSITION PRINCIPLE

The reaction of a linear filter model $\tilde{z}_t = a_t + \psi_1 a_{t-1} + \psi_2 \tilde{z}_{t-2} + \ldots$ to a single pea shot has been discussed above. However, in general we will have a sequence of random shocks bombarding the system and not just a single shock. The reaction to each shock is given by the impulse response function. But for linear time series models, the reaction to a sequence of shocks can easily be generated by the super-position principle. That means, the individual responses can be added together

TABLE 1.1 The Impulse Response Function for the AR(2) for the Pendulum

Time (t)	a_t	$\tilde{z}_t = \psi_j$
-2	0	0.00000
-1	0	0.00000
0	1	1.00000
1	0	0.98240
2	0	0.59291
3	0	0.21683
4	0	-0.00767
5	0	-0.08824
6	0	-0.08383
7	0	-0.04951
8	0	-0.01744
9	0	0.00130
10	0	0.00776
11	0	0.00715
12	0	0.00413
13	0	0.00140
14	0	-0.00016
15	0	-0.00068

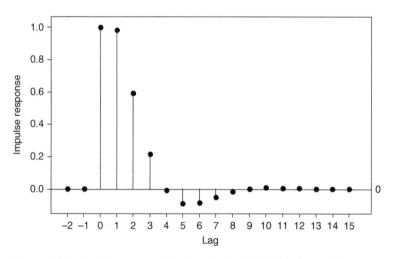

Figure 1.18 Impulse response function for the AR(2) for the pendulum.

to form the full response to a general sequence of inputs. Indeed, the impulse responses to each of the individual shocks are simply added up as they occur over time. For example, if the pendulum model $\tilde{z}_t = 0.9824\tilde{z}_{t-1} - 0.3722\tilde{z}_{t-2} + a_t$ was hit by a random sequence of 10 shocks as shown in Figure 1.20a starting

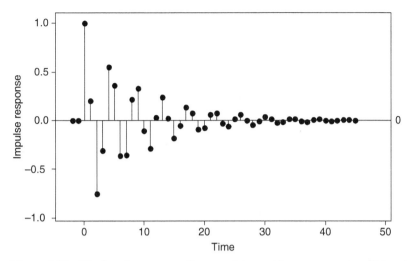

Figure 1.19 The impulse response for a pendulum with parameters $\phi_1 = 0.2$ and $\phi_2 = -0.8$.

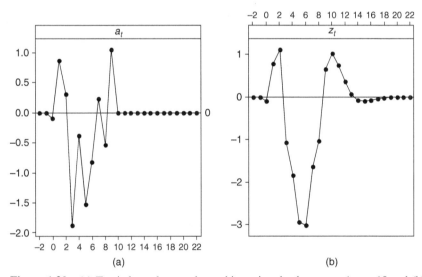

Figure 1.20 (a) Ten independent random white noise shocks $a_t, t = 1, \ldots, 10$ and (b) the superimposed responses of a linear filter generated by the AR(2) model $\tilde{z}_t = 0.9824\tilde{z}_{t-1} - 0.3722\tilde{z}_{t-2} + a_t$.

at time $t = 0$, then the pendulum's response over time would be as shown in Figure 1.20b.

The impulse response function helps us to visualize and gain an intuitive understanding of the dynamic reaction of a system. Specifically, we can consider any stationary time series model as a linear filter subject to a sequence of random shocks. How a process reacts to a single shock provides us with important information about how the noise propagates through the system and what effect it has over time. Indeed, we can always intuitively think of any stationary time series model as a system that mimics the dynamic behavior of something like a pendulum subject to a sequence of small random pea shots. Further, if the process is nonstationary, what has been said above will apply to the first or possibly higher order difference of the data. In either case, the impulse response function is still a useful tool for visualizing the dynamic behavior of a system.

1.7 PARSIMONIOUS MODELS

In any modeling effort, we should always keep in mind that the model is only an approximation of the true behavior of the system in question. One of the cardinal sins of modeling is to fall in love with the model. As George Box famously stated, "All models are wrong. Some are useful." This is particularly true in time series modeling. There is quite a bit of personal judgment when it comes to determining the type of model we would like to use for a given data. Even though this interpretation adds extra excitement to the whole time series modeling process (we might admittedly be a bit biased when we say "excitement"), it also makes it subjective. When it comes to picking a model among many candidates, we should always keep in mind Occam's razor, which is attributed to philosopher and Franciscan friar William of Ockham (1285–1347/1349) who used it often in analyzing problems. In Latin it is "Pluralitas non est ponenda sine necessitate," which means "Plurality should not be posited without necessity" or "Entities are not to be multiplied beyond necessity." The principle was adapted by many scientists such as Nicole d'Oresme, a fourteenth century French physicist, and by Galileo in defending the simplest hypothesis of the heavens, the heliocentric system, or by Einstein who said "Everything should be made as simple as possible, but not simpler." In statistics, the application of this principle becomes obvious in modeling. Statistical models contain parameters that have to be estimated from the data. It is important to employ models with as few parameters as possible for adequate representation. Hence our principle should be, "When everything else is equal, choose the simplest model (... with the fewest parameters)." Why simpler models? Because they are easier to understand, easier to use, easier to interpret, and easier to explain. As opposed to simpler models, more complicated models with the prodigal use of parameters lead to poor estimates of the parameters. Models with large number of parameters will tend to overfit the data, meaning that locally they may provide very good fits; however, globally, that is, in forecasting, they tend to produce poor forecasts and larger forecast variances. Therefore, we strongly recommend the use of Occam's razor liberally in modeling efforts and always seek the simpler model when all else is the same.

EXERCISES

1.1 Discuss why we see serial dependence in data collected in time.

1.2 In the pendulum example given in Section 1.3, what are the factors that affect the serial dependence in the observations?

1.3 Find the Wold decomposition for the AR(2) model we obtain for the pendulum example.

1.4 The impulse response function in Table 1.1 is obtained when $a_0 = 1$. Repeat the same calculations with $a_0 = 1$ and $a_1 = 1$, and comment on your results.

CHAPTER 2

VISUALIZING TIME SERIES DATA STRUCTURES: GRAPHICAL TOOLS

2.1 INTRODUCTION

Most univariate time series analysis publications, introductory and advanced textbooks, monographs, and scholarly publications project the notion that time series graphics amounts to time series plots of the raw data and the residuals as well as plots of the autocorrelation and partial autocorrelation coefficients against lags. In this chapter, we demonstrate with a few examples that a properly constructed graph of a time series can dramatically improve the statistical analysis and accelerate the discovery of the hidden information in the data. For a much detailed coverage of the topic in general, we refer our readers to the excellent books by Cleveland (1993, 1994) and Tufte (1990, 1997) from which we got the main inspiration for this chapter.

In our academic and consulting careers, our first recommendation for any data analysis exercise is to "plot the data" and try to be creative at it. Indeed, the purpose of this chapter is to stimulate a discussion of the tools, methods, and approaches for detailed time series analysis and statistical craftsmanship akin to the exacting style of data analysis demonstrated by Daniel (1976) for designed factorial experiments. We consider it important to bring this issue back in focus because we have detected a tendency toward automation of time series analysis among practitioners. This may be an unintended consequence of the success of the Box–Jenkins approach, combined with the proliferation of standard time series software packages that tend to encourage a somewhat routed and mechanical approach to time series analysis. As Anscombe (1973) pointed out, "Good statistical analysis is not a purely routine matter, and generally calls for more than one pass through the computer."

It is almost too obvious to repeat, but the distinguishing features of a time series lie in its (auto)correlation structure. Nevertheless, we think it is important to reemphasize. It is generally well recognized that a summary statistic can be highly misleading. For example, the average, without any further qualification of the underlying distributional shape, is often inappropriate. The same is the

Time Series Analysis and Forecasting by Example, First Edition. Søren Bisgaard and Murat Kulahci.
© 2011 John Wiley & Sons, Inc. Published 2011 by John Wiley & Sons, Inc.

case when we summarize the autocorrelation structure of a given time series by reporting only the linear correlation coefficients. To do so can sometimes be highly misleading. Indeed, we will show that insight can be gained by carefully scrutinizing the plots of time series data to see if patterns reveal important features of the data that otherwise would easily be missed. As has been pointed out by many, but particularly well by Cleveland (1993, 1994), the entire inference based on formal statistical modeling can be seriously misleading if fundamental distributional assumptions are violated. The models discussed in this book have much in common with ordinary linear regression. This relationship has been exploited to suggest ways to further scrutinize time series data and to check the assumptions.

2.2 GRAPHICAL ANALYSIS OF TIME SERIES

We may think of time series analysis as being primarily focused on statistical modeling using complex mathematical models. However, statistical graphics and graphical analysis of time series data are an essential aspect of time series analysis, and indeed, in many aspects, more important than the mathematical models. Graphical analysis is where we learn about what the data is trying to say! Therefore, in this chapter, we provide a discussion of statistical graphics principles for time series analysis.

In most cases, graphical and mathematical modeling go hand in hand. Careful graphical scrutiny of the data is always the first and often crucial step in any statistical analysis. Indeed, data visualization is important in all steps of an analysis and should often be the last step as a "sanity check" to avoid embarrassing mistakes. We need to develop a "feel" for the data and generate intuition about what we are dealing with. Such "feel" and intuition come from hands-on work with and manipulation of the data. Statistical graphics is perhaps the most important means we have for learning about process behavior and for discovering relationships. Graphics is also important in data cleaning. However, graphical data analysis of time series data is not necessarily obvious and trivial. It requires skills and techniques. Some methods and approaches are better than the others. There is an art and a science to data visualization. In this chapter, we discuss a number of techniques for data visualization, explain underlying principles of how our eyes and brain process graphical information, and explain the do's and don'ts in graphical data analysis. To do so, we will use a number of historical datasets known for their peculiar patterns and some new ones.

Data always includes peculiar patterns. Most are worth careful scrutiny. Of course, some patterns may be unimportant. But in many cases, outliers and strange patterns indicate something important or unusual and in some cases may be the most important part of the entire dataset. Therefore, they should not easily be glanced over or dismissed. As Yogi Berra of the New York Yankees said at a press conference in 1963, "You can observe a lot by watching." This is particularly true with time series data analysis! In later chapters, we will introduce a number

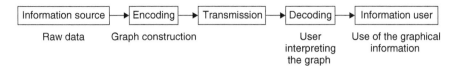

Figure 2.1   Graphical transmission of information.

of sophisticated mathematical models and methods, but graphing the data is still one of the most important aspects of time series analysis.

In English writing classes, we are taught to write, revise, rewrite, and edit text. We learn that famous authors and speech writers typically labor with the words and text, and iterate numerous times until they express precisely what they want to say clearly and succinctly. As it has been said, writing includes three steps: thinking about it, doing it, and doing it again and again. In fact, this is not even true. It is not a linear process. We do all these things at the same time. The point is that expressing ourselves concisely and succinctly is an iterative learning process. The same should be true with statistical graphics. Statistical graphics is an iterative process involving plotting, thinking, revising, and replotting the data until the graph precisely says what we need to convey to the reader. It has often been said that a graph is worth a thousand words. However, it may require hard work and several hours in front of the computer screen, rescaling, editing, revising, and replotting the data before that becomes true.

Over the years, a certain set of rules have emerged that guide good graphics. We will review a number of those that are relevant to time series analysis. The goal of statistical graphics is to *display* the data as *accurately* and *clearly* as possible. Good graphics helps highlight important patterns in the data. There are many ways of looking at data, but not all are equally good at displaying the pertinent features of the data.

When we construct a graph, we are encoding the data into some form of graphical elements such as plotting symbols, scales, areas, color, and texture. When the user of the graph reads and interprets the graph, he or she reverses the process and visually decodes the graphical elements. This process is depicted in Figure 2.1. When constructing graphs, we control this process through our own choice of graphical elements.

2.3   GRAPH TERMINOLOGY

To avoid ambiguity in discussing statistical graphics, it is helpful to develop a terminology that provides meaning to words and concepts (see Cleveland, 1994). In the following sections, we review some of the basic concepts and terms that will be used in this book. The main terms are defined graphically in Figures 2.2, 2.3, and 2.4 showing time series data from monitoring a complex manufacturing process.

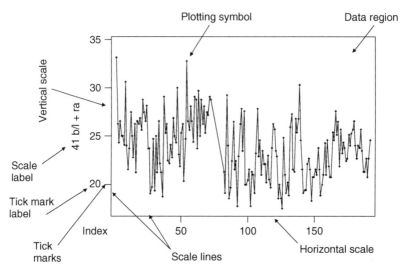

Figure 2.2 A typical time series graph of the temperature of an industrial process. Superimposed on the graph are a number of graph concepts.

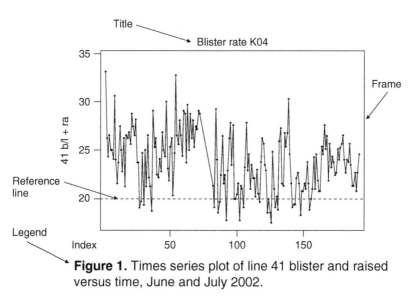

Figure 1. Times series plot of line 41 blister and raised versus time, June and July 2002.

Figure 2.3 The same plot as in Figure 2.2 with additional graphical elements such as legend, title, and reference line.

2.4 GRAPHICAL PERCEPTION

Statisticians and experimental psychologists have collaborated to investigate how the human brain processes data. For example, consider the very popular pie chart often used in the economic and business context when we want to display

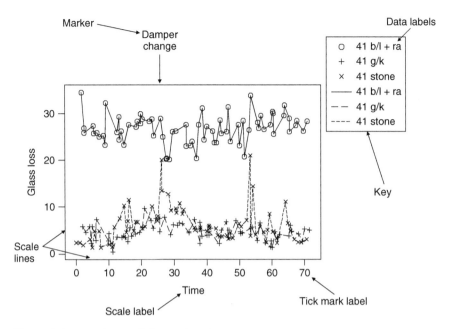

Figure 2.4   Superimposed time series plots.

fractions of the whole. In Figure 2.5, we show five categories, A, B, C, D, and E that make up the whole. In this case, the fractions are 23, 21, 20, 19, and 17%. Now looking at the pie chart, it is impossible to determine the differences in the five categories.

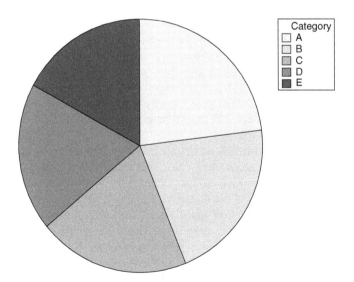

Figure 2.5   A pie chart of the five categories, A, B, C, D, E.

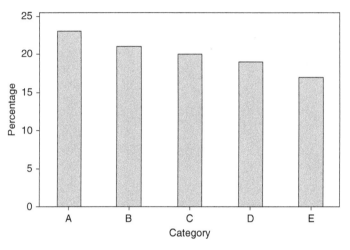

Figure 2.6   A bar chart of the same five categories.

Figure 2.6 shows a bar chart of the same five categories. If the purpose is to be able to say something about the size of the categories, we immediately see that the bar chart is much more useful. From this chart, it is very easy to see that A is larger than B, etc. In fact, we can directly read off from the graph what the sizes are. Now, this does not mean that we condemn the use of pie charts. They have their use especially when the objective is to show that one category is dominating.

With this background, let us discuss the general issues about graphic data. When making a graph, the data is encoded onto the graph by a number of mechanisms such as the shape of the graph, the selection of plotting symbols, the choice of scales, whether the points are connected or not, and the texture and color of the graph. Likewise, when the user of the graph is studying the graph to absorb the information encoded into the plotting, the person is visually decoding the graph. This decoding process is called *graphical perception*. In designing a graph, we can, for good or for worse, control and manipulate this decoding process to allow us to convey the information in the way we prefer. Our task here is to convey the information accurately, but unfortunately, sometimes graphs are also used for propaganda purposes.

Cleveland and McGill (1987) provide an outline about the elementary codes of graphs, which represent the basic geometric, color, and textural characteristics of a graph. These, in the order of how well we can judge them, are positions on a common scale, positions along identical but nonaligned scales, length, angles and slopes, areas, volume, color hue, saturation, and density. Consider, for example, the common scale issue. In Figure 2.7, it is easy to see that the two bars A1 and B1 are of unequal length, because they have a common baseline. Bars A2 and B2 are exactly the same as A1 and B1 but now that they do not share the same baseline, it is difficult to see that the two bars are of unequal length. Bars A3 and B3 are of the same length and position as A2 and B2. However, by adding

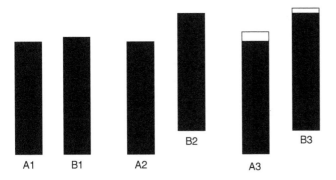

Figure 2.7   Bar charts on common and different baselines.

a reference box around black bars, which provides a scale, we can now see that A3 and B3 are unequal.

One of the main problems in graphical perception is to be able to detect differences between superimposed graphs, which are quite commonly used in time series analysis. Often, the problem is not so much of detecting the difference between the curves, but rather of our perceiving what is incorrect. We are often fooled by optical illusions. We now provide two examples and explain what is happening.

First, in Figure 2.8, we present a historical graph from William Playfair (1786) showing yearly import and export between England and East Indies from 1700 to 1780. We notice that the difference between import (the solid curve)

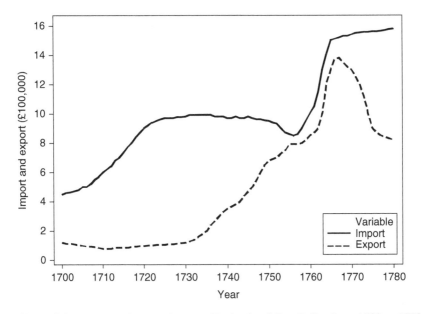

Figure 2.8   Import and export between England and East Indies from 1700 to 1780.

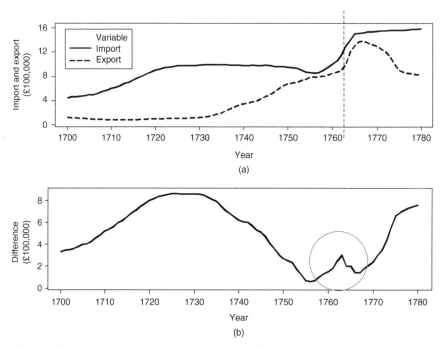

Figure 2.9 (a) Import and export between England and East Indies from 1700 to 1780 and (b) the difference between the import and the export.

and export (the dashed curve) becomes quite small between 1755 and 1770. However, this is because when it comes to comparing two curves like the ones in Figure 2.8, we tend to consider the shortest distance between the superimposed curves and not necessarily the vertical distance, which in this case corresponds to the difference between exports and imports for a given year. Now consider Figure 2.9 where we provide the difference between the import and the export in a separate panel (b). In that figure, we can clearly see that the trade deficit is not constant between 1755 and 1770 with a sudden increase around 1765.

As another example for optical illusion, consider the curves in Figure 2.10. Imagine the solid curve to represent the expenditures of a start-up company and the dashed curve as the revenues. On studying this figure, we might have the impression that while both expenditures and revenues are increasing exponentially, the difference between the two is getting smaller. In fact, the difference between these two curves for a given year remains the same. We will leave the proof to our readers to whom we recommend the use of a ruler to measure the vertical distance between the two curves.

2.5 PRINCIPLES OF GRAPH CONSTRUCTION

In this section, we adapt the principles of graph construction originally presented in Cleveland (1993, 1994). However, like any rules, the rules presented here

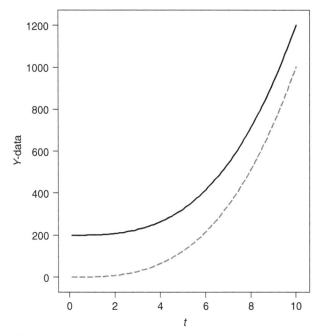

Figure 2.10 Two exponentially increasing curves. Note that the difference between the numerical values and hence the vertical distance is the same for all values of t.

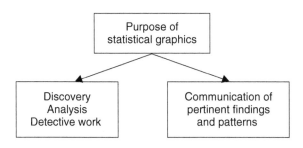

Figure 2.11 Purposes of statistical graphics.

should not be followed dogmatically, but be considered sensible guidelines that should preferably be violated only after careful consideration and trade off between expediency and clarity.

As shown in Figure 2.11, there are primarily two purposes of statistical graphics: (i) For data scrutiny, discovery, analysis, and detective work (ii) for communication of pertinent findings in the data. In both cases, we can summarize "to do's" in the following principles:

1. Large quantities of information can be communicated succinctly with a well-designed graph. We should strive to communicate key points and features of the data. For that, we should make the data stand out while trying to avoid clutter in terms of more than absolutely necessary amount of notes,

markers, and data labels in the data region. For example, keys and markers should preferably be placed outside the data region and the notes in the legend. Similarly, we should put tick marks outside the data region.

2. Compact graphs and juxtaposed panels are often useful and can convey the message clearer. But we should make sure that superposed time series in the same panel are clearly distinguishable.

3. Graph construction is an iterative process. We should iterate until the graphs tell the story we want to communicate and always proofread them.

4. Sometimes, several graphs will be necessary to highlight different aspects of the same data. So we should not shy away from having several graphical representations of the same data in order to highlight some of its features. Examples of this practice can be found throughout this book.

5. Scaling the data and the graph is crucial. We should carefully choose the axis scales of the variables, so that the data fills up the data region. If several graphs are to be compared, we should make sure that they have the same scales. We should not necessarily insist that zero is included. We should use scale breaks only when necessary. We should always consider using logarithmic scales when the data spans several orders of magnitude or when it is important to understand percentage changes.

2.6 ASPECT RATIO

One of the crucial issues in graphical representation is the aspect ratio, α, of a graph, which is defined as $\alpha = h/w$ where h is the height and w is the width of the data region. The aspect ratio has significant impact on our ability to visually decode slopes. To illustrate this, consider Figure 2.12. Which of the three plots 2.12a, 2.12b, or 2.12c provides the best discrimination between the convex and the linear part of the curve? Clearly, the difference between two segments of the curve is most apparent in Figure 2.12a. Why is that so? Our perception of a curve is based on our judgment of the relative orientation of the curve segments. Typically, our eye has the best discriminative power when the orientation (banking) of what we like to compare is at or close to a $45°$ angle. We would need to arrange the aspect ratio so that what you like to stand out is banked on average at or around a $\pm 45°$ angle, which is the case in Figure 2.12a if we would like to draw attention to the difference between the two segments of the curve.

Figure 2.13 shows the famous time series of the annual sunspot numbers from 1770 to 1869. The aspect ratio of this graph is 4/6, the typical value used as the aspect ratio for many graphs. Figure 2.14 shows exactly the same data with an aspect ratio of $6/6 = 1$. Finally, Figure 2.15 has the aspect ratio of 1/6. Figure 2.13, we clearly see the cycle in the data but it is not clear that the curve rises faster than it falls, which is a very interesting feature of this data. In Figure 2.14 with an aspect ratio of 1, it becomes even more difficult (if at

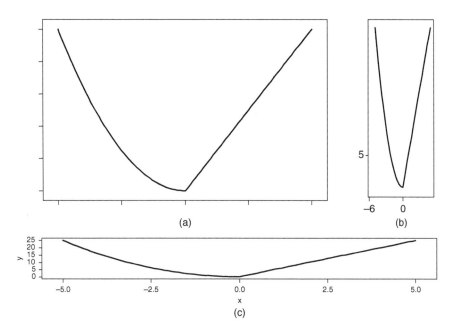

Figure 2.12   The same graph with different aspect ratios.

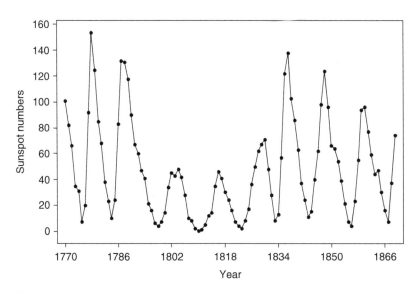

Figure 2.13   The annual sunspot numbers from 1770 to 1869. The aspect ratio is 4/6.

all possible) to notice this lack of symmetry. However, when the aspect ratio is changed to 1/6 as in Figure 2.15, we clearly see that the sunspot series is not a harmonic cycle but that the curve rises much more rapidly than it falls. This is possible because the slopes are at almost the optimal 45°. We could have missed this if we used the default settings in many statistical software packages. Indeed,

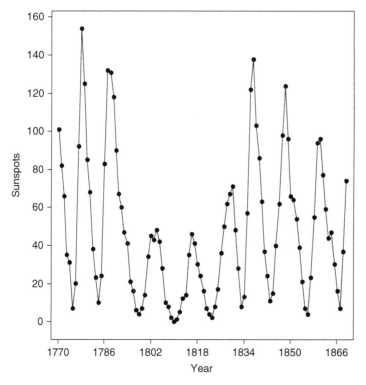

Figure 2.14 The annual sunspot numbers from 1770 to 1869. The aspect ratio is 1.

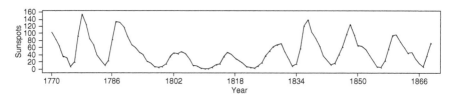

Figure 2.15 The annual sunspot numbers from 1770 to 1869. The aspect ratio is 1/6.

the eye and the brain's visual decoding of the figure is highly dependent on the graph's aspect ratio.

Let us now consider a dataset that we use in Chapter 5 and show that we can emphasize different features of the data using different aspect ratios. Consider Figure 2.16 where the number of airline passengers from 1949 to 1960 is shown. The aspect ratio of this plot is 4/3. When we look at the general trend and try to ignore the seasonality, the eye connects the peaks. Because the aspect ratio is chosen so that the trend is about the optimal 45°, we can see a bit of a nonlinear (exponential) trend in the overall behavior of the data.

Now, consider Figure 2.17 where the same data as in Figure 2.16 is shown with an aspect ratio of 5/16. Now the trend looks more or less linear. The trick is

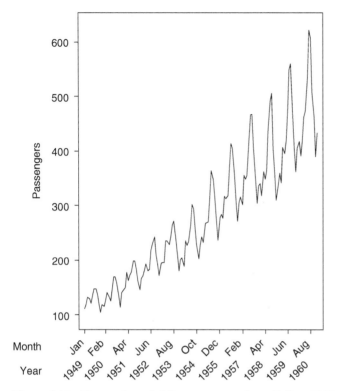

Figure 2.16 The number of airline passengers from 1949 to 1960. The aspect ratio is 4/3.

that the eye looks at and connects the peaks and since the angle now is far from the optimal 45°, we do not see the nonlinearity in the trend so well. However, we can clearly see that the drop after each peak is sharper than the rise to the peak. This asymmetry around the peaks was not so obvious in Figure 2.16. So what is the conclusion? The best aspect ratio depends on what we are looking

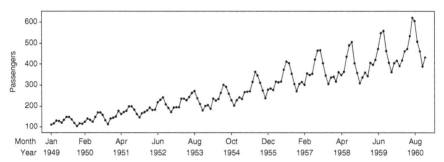

Figure 2.17 The number of airline passengers from 1949 to 1960. The aspect ratio is 5/16.

for or what message we are trying to convey and is case specific. Therefore, it could be quite beneficial to try a few different aspect ratios before deciding on one.

2.7 TIME SERIES PLOTS

In time series plots, the dependent variable is plotted against the independent variable time. There is usually a single value of the dependent variable for each value of the independent variable and those values are typically equally spaced. There are various ways of plotting time series data and there are pros and cons for each.

2.7.1 Connected Symbols Graph

The connected symbols graph is the most common type of time series graph. The advantage is that each individual data point as well as the ordering of the points can be seen clearly. Figure 2.18 shows the connected symbols graph of the number of airline passengers data we discussed earlier. Notice that the annual cycle is clearly visible in this plot.

2.7.2 Connected Lines Graph

The connected lines graph for the airline passenger data is given in Figure 2.19. It does provide clarity about the flow of the data and the cyclic pattern. Sometimes, especially with large number of observations, the symbols can get in the way.

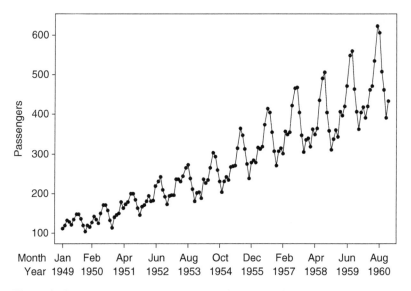

Figure 2.18 The connected symbols graph for the number of airline passengers data.

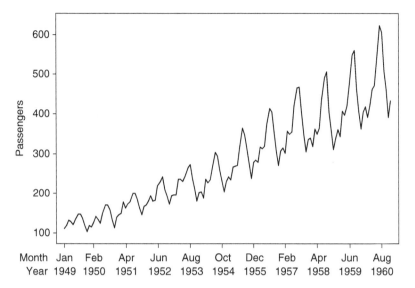

Figure 2.19   The connected lines graph for the number of airline passengers data.

However, without the symbols, it is not possible to see how many individual values are involved, for example, around peak values such as around October 1957—is it a single point or more?

2.7.3   Graph with Symbols Only

In this graph, the symbols are not connected. This may be of value when considering the general trend in a time series. However, since ordering is no longer obvious, it will be difficult to see short- to mid-term patterns. In Figure 2.20, without the connecting lines, it is now harder to see the systematic yearly cycles.

2.7.4   Graph with Projected Bars and Symbols

Projected bar charts are popular in economics and are often used by newspapers. The symbols may make them a bit too busy as in Figure 2.21. It is also hard to see troughs compared to the peaks in the data. These plots can be useful if the bars are projected to a reference horizontal line in the middle of the graph, for example, as in autocorrelation plots that are shown in the following chapters.

2.7.5   Graph with Projected Bars (Vertical Line Plot)

Figure 2.22 shows the vertical line plot as in Figure 2.21 without the symbols. It certainly looks less busy. However, projected bars create an asymmetry in the appearance that makes peaks stand out more than troughs. Therefore, when it comes to time series plots, projected bars should be used only after careful consideration and sparingly.

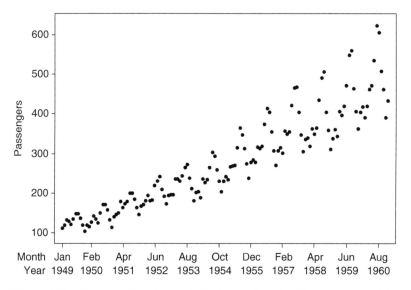

Figure 2.20 The symbols only graph for the number of airline passengers data.

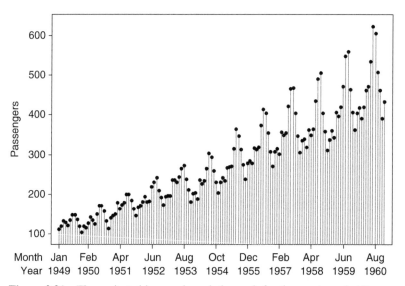

Figure 2.21 The projected bars and symbols graph for the number of airline passengers data.

2.7.6 Area Graph

This type of graph is also used somewhat commonly, particularly with superimposed time series. Figure 2.23 is worse, however, than the vertical line graph in Figure 2.22. The asymmetry is even more pronounced; peaks stand out more than troughs. We recommend using this type of graph very sparingly.

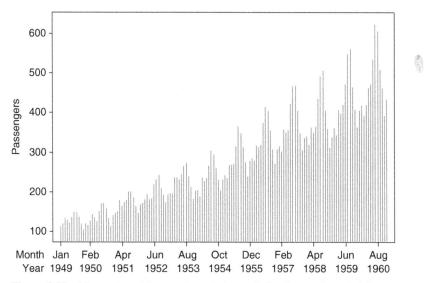

Figure 2.22   The projected bars and symbols graph for the number of airline passengers data.

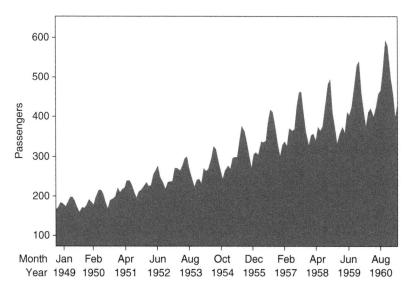

Figure 2.23   The area graph for the number of airline passengers data.

2.7.7   Cut and Stack Graph

In time series analysis, it is quite usual to get a dataset with a large number of observations. Consider 2195 hourly temperature observations from a cooling tower shown in Figure 2.24. It is nearly impossible to identify any short- to

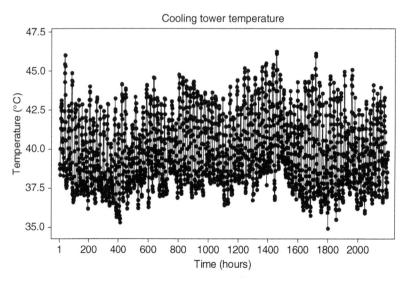

Figure 2.24 The hourly temperature observations from a cooling tower.

mid-term oscillations, even if we want to emphasize certain known features of the data that will most likely require unfeasibly low aspect ratios. Instead, we can cut the time series in segments, make appropriate plots of those segments, and stack them—hence the name "cut and stack graph."

Figure 2.25 gives 11 panels representing segments of the temperature data. Those panels are stacked on the same graph. We can now clearly see the seasonal pattern in the data that was impossible to see in Figure 2.24.

2.8 BAD GRAPHICS

So far, we have mainly focused on what to do in graphical analysis. Here are a few examples of what not to do.

Case 1. Defect Report to Upper Management. Figure 2.26 shows the number of daily defect counts (in percentage) for 6 months of production in a ceramic production process. We observe some variation from month to month. Before we go on with bad representation of the same data, consider the box plot in Figure 2.27. This graph helps in summarizing the data succinctly and is sometimes preferred as a summary for upper management.

In Figure 2.28, we adjust the x-axis to include the entire year even though the data is not available yet. Similarly, for the y-axis we include 0, a very common practice. Both changes in the scale caused the data to get compressed and subsequently some of its features are no longer observed as easily as in Figure 2.26. In the case below, we show an even more dramatic effect of changing the scale of the y-axis by including 0.

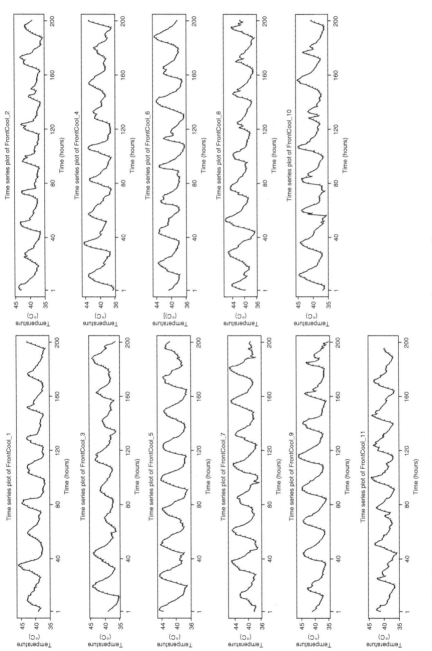

Figure 2.25 Eleven segments of the hourly temperature observations from a cooling tower.

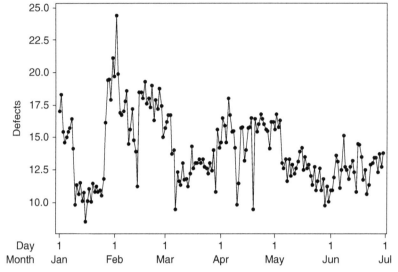

Figure 2.26 Time series plot of the daily defect count in percentage for 6 months of production in a ceramic production process.

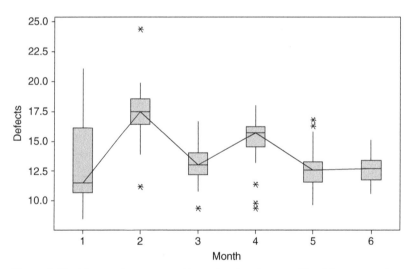

Figure 2.27 Time series connected box plot showing monthly defect counts in percentage for 6 months of production in a ceramic production process.

Case 2. Temperature of an Industrial Furnace. The temperature of industrial furnaces can get quite high. Figure 2.29 shows 200 observations of the temperature of such a furnace. From that figure, it is hard not to conclude that the temperature of the furnace was stable during that period.

We can further "hide" the data by including 0 on the y-axis as in Figure 2.30. As discussed earlier, in time series plots, it is quite common to

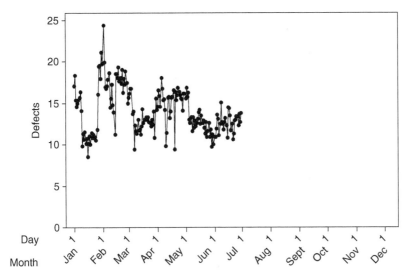

Figure 2.28 Time series plot of daily defect count in percentage for 6 months of production in a ceramic production process.

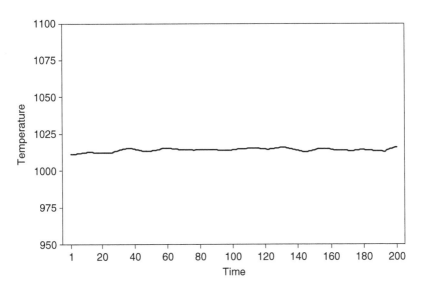

Figure 2.29 Temperature measurements for an industrial furnace.

include 0 on the y-axis even if the minimum value is far from 0. To make things even worse, we add horizontal gridlines in Figure 2.31. This is in fact a very effective way of hiding the data. This becomes quite obvious when we use more appropriate limits for the y-axis in Figure 2.32. The variation in the temperature for the given time period is now visible. One might argue that the range of the variation is quite small (5°C) compared to the average value of

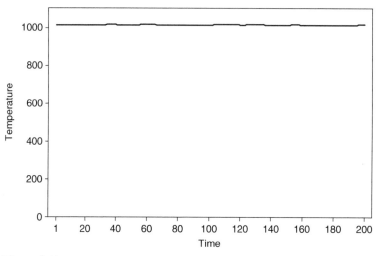

Figure 2.30   Temperature measurements for an industrial furnace with 0 included in the
y-axis.

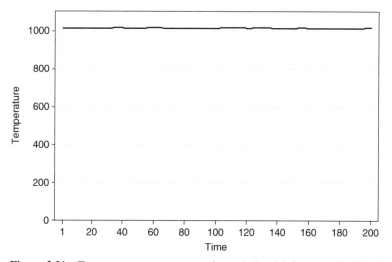

Figure 2.31   Temperature measurements for an industrial furnace with 0 included in the
y-axis and horizontal gridlines.

1014°C. However, this particular furnace is controlled very tightly to maintain a
target temperature value since even small changes in temperature can potentially
cause a lot of defects. Therefore, studying the behavior of the temperature is a
very crucial issue in controlling the quality, and a variation over a range of 5°C
may very well be unacceptable. When this variation is hidden in the scale as in
Figure 2.30, the focus might erroneously shift to other areas of production to
maintain the quality, which will result in wasted resources, energy, and money.

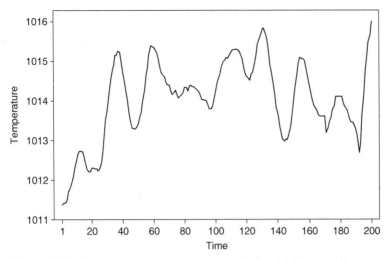

Figure 2.32   Temperature measurements for an industrial furnace with an appropriate scale for y-axis.

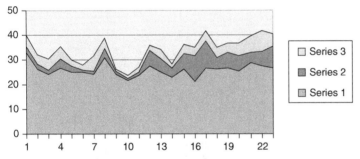

Figure 2.33   Area graph for stacked proportions of causes of glass loss where Series 1, 2, and 3 are blisters, stones, and glass knots respectively.

Case 3. Type of Defects in Glass Manufacturing.   Figure 2.33 shows an area graph with stacked proportions of the three causes of glass loss during production—blisters, stones, and glass knots. From that graph, it is not quite clear if there is a ranking between these three causes. That is, it is hard to see if any one of the causes is worse than the others in terms of glass loss. This is a common problem with stacked area graphs.

Figure 2.34 shows simple time series plots of percentage glass loss due to each cause. We can now clearly see that blisters are the number one cause for percentage glass loss.

Case 4. Auto and Truck Production in the United States.   Figure 2.35 shows the monthly production of autos and trucks in the United States from January 1986 to September 2007. This is another example of a stacked area graph. It is

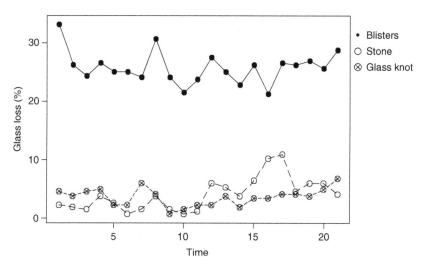

Figure 2.34   Time series plots of the causes of glass loss.

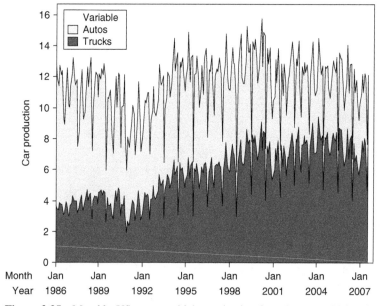

Figure 2.35   Monthly US motor vehicle production from January 1986 to September 2007.

very hard to make a comparison between auto and truck production during any one year. A better graph is given in Figure 2.36 where the time series plots of the production number for both types of cars are shown. Now both graphs are sharing the same baseline and are not stacked. It is now possible to see that auto production is declining, whereas truck production is increasing. However, due to

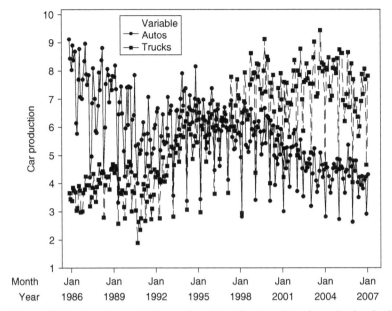

Figure 2.36 Superimposed time series plots of auto and truck production in the United States.

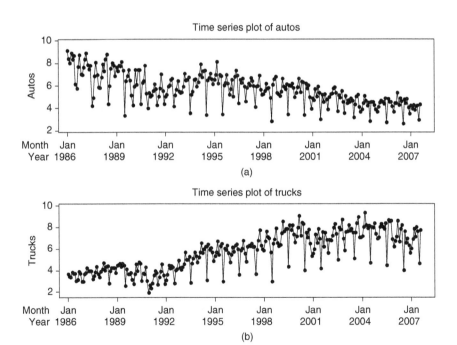

Figure 2.37 Time series plots of auto and truck production in the United States in separate panels.

the large number of observations and the variation both time series exhibit, it is still quite difficult to make any comparisons. This could of course be improved using different colors. Instead, we decide to plot these time series in two separate panels on the same plot as in Figure 2.37. It is now easier to notice the decline in auto production as truck production increases over the years. Note that both graphs have the same y-axis to ease comparison.

Proper graphical display of the data can greatly improve the subsequent statistical analysis and help in conveying the analyst's message in a clearer and more effective way. While we encourage our readers to start any statistical analysis by plotting their data, we also recommend that they be creative and to keep in mind that with today's powerful hardware and software, we are in many cases only bound by our imagination.

EXERCISES

2.1 Table B.17 contains the quarterly dollar sales (in $1000) of Marshall Field & Company for the period 1960 through 1975. Use different aspect ratios to plot this time series and comment on which aspect ratio is preferable.

2.2 Repeat Exercise 2.1 for the outstanding consumer credits data provided in Table B.9.

2.3 Repeat Exercise 2.1 for the monthly data on sea levels in Los Angeles, CA, given in Table B.4.

CHAPTER 3

STATIONARY MODELS

3.1 BASICS OF STATIONARY TIME SERIES MODELS

3.1.1 Deterministic Versus Stochastic Models

In many application areas such as in engineering, natural, and social sciences, processes (or systems) are often described through mathematical models based on the governing physics of such processes. These models can then be used to predict the output of these processes at a given time in the future. If the mathematical model of a physical process can predict the output of this process exactly, it is called a *deterministic model*. Consider for example, Figure 3.1a, the simple harmonic function as the mathematical model describing the behavior of Yule's simple pendulum mentioned in Chapter 1. How realistic is the deterministic model though? Yule suggests two possible problems. The first one he calls superposed fluctuations may simply be some recording errors due to some inaccuracies in the measurement system. In that case, for example, the time series we observe would be of the kind presented in Figure 3.1b. We generated this time series by simply sprinkling some random noise on top of the simple harmonic function in Figure 3.1a. This type of disturbance would of course cause our observations to deviate from what is predicted by the deterministic model. While this is indeed a legitimate and often overlooked issue in data collection, the disturbances caused by the recording errors do not, by definition, affect the physical process itself.

There could however be, as Yule calls them, true disturbances caused by unknown and/or uncontrollable factors that have direct impact on the process. The example given by Yule has boys barging "into the room and start pelting the pendulum with peas, sometimes from one side and sometimes from the other" as the true disturbance. It could be argued that this disturbance can in fact be incorporated into the model if, for example, it could be generated in a more systematic manner. Even if that could be achieved, there would still be many potential sources of disturbances; for example, when each time the boys barge into the room, the air circulation, changes in the temperature, and the humidity of the room may very well affect the movement of the pendulum. Unfortunately, it is impossible to come up with a comprehensive deterministic model to account for all these possible disturbances, since by definition they are "unknown." In these cases, a probabilistic or stochastic model will be more appropriate to describe the behavior of the process. This model will be based on

Time Series Analysis and Forecasting by Example, First Edition. Søren Bisgaard and Murat Kulahci.
© 2011 John Wiley & Sons, Inc. Published 2011 by John Wiley & Sons, Inc.

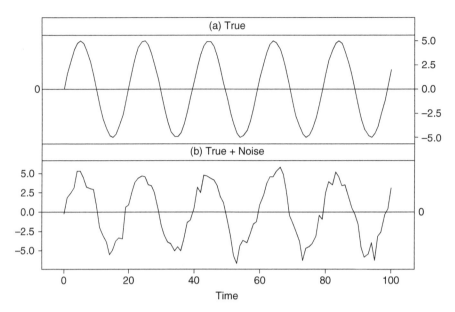

Figure 3.1 (a) Simple harmonic motion and (b) simple harmonic motion plus noise.

the fact that the observed data for a given period simply reflects the "reaction" of the process to the disturbances that happen during that period. But we should consider the fact that for the same period we could have had a different set of disturbances resulting in another set of observed data. Hence as stated in BJR, we should treat the observed data as "a realization from an infinite population of such time series that could have been generated by the stochastic process." Therefore, it is not possible to predict exactly the output of the process in the future. Instead, we may provide a prediction interval and the probability with which the future observation will lie in that interval. However, in order to do that we need to assume that certain properties of our process remain constant in time. As Brockwell and Davis (2002) (BD hereafter) put it "If we wish to make predictions, then clearly we must assume that *something* does not vary with time." This brings us to the notion of stationarity that will be discussed in the next section.

3.1.2 Stationarity

The basis for any time series analysis is stationary time series. It is essentially for stationary time series that we can develop models and forecasts. However, it is the nonstationary time series that is most interesting in many applications, especially in business and economics. Similarly in industrial applications, when processes are left alone, they are expected to show nonstationary behavior simply following the second law of thermodynamics. In fact, stationarity in most processes is only ensured by control actions taken at regular intervals and/or continuous maintenance of the components of the system. Without such deliberate actions to make the processes stationary, one solution is not to focus on the original data that exhibits nonstationarity but instead, for example, on the changes in the successive observations. In many applications with nonstationary data, the changes from time $t - 1$ to time t of the time series z_t, denoted by $\nabla z_t = z_t - z_{t-1}$ may instead be stationary. If that is the case, we can then model the changes, make forecasts about the future values of these changes, and from the model of the changes build models and create forecasts of the original nonstationary time series. Therefore while in real life applications it happens only under specific situations, the stationary time series play a key role as the foundation for time series analysis. In this chapter, we primarily focus on how to model stationary time series.

In the strict sense, a time series is called *stationary* if the joint probability distribution of any n observations, $\{z_{t+1}, z_{t+2}, \ldots, z_{t+n}\}$ (i.e., realizations of the time series at times $t + 1, t + 2, \ldots, t + n$) of this series remain the same as another set of n observations shifted by k time units, that is, $\{z_{t+1+k}, z_{t+2+k}, \ldots, z_{t+n+k}\}$. For further details, see BD, Chapter 1. For practical purposes, however, we define a stationary time series as a finite variance process of which the mean and the variance are constant in time, and the correlation between observations from different points in time is only lag dependent. In other words, we deem a time series z_t stationary, if $E(z_t) = \mu$ is independent of t and the autocovariances $\text{Cov}(z_{t+k}, z_t)$ depend only on k for all t. This is sometimes referred to as *weak stationarity* (BJR and BD). Unless otherwise specified, in this book whenever we call a times series stationary, we in fact refer to weak stationarity as defined above. Although there are some rigorous statistical tests that can be performed, often a visual inspection of the time series plot will provide important information whether the process is stationary or not. If arbitrary snapshots of the time series we study exhibit similar behavior in central tendency and spread, we will proceed with our analysis assuming that the time series is indeed stationary. As we will show in Section 3.3, once the model is obtained, we can check this assumption based on the parameter estimates of the proposed model. In fact, many modern statistical software packages nowadays provide an automatic warning if the fitted model is nonstationary, making it even easier to confirm the assumption. In Figure 3.2, we provide examples of stationary and nonstationary time series. We can see that the visual inspection is indeed sufficient to confirm or deny the tentative assumption of stationarity of these time series.

We need to be a bit cautious, however, with the visual inspection as the time span considered can be misleading. For a short period of time, a nonstationary time series may give the impression of being stationary. Therefore besides the visual inspection, we also highly recommend for the practitioners to use their intuition and knowledge about the process in question. Does it make sense for a tightly controlled chemical process to exhibit similar behavior in mean and variance in time? Should we expect the stock market as BJR put it "to remain in equilibrium about a constant mean level" and if your answer is yes, would you invest in it? In practice, in most cases the visual inspection and the answer to the

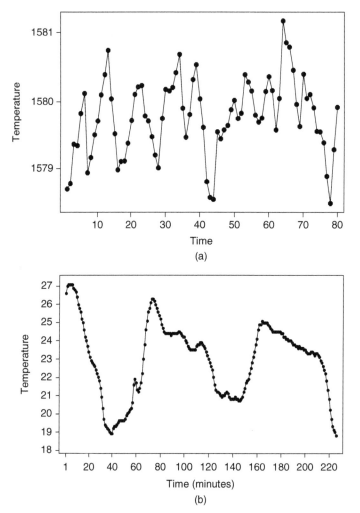

Figure 3.2 Temperature measurements for two different chemical processes. The process in (a) is tightly controlled to be around a certain target value (stationary behavior). The process in (b) is not controlled so that the temperature measurements do not vary around a target value (nonstationary behavior).

question "Does it make sense ...?" are often enough to tentatively decide on the stationarity of a time series. As mentioned earlier, once the tentative model is obtained, we can further check the stationarity of the model through the parameter estimates.

3.1.3 Backshift Operator

In our model representations, we will often use the backshift operator B. When applied to a time series, it basically shifts the time by one unit. That is, $Bz_t = z_{t-1}$.

Similarly when we consider the differences between successive observations, we will simply have $z_t - z_{t-1} = z_t - Bz_t = (1 - B)z_t$. It should be noted that the operators $(1 - B^2)$ and $(1 - B)^2$ are not equivalent, since $(1 - B^2)z_t = z_t - z_{t-2}$ and $(1 - B)^2 z_t = z_t - 2z_{t-1} + z_{t-2}$.

3.1.4 Autocorrelation Function (ACF)

As mentioned in Section 3.1.2, a (weakly) stationary time series, z_t, has finite variance with constant mean and variance over time t. Hence we can write

$$E(z_t) = \mu_t = \mu \tag{3.1}$$

The sample estimate for the mean is given by

$$\bar{z} = \frac{1}{T} \sum_{t=1}^{T} z_t \tag{3.2}$$

Similarly, the sample variance can be calculated using

$$\hat{\sigma}_z^2 = \frac{1}{T} \sum_{t=1}^{T} (z_t - \bar{z})^2 \tag{3.3}$$

To further study the stationary time series, we introduce the autocovariance and autocorrelation. In general, we define the covariance between two random variables, X and Y, as

$$\text{Cov}(X, Y) = E[(X - E(X))(Y - E(Y))] \tag{3.4}$$

and their correlation is simply their covariances scaled for their standard deviations.

$$\text{Corr}(X, Y) = \frac{\text{Cov}(X, Y)}{\sqrt{\sigma_X^2}\sqrt{\sigma_Y^2}} \tag{3.5}$$

We can also visually investigate how correlated X and Y are by simply having a scatterplot of Y versus X. However, so far we have only talked about a single time series. So how can we define covariance or correlation for a single variable? Here is what we mean by that: Consider a time series data with 100 observations, take the 2nd to 100th observations and call these 99 observations z_t; then take the 1st to 99th observations and call these 99 observations z_{t-1}. Now from a set of 100 observations, we managed to create two variables z_t and z_{t-1} with 99 observations each. We can then consider the covariance and correlation between these two variables as we did earlier. The covariance between z_t and z_{t-1} can then be written as $\text{Cov}(z_t, z_{t-1}) = E[(z_t - \mu)(z_{t-1} - \mu)]$ (Note that owing to stationarity, $E(z_{t-1}) = E(z_t) = \mu$). We can also have a scatterplot of these new variables to see how successive observations in our time series data are correlated. We can similarly find the covariance between observations of k lags apart from $\text{Cov}(z_{t+k}, z_t) = E[(z_{t+k} - \mu)(z_t - \mu)]$, also called *autocovariance* since we are actually dealing with the same data set. Once again, owing to stationarity we have $\text{Cov}(z_{t_1+k}, z_{t_1}) = \text{Cov}(z_{t_2+k}, z_{t_2})$, making the autocovariance only a function of the time lag k. Therefore, we define the autocovariance function as

$$\gamma(k) = E[(z_{t+k} - \mu)(z_t - \mu)] \tag{3.6}$$

Note that the variance of the time series is $\gamma(0)$. Following Equation (3.5), we define the *autocorrelation function* (ACF) for a stationary time series as

$$\rho(k) = \frac{\gamma(k)}{\sqrt{\gamma(0)}\sqrt{\gamma(0)}} = \frac{\gamma(k)}{\gamma(0)} \tag{3.7}$$

The ACF plays an extremely crucial role in the identification of time series models as it summarizes as a function of k how correlated the observations that are k lags apart are. Of course in real life we cannot know the true value of ACF, but instead we will estimate it from the data at hand using

$$\hat{\gamma}(k) = \frac{1}{T} \sum_{t=1}^{T-k} (z_{t+k} - \bar{z})(z_t - \bar{z}) \tag{3.8}$$

and

$$\hat{\rho}(k) = \frac{\hat{\gamma}(k)}{\hat{\gamma}(0)} \tag{3.9}$$

It should be noted that as k increases, that is, we are considering the autocorrelation between observations further and further apart, the two variables we create, z_{t+k} and z_t, will have fewer and fewer observations since each will have $T - k$ observations. Therefore after a while, the estimates of the autocovariance and autocorrelation will get more and more unreliable. BJR, as a rule of thumb, suggest that the total number of observations T should be at least 50 and in the estimation of $\gamma(k)$ and $\rho(k)$, we should have $k \leqslant T/4$.

Also note that in Equation (3.8), the summation on the left hand side is divided by T and not $T - k$ as one may expect. In both cases, the estimates will be biased. But the former ensures that the estimate of the covariance matrix is a nonnegative definite matrix. See BD for details.

When dealing with point estimates we should also be aware of the uncertainty in the estimation and quantify it in the form of the standard error of the estimate or a confidence interval. For a general formula for the variance of the sample ACF, see Brockwell and Davis (1991). We will focus on a special case where z_t is independent and identically distributed (i.i.d.) with 0 mean and constant variance. In that case, $\rho(k)$ for $k > 0$ is equal to 0. However, we do not expect the sample ACF to be exactly 0. But if z_t is i.i.d., its sample ACF is expected to be quite small. In fact in this case, for large T, it turns out that $\hat{\rho}(1), \hat{\rho}(2), \ldots, \hat{\rho}(k)$ are approximately normally distributed with mean zero and standard deviation $1/T$. Therefore, $\pm 2/T$ is often used as the 95% confidence interval. If ACF at any lag k plots beyond these confidence limits, we suspect that the assumption of independent time series z_t is violated. This will be the primary tool that we will use in our modeling efforts in checking for the autocorrelation in the data as well as in the residuals obtained after a tentative model is employed.

3.1.5   Linear Processes

In Chapter 1, we briefly discussed the representation of a stationary process as the infinite weighted sum of random shocks. This representation constitutes the basis for the linear time series models that will be discussed in this book. In this section, we give the definition of a linear process as follows: if a (weakly) stationary process has the following representation

$$Z_t = \sum_{i=-\infty}^{\infty} \psi_i a_{t-i} \tag{3.10}$$

where a_t is white noise, that is, uncorrelated with zero mean and constant variance σ^2, and the weights $\{\psi_i\}$ are absolutely summable, that is, $\sum_{i=-\infty}^{\infty} |\psi_i| < \infty$, then it is called a *linear process*. It can further be shown that in general

$$Z_t = \sum_{i=-\infty}^{\infty} \psi_i x_{t-i} \tag{3.11}$$

is stationary, if the time series $\{x_t\}$ is stationary. In the case of a linear process where $\{x_t\}$ is white noise with zero mean and σ^2 variance, then z_t has the zero mean and autocovariance function

$$\gamma(k) = \sum_{j=-\infty}^{\infty} \psi_j \psi_{j+k} \sigma^2 \tag{3.12}$$

since a white noise sequence is stationary. We will leave the proof to the readers. We will frequently refer to this definition and its implications in the following sections.

It should be noted that there is a difference between independence and correlation. Two random variables are said to be independent if their joint probability density function (pdf) is equal to the product of the marginal pdfs of the individual variables, that is,

$$f(X, Y) = f(X) f(Y) \tag{3.13}$$

The more intuitive interpretation is that if two variables are independent, knowing the value of one does not give us any information about the possible value of the other one.

As for the correlation of two variables, we have

$$\text{Cor}(X, Y) = \frac{\text{Cov}(X, Y)}{\sigma_X \sigma_Y} \tag{3.14}$$

where the covariance is defined as

$$
\begin{aligned}
\mathrm{Cov}(X, Y) &= E[(X - \mu_X)(Y - \mu_Y)] \\
&= E[XY - X\mu_Y - Y\mu_X + \mu_X\mu_Y] \\
&= E[XY] - E[X\mu_Y] - E[Y\mu_X] + E[\mu_X\mu_Y] \\
&= E[XY] - \mu_X\mu_Y - \mu_X\mu_Y + \mu_X\mu_Y \\
&= E[XY] - \mu_X\mu_Y \\
&= E[XY] - E[X]E[Y] \quad\quad\quad\quad\quad\quad (3.15)
\end{aligned}
$$

Hence, if two variables are uncorrelated we have $E[XY] - E[X]E[Y] = 0$ or $E[XY] = E[X]E[Y]$. We can now show that if two variables are independent, they are also uncorrelated but the opposite is not always true. Assuming that X and Y are independent, then

$$
\begin{aligned}
E[XY] &= \iint xy f(x, y)\,dx\,dy \\
&= \iint xy f(x)f(y)\,dx\,dy \\
&= \left\{\int xf(x)\,dx\right\}\left\{\int yf(y)\,dy\right\} \\
&= E[X]E[Y] \quad\quad\quad\quad\quad\quad\quad (3.16)
\end{aligned}
$$

Therefore if X and Y are independent they are also uncorrelated. To show that the reverse is not always true, let X be a random variable with a symmetric pdf around 0, $E[X] = \mu_X = 0$, and $Y = |X|$. Clearly X and Y are not independent since knowing the value of X also determines the value of Y. However, $E[Y] = 2\int_0^\infty xf(x)dx$ and $E[XY] = \int_0^\infty x^2 f(x)dx - \int_{-\infty}^0 x^2 f(x)dx = 0$. Therefore $E[XY] = E[X]E[Y]$, which implies that the dependent X and Y are uncorrelated.

The Wold decomposition theorem covered in Chapter 1 requires the random shocks to be only uncorrelated and not necessarily independent. This decomposition is the foundation of many models we discuss in this book. Therefore, independence of the random shocks is not needed unless otherwise specified as in the calculations of the forecasts. Hence we generally assume that random shocks are *white noise*, which is defined as a sequence of uncorrelated random variables with zero mean and constant variance σ^2.

3.2 AUTOREGRESSIVE MOVING AVERAGE (ARMA) MODELS

3.2.1 Example: Temperature Readings from a Ceramic Furnace

In Figure 3.3, we provide 80 consecutive hourly temperature readings coming from a ceramic furnace. A keen observer will notice that Figure 3.3 shows a slow

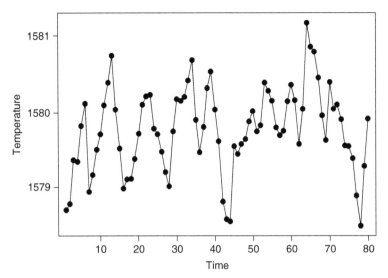

Figure 3.3 A times series plot of 80 consecutive hourly temperature observations from a ceramic furnace.

moving wavy pattern rather than a completely random one. Given the physics of the circumstances, this is to be expected. With a sampling rate of one observation per hour of the temperature of something extremely large such as a furnace, the temperature at time t and 1 h later at time $t + 1$ are obviously going to be related. Things do not change that fast!

In the following section, we present preliminary analysis techniques to investigate whether a process exhibits serial correlation.

3.2.2 Serial Correlation

For the present process, we can readily see that the observations are correlated over time. Figure 3.4 shows a panel of four plots (a) the lag one temperatures z_{t-1} versus the temperatures z_t (b) the lag 2 temperatures z_{t-2} versus the temperatures z_t, (c) the lag 3 temperatures z_{t-3} versus the temperatures z_t, and (d) the lag 4 temperatures z_{t-4} versus the temperatures z_t. From these four plots we see that observations one time unit apart are highly positively correlated, observations two time units apart are also correlated but less so, and after three and four time lags, they are more or less uncorrelated. As we mentioned earlier, since in essence there is only one variable in question, this kind of correlation is called *autocorrelation*.

An alternative representation of the autocorrelation is provided in Figure 3.5. In this plot, only the correlation coefficients for different lags, that is, ACF, are presented. Specifically, the correlation shown as scatter plots in Figure 3.4a between z_{t-1} and z_t is about 0.7, and the correlation between z_{t-2} and z_t shown in Figure 3.4b is about 0.3. These lag 1 and lag 2 autocorrelations are shown in Figure 3.5 as the two first bars of the length

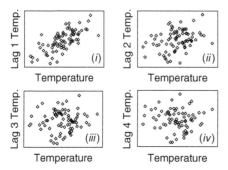

Figure 3.4 The correlation between observations (a) one time unit apart, (b) two time units apart, (c) three time units apart, and (d) four time units apart.

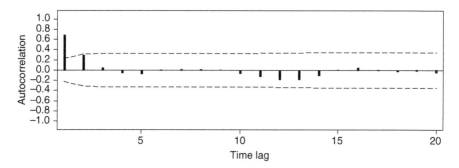

Figure 3.5 The autocorrelation function of the furnace temperature.

0.7 and 0.3, respectively. Further the two dotted horizontal lines in Figure 3.5 indicate 95% confidence intervals around 0 as discussed in Section 3.1.4. Thus, the lag 1 autocorrelation is clearly significantly different from 0, and the lag 2 autocorrelation is borderline significant.

3.2.3 Time Series Model for the Furnace Data

Stationary autocorrelated process data can often be modeled via an autoregressive moving average (ARMA) time series model. The identification of the particular model within this general class of models is determined by looking at the ACF and the partial autocorrelation function (PACF). We have already discussed ACF. In the following, we provide a brief description of PACF.

To appreciate what PACF is, consider a pth order autoregressive process, AR(p), which is the generalization of the AR(2) model discussed in Chapter 1. This model is like a regular regression equation except that we regress the current observations z_t on the past p-values, $z_{t-1}, \ldots, z_{t-p}$. That is, if $\tilde{z}_t = z_t - \mu$ is the current observation's deviation from the process mean, μ, then

$$\tilde{z}_t = \phi_1 \tilde{z}_{t-1} + \phi_2 \tilde{z}_{t-2} + \cdots + \phi_p \tilde{z}_{t-p} + a_t \tag{3.17}$$

where a_t is assumed to be "white noise" errors. The kth order partial autocorrelation measures the additional correlation between $\tilde{z}_t$ and $\tilde{z}_{t-k}$ after adjustments

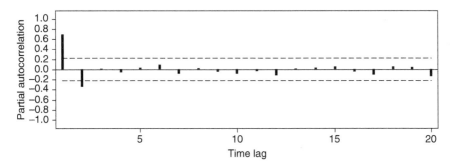

Figure 3.6   The partial autocorrelation function for the furnace data.

have been made for the intermediate observations $\tilde{z}_{t-1}, \tilde{z}_{t-2}, \ldots, \tilde{z}_{t-k+1}$. In other words, the lag k partial autocorrelation can be thought of as the last regression coefficient ϕ_{kk} if we fit regression equations for $k = 1, 2, \ldots$

$$\tilde{z}_t = \phi_{k1}\tilde{z}_{t-1} + \cdots + \phi_{kk}\tilde{z}_{t-k} + a_t \qquad (3.18)$$

to the data. Thus, by the time we fit too many terms, the partial autoregression coefficients ϕ_{kk} will approximately be 0. For example, if we fit an AR(2) model to data that truly follows an AR(2) model, then the regression coefficients ϕ_{kk} for lag $k = 3, 4, \ldots$ will be 0. It should be noted that for a given data we obtain the sample estimates for PACF at different lags. For an AR(p) process, the sample PACF values for $k > p$ are approximately independent and normally distributed with zero mean and variance $1/T$, where T is the number of observations. Therefore as in ACF, the approximate 95% confidence interval for sample PACF will be $\pm 2/\sqrt{T}$.

For the furnace data, the ACF in Figure 3.5 can be to some extent interpreted as a damped sine function. The PACF in Figure 3.6 shows that the first two partial autocorrelations are larger than the two standard error limits and hence are deemed significant. After that the PACF cuts off. To identify the particular type of ARMA model we use Table 3.2. From that table we see that a pattern of an exponentially decaying or sine wave decaying ACF and a PACF that cuts off after lag 2 suggest an AR (2) model. Therefore, we tentatively fit the following AR (2) model

$$\tilde{z}_t = \phi_1\tilde{z}_{t-1} + \phi_2\tilde{z}_{t-2} + a_t \qquad (3.19)$$

This model can also be written directly in terms of the data z_t. Thus, if we use the substitution $\tilde{z}_t = z_t - \mu$ we get

$$z_t - \mu = \phi_1(z_{t-1} - \mu) + \phi_2(z_{t-2} - \mu) + a_t$$
$$z_t = \mu - \phi_1\mu - \phi_2\mu + \phi_1 z_{t-1} + \phi_2 z_{t-2} + a_t$$
$$z_t = \text{constant} + \phi_1 z_{t-1} + \phi_2 z_{t-2} + a_t$$

where the constant $= \mu - \phi_1\mu - \phi_2\mu$ or $\mu = \text{constant}/(1 - \phi_1 - \phi_2)$.

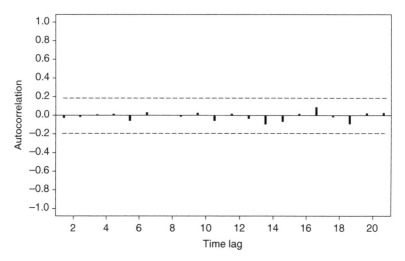

Figure 3.7 The ACF of the residuals after fitting an AR (2) model to the furnace data.

For this AR(2) model to be stationary, it is required that the coefficients satisfy the relations

$$\phi_2 + \phi_1 < 1$$
$$\phi_2 - \phi_1 < 1$$
$$-1 < \phi_2 < 1$$

Thus with $\hat{\phi}_2 + \hat{\phi}_1 = 0.61, \hat{\phi}_2 - \hat{\phi}_1 = -1.35$, and $\hat{\phi}_2 = -0.3722$ from Table 3.1, we conclude that the process is indeed stationary. We will come back to this in the next section.

In the AR(2) model in Equation (3.19), we assumed that the errors were uncorrelated with zero mean and constant variance σ_a^2. Hence, if AR(2) is an appropriate model for our data, we do not expect any remaining autocorrelations after the model is taken out of the data, that is, in the residuals. Therefore, we will consider ACF and PACF of the residuals to check for any remaining auto-correlation that was not taken out adequately by the model. Figures 3.7 and 3.8 show the ACF and the PACF for the residuals after fitting the AR (2) model to the furnace data. Both the ACF and the PACF are essentially zero for all lags.

Now that we have identified the furnace temperature to follow a stationary AR(2) process and confirmed that the model fits the data well, we can compute the process' variance. The variance (see BJR, p. 64) for a stationary AR(2) process with $\phi_1 = 0.9824$ and $\phi_2 = -0.3722$ is

$$
\begin{aligned}
\sigma_z^2 &= \left(\frac{1 - \phi_2}{1 + \phi_2}\right) \frac{\sigma_a^2}{\{(1 - \phi_2)^2 - \phi_1^2\}} \\
&= \left(\frac{1 + 0.3722}{1 - 0.3722}\right) \frac{\sigma_a^2}{\{(1 + 0.3722)^2 - 0.9824^2\}} \\
&= 2.38146\sigma_a^2
\end{aligned}
\tag{3.20}
$$

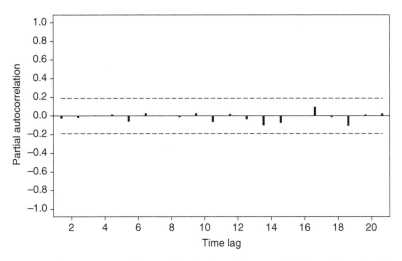

Figure 3.8   The PACF of the residuals after fitting an AR (2) model to the furnace data.

TABLE 3.1   Estimated Coefficients for an AR(2) Process

Coefficient	Estimate	Standard error	t-value	p-value
$\hat{\phi}_1$	0.9824	0.1062	9.25	0.000
$\hat{\phi}_2$	−0.3722	0.1066	−3.49	0.001
Constant	615.836	0.042	—	—
$\hat{\mu}$	1579.79	0.11	—	—
$\hat{\sigma}_a^2$	0.1403	—	—	—

Thus the variance of the furnace temperature process is approximately 2.4 times larger than the residual variability, σ_a^2. Further, from Table 3.1 we have that $\hat{\sigma}_a^2 = 0.1403$. Thus, $\hat{\sigma}_z = \sqrt{2.38146 \times 0.1403} \cong 0.5780$. This compares well with the estimate of the overall sample standard deviation, $s = \sqrt{\sum (z_t - \overline{z})^2/(80 - 1)} \cong 0.5685$.

3.2.4   ARMA(p, q) Models

Typical stationary time series models come in three general classes, autoregressive (AR) models, moving average (MA) models, or a combination of the two, the ARMA models. Let us denote the stationary time series by w_t. It should be noted that w_t can be the first difference of a nonstationary time series z_t, that is, $w_t = \nabla z_t = z_t - z_{t-1}$. We will discuss the nonstationary time series and the importance of the differencing operation in the next chapter. But for now, we will simply focus on w_t being a stationary time series. Autoregressive models relate current observations w_t to previous observations $w_{t-1}, w_{t-2}, \ldots, w_{t-p}$ the same way a regular regression model does. Hence, an AR(p) model is written as

$$w_t = \phi_1 w_{t-1} + \phi_2 w_{t-2} + \cdots + \phi_p w_{t-p} + a_t \qquad (3.21)$$

where a_t is an error term assumed to be uncorrelated with mean $E\{a_t\} = 0$ and constant variance, $\text{Var}\{a_t\} = \sigma_a^2$. The regression coefficients, $\phi_i, i = 1, \ldots, p$, are parameters to be estimated from the data.

The MA models are "averages" of past and present noise terms. An MA(q) model is written as

$$w_t = a_t - \theta_1 a_{t-1} - \theta_2 a_{t-2} - \cdots - \theta_q a_{t-q} \qquad (3.22)$$

where the coefficients $\theta_i, i = 1, \ldots, q$ are parameters to be determined from the data. Note that it is standard convention to use minus signs in front of the θs.

Finally, the combined ARMA(p, q) process is written as

$$w_t = \phi_1 w_{t-1} + \phi_2 w_{t-2} + \cdots + \phi_p w_{t-p} + a_t - \theta_1 a_{t-1}$$
$$-\theta_2 a_{t-2} - \cdots - \theta_q a_{t-q} \qquad (3.23)$$

The summary of the behaviors of the ACF and PACF for ARMA models is provided in Table 3.2. This table will play a crucial role in the model identification process. Note that ACF of an MA(q) process cuts off after lag q. In fact, ACF of an MA(q) process is given as

$$\rho_y(k) = \frac{\gamma_y(k)}{\gamma_y(0)} = \begin{cases} \dfrac{-\theta_k + \theta_1\theta_{k+1} + \cdots + \theta_{q-k}\theta_q}{1 + \theta_1^2 + \cdots + \theta_q^2} & k = 1, 2, \ldots, q \\ 0 & k > q \end{cases} \qquad (3.24)$$

This means that the sample ACF will be significant and hence plot outside the approximate 95% confidence interval given by $\pm 2/\sqrt{T}$ up to and including lag q. After lag q, sample ACF values are expected to be within the confidence limits for an MA(q) process.

TABLE 3.2   Summary of Properties of Autoregressive (AR), Moving average (MA), and Mixed Autoregressive moving average (ARMA) processes

	AR(p)	MA(q)	ARMA(p, q)
Model	$w_t = \phi_1 w_{t-1} + \ldots + \phi_p w_{t-p} + a_t$	$w_t = a_t - \theta_1 a_{t-1} - \ldots - \theta_q a_{t-q}$	$w_t = \phi_1 w_{t-1} + \cdots + \phi_p w_{t-p} - \theta_1 a_{t-1} + \ldots - \theta_q a_{t-q} + a_t$
Autocorrelation function (ACF)	Infinite damped exponentials and/or damped sine waves; Tails off	Finite; cuts off after q lags	Infinite damped exponentials and/or damped sine waves; Tails off
Partial autocorrelation function (PACF)	Finite; cuts off after p lags	Infinite; damped exponentials and/or damped sine waves; Tails off	Infinite damped exponentials and/or damped sine waves; Tails off

Source: Adapted from BJR.

ACF of an AR(p) model on the other hand satisfies the so-called Yule–Walker equations given as

$$\rho(k) = \sum_{i=1}^{p} \phi_i \rho(k - i) \quad k = 1, 2, \dots \tag{3.25}$$

It should be noted that the equations in Equation (3.25) are pth order linear difference equations. We will discuss linear difference equations in Chapter 7. It can be shown that the solution to Equation (3.25) is simply a mixture of exponential decay and damped sinusoid expressions depending on the ϕ_i's. But PACF of an AR(p) process will cut off after lag p causing the sample PACF values to plot within the confidence limits for lags greater than p. In Figure 3.9, we provide some simulated examples of ARMA models and their ACFs and PACFs. For further technical derivations and explanations about the behaviors of ACF and PACF of ARMA models, we refer the readers to BJR, BD, and Montgomery *et al.* (2008) (MJK hereafter).

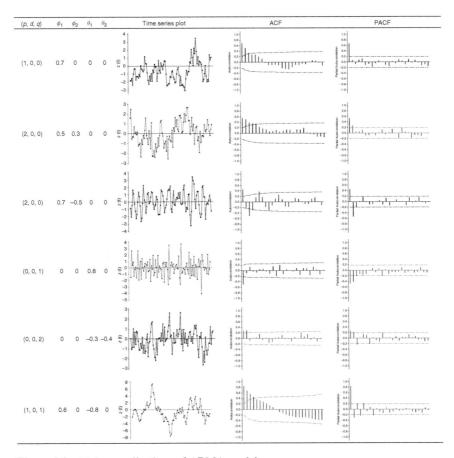

Figure 3.9 Various realizations of ARMA models.

From Table 3.2 and Figure 3.9, we can see that ACF and PACF are excellent tools in identifying the orders of MA and AR models, respectively. However, for ARMA models picking the right orders for MA and AR components are not that straightforward and ACF and PACF offer little help except for potentially revealing that the model we should entertain is not a pure MA or AR model. There are other tools besides ACF and PACF such as extended sample autocorrelation function (ESACF), generalized sample partial autocorrelation function (GPACF), and inverse autocorrelation function (IACF) that can be of help in determining the order of the ARMA model [see BJR and Wei (2006)]. However, since the model we initially entertain is a tentative model to start with, its residual analysis will often reveal if there is still autocorrelation not picked up by the model. It may also be possible to identify which term should be added to the model from the ACF and PACF plots of the residuals. In many respects, with currently available statistical software packages, fitting time series models requires only a few seconds, making it very easy to consider various models at once. However, the question of which model should be picked remains. We will address this issue in Chapter 6 where we discuss model identification in general.

3.3 STATIONARITY AND INVERTIBILITY OF ARMA MODELS

In Figure 3.10, we show four realizations of the AR(1) process,

$$z_t = \phi z_{t-1} + a_t \tag{3.26}$$

with $\phi = 0, 0.75, 1$, and 1.25. Recall that in Equation (3.26), $\{a_t\}$ is white noise and a_j and z_i are uncorrelated for $i < j$.

For $\phi = 0$, z_t is the same as a_t, making it white noise and hence stationary. For $\phi = 0.75$, we can see that the time series exhibits short sequences of up and down trends but always returns back to an equilibrium. The tendency of the time series to return back to a state of equilibrium indicates that we may have a stationary time series. We will now show this more explicitly. Consider the AR(1) model in Equation (3.24); we can write it for z_{t-1} as

$$z_{t-1} = \phi z_{t-2} + a_{t-1} \tag{3.27}$$

Combining Equations (3.26) and (3.27), we have

$$\begin{aligned} z_t &= \phi z_{t-1} + a_t \\ &= \phi(\phi z_{t-2} + a_{t-1}) + a_t \\ &= \phi^2 z_{t-2} + a_t + \phi a_{t-1} \end{aligned} \tag{3.28}$$

We can show that by replacing z_{t-2} with z_{t-3}, z_{t-3} with z_{t-4}, and so on, we will arrive at the following representation

$$\begin{aligned} z_t &= a_t + \phi a_{t-1} + \phi^2 a_{t-2} + \ldots \\ &= \sum_{i=0}^{\infty} \phi^i a_{t-i} \end{aligned} \tag{3.29}$$

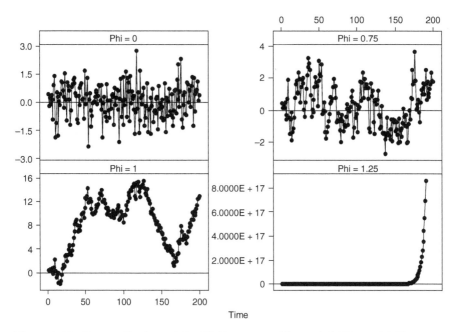

Figure 3.10 Four realizations of the AR(1) process with $\phi = 0, 0.75, 1$ and 1.25.

The solution for z_t given in Equation (3.29) shows that it can be written as the weighted sum of past disturbances defined by $\{a_t\}$. The weights in the summation follow an exponentially decaying pattern for $|\phi| < 1$, making z_t a linear process as described in Section 3.1.5 and hence stationary.

Now for $|\phi| > 1$, it is obvious that the time series becomes explosive as seen in Figure 3.10 for $\phi = 1.25$. BJR argue that except for some natural processes such as bacteria growth, this type of explosive or evolutionary behavior is not of practical interest and therefore they refrain from further discussing them and focus on cases where $|\phi| = 1$ and $|\phi| < 1$. From a theoretical stand point, the solution in Equation (3.29) will not converge for $|\phi| > 1$. However, consider for example z_{t+1} for an AR(1) process

$$z_{t+1} = \phi z_t + a_{t+1} \tag{3.30}$$

With little rearranging, we have

$$
\begin{aligned}
z_t &= \phi^{-1} z_{t+1} - \phi^{-1} a_{t+1} \\
&= \phi^{-1}(\phi^{-1} z_{t+2} - \phi^{-1} a_{t+2}) - \phi^{-1} a_{t+1} \\
&= \phi^{-2} z_{t+2} - \phi^{-1} a_{t+1} - \phi^{-2} a_{t+2} \\
&\ \ \vdots \\
&= -\phi^{-1} a_{t+1} - \phi^{-2} a_{t+2} - \phi^{-3} a_{t+3} - \ldots \\
&= -\sum_{i=1}^{\infty} \phi^{-i} a_{t+i} \tag{3.31}
\end{aligned}
$$

Since $|\phi| > 1 \rightarrow |\phi^{-1}| < 1$, the solution for z_t given in Equation (3.31) is in fact stationary. There is one caveat though. The solution in Equation (3.31) defines the current value of z_t as the weighted sum of future disturbances. This is practically quite useless since, for example, to make forecasts about the future, we need to know what will happen in the future. These types of future-dependent models are called *noncausal models*. So even though we have a stationary solution for an AR(1) process when $|\phi| > 1$, it is noncausal. Throughout our discussions in this book unless specified otherwise when we use the term stationary, we will refer to causal, stationary processes. In fact, it can be shown that an AR(1) process is nonstationary if and only if $|\phi| = 1$. Therefore for our discussions, the only stationary solution (worth considering) for an AR(1) process is when $|\phi| < 1$.

What is then the requirement for an AR(p) model to be stationary? Consider, for example, the AR(2) model

$$z_t = \phi_1 z_{t-1} + \phi_2 z_{t-2} + a_t \tag{3.32}$$

After rearranging Equation (3.32), we have

$$z_t - \phi_1 z_{t-1} - \phi_2 z_{t-2} = a_t \tag{3.33}$$

We will call the following second order equation in m the associated polynomial of Equation (3.33)

$$m^2 - \phi_1 m - \phi_2 = 0 \tag{3.34}$$

Clearly, the associated polynomial in Equation (3.34) has two roots. It can be shown that AR(2) process in Equation (3.32) is stationary, if the roots of the associate polynomial in Equation (3.34) are both <1 in absolute value. The roots can be complex conjugates, for example, $m_1, m_2 = a \pm ib$. In that case the magnitude of the complex conjugates, that is, $\sqrt{a^2 + b^2}$ should be <1 for the AR(2) process to be stationary. Hence an AR(2) process is stationary if the roots of the associated polynomial are within the unit circle in the complex plane. We will leave it to the reader to show that this condition is the same as the set of conditions we provided in Section 3.2.2 for the stationarity of an AR(2) process, namely,

$$\phi_2 + \phi_1 < 1$$
$$\phi_2 - \phi_1 < 1$$
$$-1 < \phi_2 < 1 \tag{3.35}$$

To generalize the above results for AR(p) process, an AR(p) process

$$z_t - \phi_1 z_{t-1} - \phi_2 z_{t-2} - \cdots - \phi_p z_{t-p} = a_t \tag{3.36}$$

is stationary (and causal) if the roots of the associate polynomial

$$m^p - \phi_1 m^{p-1} - \phi_2 m^{p-2} - \cdots - \phi_p = 0 \tag{3.37}$$

are all within the unit circle.

What about MA processes? From Section 3.1.5, it is easy to show that all finite order MA processes, that is, MA(q), are stationary. However, similar to the causality concept we discussed for the AR models, we have a similar situation for the MA processes. Consider for example, the MA(1) model

$$z_t = a_t - \theta a_{t-1} \tag{3.38}$$

By rearranging Equation (3.38), we have

$$a_t = z_t + \theta a_{t-1} \tag{3.39}$$

As in the AR(1) process, we can successively substitute a_{t-1} with $a_{t-1} = z_{t-1} + \theta a_{t-2}, a_{t-2}$ with $a_{t-2} = z_{t-2} + \theta a_{t-3}, \ldots$ to have the current disturbance a_t in terms of present and past observations

$$a_t = \sum_{i=0}^{\infty} \theta^i z_{t-i} \tag{3.40}$$

Hence a_t in Equation (3.40) can be presented as a *convergent* series of present and past observations if $|\theta| < 1$. The MA(1) process is then called *invertible*. As in the causality argument of the AR(1) process, for $|\theta| > 1, a_t$ can be expressed as a convergent series of future observations and therefore it will be called *noninvertible*. For $|\theta| = 1$, the MA(1) process is considered noninvertible in a more restricted sense. For technical details, see Brockwell and Davis (1991, Chapter 4). Another important implication of invertibility becomes evident in model identification. Consider the first lag autocorrelation of an MA(1) process given as

$$\rho(1) = \frac{-\theta}{1 + \theta^2} \tag{3.41}$$

Now when it comes to an MA(1) process this is all we got in terms of autocorrelation since for an MA(1) process $\rho(2) = \rho(3) = \ldots = 0$. If we are asked to calculate θ for a given $\rho(1)$, we rearrange Equation (3.41) as

$$\theta^2 - \frac{\theta}{\rho(1)} + 1 = 0 \tag{3.42}$$

and solve for θ. Clearly, there will be two solutions that will satisfy Equation (3.42), leaving the analyst in a dilemma. To make it a little less abstract, let $\rho(1) = 0.4$. Then both $\theta = -0.5$ and $\theta = -2$ satisfy Equation (3.42). However, only the MA(1) model with $\theta = -0.5$ gives us the invertible process, and hence that value is chosen. It can be shown that when there are multiple solutions as in Equation (3.42), there is only one solution that will satisfy the invertibility condition; see BJR Section 6.4.1.

We can extend the invertibility condition to MA(q) models as in the stationarity condition for the AR(p) models. An MA(q) process

$$z_t = a_t - \theta_1 a_{t-1} - \theta_2 a_{t-2} - \cdots - \theta_q a_{t-q} \tag{3.43}$$

is invertible if the roots of the associate polynomial

$$m^q - \theta_1 m^{q-1} - \theta_2 m^{q-2} - \cdots - \theta_q = 0 \tag{3.44}$$

are all within the unit circle. In general, an ARMA(p, q) process

$$z_t = \phi_1 z_{t-1} + \phi_2 z_{t-2} + \cdots + \phi_p z_{t-p} + a_t - \theta_1 a_{t-1}$$
$$-\theta_2 a_{t-2} - \cdots - \theta_q a_{t-q} \tag{3.45}$$

is stationary (and causal) if the roots of the associate polynomial

$$m^p - \phi_1 m^{p-1} - \phi_2 m^{p-2} - \cdots - \phi_p = 0 \tag{3.46}$$

are all within the unit circle and invertible if the roots of the associate polynomial

$$m^q - \theta_1 m^{q-1} - \theta_2 m^{q-2} - \cdots - \theta_q = 0 \tag{3.47}$$

are all within the unit circle.

3.4 CHECKING FOR STATIONARITY USING VARIOGRAM

Modeling the process and checking that the parameter estimates satisfy stationarity conditions is one way to test for stationarity. However, it would be desirable to have a simple exploratory tool to investigate whether a process is stationary without having to model it first. The variogram is such a tool.

3.4.1 Periodicities in Stationary Processes

Before we introduce the variogram, let us first discuss why data may exhibit cycles and specifically the reason for the apparent cyclic pattern in the data shown in Figure 3.3. We believe it is not as unusual as one might think. For that we refer back to Yule's article on Wölfers sunspot numbers that we discussed in Chapter 1. There Yule wanted to explain how periodicities in times series data may occur. To this end, he provided a nice physical illustration that may also be instructive in our case. First he reminded the reader that the horizontal movement of a pendulum "when left to itself" in a room can be modeled in continuous time as a second order homogenous differential equation. It is well-known that the solution to this second order differential equation is a damped sinusoidal curve. Yule then proceeded to explain that a second order autoregressive model is the discrete time equivalent of a second order linear differential equation with constant coefficients. But, as he explained, suppose "boys get into the room and start pelting the pendulum with peas, sometimes from one side and sometimes from the other." The pendulum, he argued, will then exhibit horizontal movements that will follow a second order autoregressive model with random inputs or shocks, a_t. In other words, the system will continue to exhibit second order dynamic behavior. But, rather than vary as a regular sinusoidal curve, "the graph will remain surprisingly smooth, but amplitude and phase will vary continuously."

It is often possible to approximate the dynamic behavior of a system with second order differential equations. Indeed, it does not seem far-fetched to imagine that our ceramic furnace with its very considerable inertia approximately

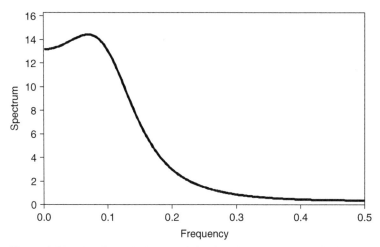

Figure 3.11 The theoretical spectral density of an AR (2) model with parameters similar to the ceramic furnace.

resembles the dynamic behavior of a second order system. Therefore, we can expect an oscillatory behavior analogous to a pendulum disturbed by naughty boys shooting peas at it.

Let us further explore the issue of inherent periodicities of a stationary time series. The frequency spectrum is a graph that shows the amplitude of the sinusoidal components for a range of frequencies. For the present set of parameters of the AR (2) model fitted to the furnace data, it can be shown (see BJR for details) that the theoretical frequency spectrum is as shown in Figure 3.11. In the present case, the spectrum shows a peak at a frequency around $f_0 \approx 0.07$ corresponding to a period $p = 1/f$ of approximately 14 h. Thus the prevalent cycle is about 14 h. This matches well with the pattern seen in the time series plot in Figure 3.3.

3.4.2   Stationary or Not Stationary?

In practice, there is no clear demarcation line between a stationary and a non-stationary process. A process that is nearly nonstationary will wander randomly away from its mean and not return for considerable time lags. On the basis of a finite record, it will therefore be difficult to distinguish such a process from one that is truly nonstationary.

The most basic assessment of stationarity is simply to plot a sufficiently long series of data to see whether the process appears stationary. Another approach is to plot the ACF. If the ACF does not dampen out within, say 15 to 20 lags, then the process is likely not stationary. However, the autocorrelation is, strictly speaking, not defined for nonstationary processes.

An alternative and more flexible tool is the variogram; for a detailed discussion, see BJR and Box and Luceño (1997). This tool is well defined and applicable to stationary as well as nonstationary processes. To understand how it works suppose $\{z_t\}$ is a time series. The variogram G_k measures the variance of differences k time units apart relative to the variance of the differences one time unit apart. Specifically, the variogram is defined as

$$G_k = \frac{V\{z_{t+k} - z_t\}}{V\{z_{t+1} - z_t\}}, \qquad k = 1, 2, \ldots \tag{3.48}$$

where G_k is plotted as a function of the lags, k.

For a stationary process,

$$G_k = (1 - \rho(k))/(1 - \rho(1)). \tag{3.49}$$

But for a stationary process, as k increases, $\rho(k) \to 0$. Hence G_k when plotted as a function of k will reach an asymptote $1/(1 - \rho(1))$. However, if the process is nonstationary, G_k will instead increase monotonically. For a stationary process with positive autocorrelation and no seasonality, the intuitive interpretation of the variogram is as follows: because of the positive autocorrelation, consecutive observations are expected to be similar. The variance of the first differences will therefore be less than the variance of the differences two lags apart, three lags apart, and so on. However, as k gets large the differences k lag and $k + 1$ lag apart will eventually be very similar. Thus, while the variogram for small k's tends to get larger as k increases, it will eventually reach an asymptote.

Figure 3.12 provides a plot of the theoretical variogram for the AR(2) model with parameter estimates $\hat{\phi}_1 = 0.9824$ and $\hat{\phi}_2 = -0.3722$. We see that the variogram after 8 or 9 lags seems to settle down, implying that the process is stationary.

In practice, we need to estimate the variogram directly from the time series data. Thus, we need an expression for the sample estimate of G_k. The literature is a bit ambiguous about how the sample variogram should be computed. However, Haslett (1997) has shown that a good estimate can be obtained by simply using the usual sample squared standard deviation applied to the differences with

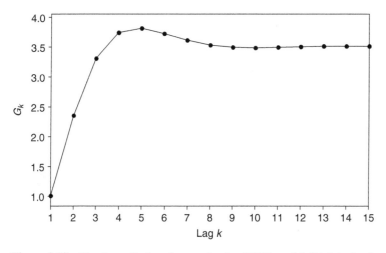

Figure 3.12   The theoretical variogram for the AR(2) model fitted to the furnace data.

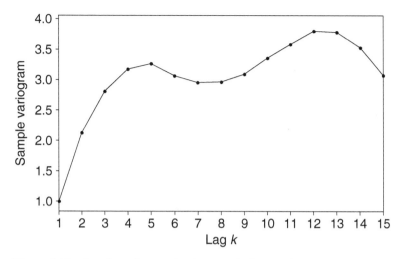

Figure 3.13 Sample variogram for the ceramic furnace data.

appropriate modifications for changing sample sizes that occur when we take differences. Specifically, he suggests using

$$\hat{V}\{z_{t+k} - z_t\} = s_k^2 = \frac{\sum_{t=1}^{n-k} (d_t^k - \overline{d}^k)^2}{n - k - 1} \tag{3.50}$$

where $d_t^k = z_{t+k} - z_t$ and $\overline{d}^k = (n - k)^{-1} \sum d_t^k$. Hence the sample variogram is given by

$$\hat{G}_k = \frac{s_k^2}{s_1^2}, m = 1, 2, \dots. \tag{3.51}$$

Figure 3.13 shows the sample variogram for the furnace data. As we should expect, the sample estimate differs somewhat from the theoretical variogram. However, aside from sample variations, the general appearance of the sample variogram does seem to indicate that it converges to a stable level confirming our previous assessment that the process most likely is stationary.

It should be noted that the sample variogram typically is not provided as a standard pull-down menu item by statistical software packages. However, it is relatively simple to compute. All that is needed is to compute successive differences for a number of lags and their sample variances. The ratios of these sample variances to the sample variance of the first differences produced the sample variogram given in Figure 3.13.

3.5   TRANSFORMATION OF DATA

An important preliminary step in any data analysis is to consider the possibility of a (nonlinear) data transformation. It is often the case that the scale in which the data naturally arrives to the data analyst is not necessarily the best scale to analyze the data in. For example, in some countries the mileage of a car is measured in miles per gallon and in others it is in liters per kilometer. Besides the obvious differences in the scale of measurement from gallons to liters and from miles to kilometers, we also need an inverse transformation $y = 1/x$ to convert one measure to the other. Which of the two scales is better? The answer depends on such things as the use and the structure of the data. Other transformations in common use are the Richter scale for measuring the strength of earthquakes and thermonuclear explosions where the amplitude of the needle's movement is converted to the base 10 logarithmic scale. The scale for measurement of sound in decibels is also the log scale.

A primary goal of the transformation is to identify a scale where the residuals after fitting a model will have homogeneous variability and be independent of the level of the time series. Another is a simplified data structure. The most common transformation used is the logarithmic transformation. In this section, we will first discuss the log transformation and then turn to a more general discussion of other useful transformations.

To appreciate the use of log transformations let us consider the US quarterly gross national product (GNP) for the period 1947–1966 plotted in Figure 3.14 as a time series plot. We see that the data trends upward and that the trend is not completely linear but somewhat concave sloping increasingly upward. Most people associate this curve shape with exponential growth. We also notice that the data spans a large range, from about 200 to about 800, relative to the variability

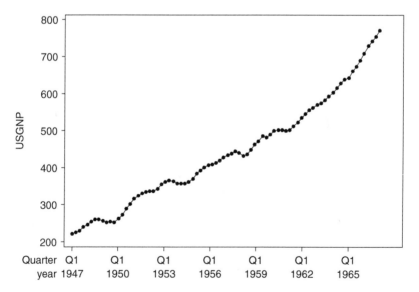

Figure 3.14 US quarterly gross national product (GNP) from 1947 to 1966.

around some imaginary slope line. This scenario is typical for data that may be better analyzed on a transformed scale.

In some cases, we may consider transformations that are suggested because theory or a logical argument informs us about how the variability in certain cases may behave. In other cases, we may apply transformations that are entirely empirically based and based on a trial and error of what seems to work in a given set of data. In fact practically in many cases, the two approaches lead to more or less the same transformation. For the GNP data we will proceed with a theory-based argument that leads to the logarithmic transformation.

As we indicated above, the scale the data arrive in is not necessarily the best scale for analysis. In the GNP case we get the absolute data but we could as well ask for the increases in GNP or more generally the changes in GNP. In some sense the change in GNP is really the interesting number, not the absolute number. The change in GNP tells us directly about productivity change in society and hence about how well the economy is doing. The absolute number is in some sense almost noninformative. For example, suppose you were told that the GNP in the third quarter of 1966 was \$756 billions (adjusted for 1973 dollars); you would probably have no idea what that meant and have no emotional reaction to that number. However, if you were told that the increase from the third to the fourth quarter of 1966 was 15, then you would know that the GNP went up and that was a good thing. However, even better would be if you were told that it went up by a proportion of 0.01984 or almost 2% from the third to the fourth quarter; you would naturally conclude that the economy was doing well. What we just did here was to transform the data into percentages and all of a sudden they made a lot more sense. We will now show how this transformation is related to the log transformation.

Now suppose the GNP y_t increases by a certain stationary percentage p_t over time. The percentage growth in GNP from time t to time $t + 1$ is then computed as

$$p_t = \frac{y_{t+1} - y_t}{y_t} \times 100. \tag{3.52}$$

Rewriting this relationship, we get

$$y_{t+1} = y_t + y_t(p_t/100) = (1 + p_t^*)y_t \tag{3.53}$$

where $p_t^* = p_t/100$ is the growth rate. This means the change in the successive observations is not constant but proportional to the level of the previous observation, $y_{t+1} - y_t = p_t^* y_t$. If the observations are trending upward in time, the differences between successive observations will also increase in time. This in turn will lead to an increase in the variability over time. Now suppose we transformed the data y_t using a logarithmic transformation $z_t = \log(y_t)$. Then, using the relationship $y_{t+1} = (1 + p_t^*)y_t$, the difference on the log scale between successive observations can be written as

$$z_{t+1} - z_t = \log(y_{t+1}) - \log(y_t)$$
$$= \log\left(\frac{y_{t+1}}{y_t}\right)$$
$$= \log\left(\frac{(1 + p_t^*)y_t}{y_t}\right)$$
$$= \log(1 + p_t^*) \qquad (3.54)$$

However, for small values of p_t^*, the logarithmic function is approximately a linear function. In other words, $\log(1 + p_t^*) \approx p_t^*$ and hence $z_{t+1} - z_t \approx p_t^*$. On the log scale the difference between successive observations therefore remains stable in time resulting in a more stable variance. Whenever it seems reasonable to think that a time series exhibits a more or less stationary percentage growth over time, it is often a good bet that a log transformation will make the variability more stable. We therefore proceed to take the log of the data. A word of caution: If not all data values are positive, an arbitrary positive number can be added to all data points to make them all positive.

The log transformed GNP data is given in Figure 3.15. Now the upward trend is more or less linear. To assess the changes from one year to another, we plot the first differences between successive log-transformed observations in Figure 3.16. Note how we can, by a simple intraocular test, see that the average growth rate was about 0.02 or 2% over the entire period. Moreover, Figure 3.16 tells us about recessions and high growth periods.

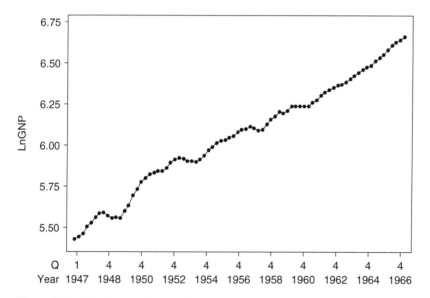

Figure 3.15 The log transformed GNP data.

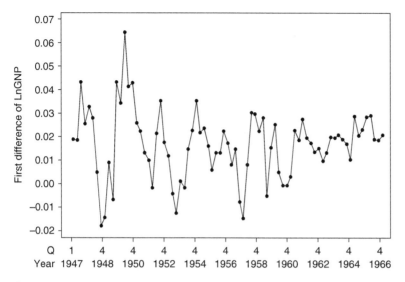

Figure 3.16 The first differences of the log transformed GNP data.

The log transformation is often an appropriate transformation, especially for economic and financial time series. However, sometimes we may want to consider a more general class of transformations. Box–Cox power transformations provide such an option and are given by

$$y^{(\lambda)} = \begin{cases} \dfrac{y^\lambda - 1}{\lambda \dot{y}^{\lambda - 1}}, \lambda \neq 0 \\ \dot{y} \ln y, \lambda = 0 \end{cases} \tag{3.55}$$

where $\dot{y} = \exp\left[(1/T) \sum_{t=1}^{T} \ln y_t\right]$ is the geometric mean of the observations and it is needed as a scale factor in order to be able to compare in a meaningful way different models obtained using different λ values based on their residual sum of squares. Then the λ value that yields the smallest residual sum of squares is chosen. We reconsider transformation of data in greater detail in Chapter 6 where we discuss seasonal models.

EXERCISES

3.1 Determine whether the following AR processes are stationary

    a. $z_t = 20 + 0.75 z_{t-1} + a_t$

    b. $z_t = 50 + 0.2 z_{t-1} - 0.4 z_{t-2} + a_t$

    c. $z_t = -10 + 0.3 z_{t-1} + 0.5 z_{t-2} + a_t$

    d. $z_t = 35 + 0.3 z_{t-1} + 0.7 z_{t-2} + a_t$

3.2 Determine the mean and the variance of the following AR processes, where a_t is the uncorrelated noise with zero mean and σ_a^2 variance

    a. $z_t = 40 + 0.5z_{t-1} + a_t$

    b. $z_t = 20 + 0.2z_{t-1} + 0.6z_{t-2} + a_t$

3.3 Determine whether the following ARMA processes are stationary and invertible

    a. $z_t = 10 + 0.7z_{t-1} + a_t - 0.5a_{t-1}$

    b. $z_t = -15 + a_t - 0.5a_{t-1} + 0.4a_{t-2}$

    c. $z_t = 75 - 0.2z_{t-1} + 0.5z_{t-2} + a_t + 0.6a_{t-1}$

    d. $z_t = 25 + 0.4z_{t-1} - 0.3z_{t-2} + a_t - 0.2a_{t-1} + 0.4a_{t-2}$

3.4 For the time series plot and corresponding ACF and PACF plots below, determine the orders p and q of a tentative ARMA(p, q) model that can be used for this data

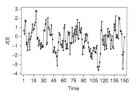 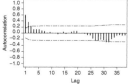 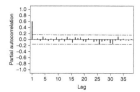

3.5 For the time series plot and corresponding ACF and PACF plots below, determine the orders p and q of a tentative ARMA(p, q) model that can be used for this data

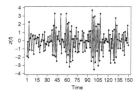 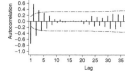 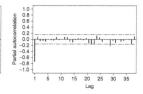

3.6 For the time series plot and corresponding ACF and PACF plots below, determine the orders p and q of a tentative ARMA(p, q) model that can be used for this data

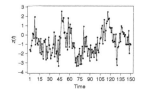 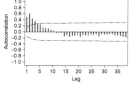 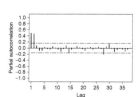

3.7 For the time series plot and corresponding ACF and PACF plots below, determine the orders p and q of a tentative ARMA(p, q) model that can be used for this data

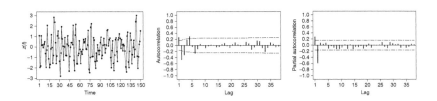

3.8 For the time series plot and corresponding ACF and PACF plots below, determine the orders p and q of a tentative ARMA(p, q) model that can be used for this data

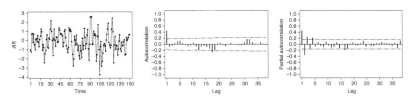

3.9 For the time series plot and corresponding ACF and PACF plots below, determine the orders p and q of a tentative ARMA(p, q) model that can be used for this data

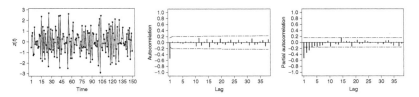

3.10 For the time series plot and corresponding ACF and PACF plots below, determine the orders p and q of a tentative ARMA(p, q) model that can be used for this data

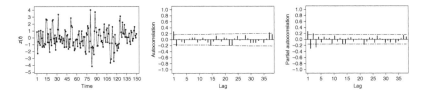

3.11 For the time series plot and corresponding ACF and PACF plots below, determine the orders p and q of a tentative ARMA(p, q) model that can be used for this data

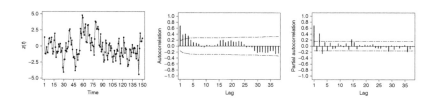

3.12 In a soft beverage bottling company, the data on the amount of beverage in each bottle is collected. Table B.1 contains 250 observations given in milliliters. Fit an appropriate ARMA model to this data and comment on your results.

3.13 In an oil refinery, the pressure of steam fed to a distillation column is tightly controlled and recorded. Table B.2 contains 120 pressure readings in bar taken every 10 min from a distillation column. Fit an appropriate ARMA model to this data and comment on your results.

3.14 A national bank started accepting electronic checks over the Internet in January 2006. Prior to that date only paper checks were accepted. A local branch collected the data on weekly number of paper checks processed at that branch from January 2004 to January 2008. The data is given in Table B.3. Consider only the first 2 years of that dataset and fit an appropriate ARMA model.

3.15 The monthly data on sea levels in Los Angeles, California, from 1889 to 1974 is given in Table B.4. Convert the data to yearly observations by taking the average sea level measurements of the 12 months of each year. Fit an appropriate ARMA model to the yearly data.

3.16 Table B.5 contains a dataset of 200 temperature readings from a chemical process. Fit (1) an ARMA(1, 1) model and (2) an appropriate AR(p) model to this data. Compare the two models.

3.17 For the models in Exercise 3.16, plot the impulse response functions and comment.

3.18 Show that the first lag autocorrelation of an MA(1) process is given as

$$\rho(1) = \frac{-\theta}{1 + \theta^2}$$

3.19 Show that the AR(2) process is (causal) stationary if

$$\phi_2 + \phi_1 < 1$$
$$\phi_2 - \phi_1 < 1$$
$$-1 < \phi_2 < 1$$

3.20 Show that the variance of a stationary AR(2) process is

$$\sigma_z^2 = \left(\frac{1 - \phi_2}{1 + \phi_2}\right) \frac{\sigma_a^2}{\{(1 - \phi_2)^2 - \phi_1^2\}}$$

3.21 Show that ACF of an MA(q) process is given as

$$\rho_y(k) = \frac{\gamma_y(k)}{\gamma_y(0)} = \begin{cases} \dfrac{-\theta_k + \theta_1\theta_{k+1} + \cdots + \theta_{q-k}\theta_q}{1 + \theta_1^2 + \cdots + \theta_q^2} & k = 1, 2, \ldots, q \\ 0 & k > q \end{cases} \qquad (3.56)$$

3.22 Show that an MA(q) process is always stationary.

3.23 Show that an AR(1) process is nonstationary when $|\phi| = 1$.

CHAPTER 4

NONSTATIONARY MODELS

4.1 INTRODUCTION

4.1.1 Example 1: Temperature Readings from a Chemical Pilot Plant

As mentioned in Chapter 3, the stationarity of most real life time series data can be observed for series with few observations and under specific conditions such as tight control actions taken on the process. Consider, for example, the well-known series C from BJR, which consists of 226 observations of the temperature of a chemical pilot plant sampled every minute. Typically, such a process will be equipped with an automatic feedback control system to keep the temperature more or less constant. However, if we observe a controlled process, we will not learn much about how the process functions. The controller will cancel out the natural dynamics of the process. Therefore, to study the process, the investigators in this case disconnected the control system for a few hours. The data is therefore sampled from the *uncontrolled* process.

A plot of the raw data is always the first step in any time series analysis. Figure 4.1 shows a time series plot of the temperature. To the trained eye, a number of interesting features are apparent. First, the data is clearly positively autocorrelated—subsequent temperature readings are close together. Second, we see that the process seems to be nonstationary. Indeed, it exhibits some large swings up and down. However, as George Box has often pointed out, the second law of thermodynamics dictates that if left alone, a process will eventually wander off and become nonstationary. In other words, stationarity is not natural! In the present case, the investigators disconnected the control system. We should therefore not be surprised that the process exhibits this natural nonstationary behavior. As a further feature, we notice a peculiar pattern around observations 58–60. We will return to this later.

4.2 DETECTING NONSTATIONARITY

Standard autoregressive moving average (ARMA) time series models as discussed in Chapter 3 apply only to *stationary* time series. Stationary processes are in

Time Series Analysis and Forecasting by Example, First Edition. Søren Bisgaard and Murat Kulahci.
© 2011 John Wiley & Sons, Inc. Published 2011 by John Wiley & Sons, Inc.

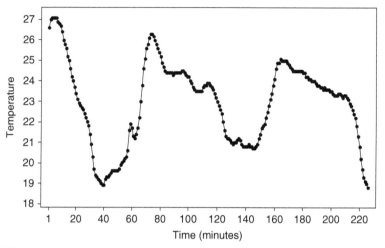

Figure 4.1 Time series plot of the temperature from a pilot plant observed every minute (series C from BJR).

equilibrium around a constant mean. However, we already noted that the process temperature, $\{z_t, t = 1, \ldots, 220\}$, does not appear to be stationary. We therefore need to investigate whether we can transform the data to a stationary form.

4.2.1 Transformation to Stationarity

There are many ways a time series can be nonstationary. The type of nonstationary behavior we typically encounter in many applications is of the type where the level changes, but the process nevertheless exhibits homogeneity in the variability. In such cases, the (first) difference, $\nabla z_t = z_t - z_{t-1}$, may be stationary. If so, we refer to the process as being (first order) homogenous nonstationary.

Another type of nonstationarity encountered in practice is when both the level and the slope of a time series are nonstationary, but the variability otherwise exhibits homogeneity. In that case, we may difference the data twice. The second difference is defined as $\nabla^2 z_t = \nabla(z_t - z_{t-1}) = (z_t - z_{t-1}) - (z_{t-1} - z_{t-2}) = z_t - 2z_{t-1} + z_{t-2}$. If the second difference is stationary and homogeneous, we refer to the process as a homogeneous nonstationary process of the second order. Of course, if second order differencing does not produce a stationary process, we can proceed to difference again. However, in practice, we seldom need to go beyond second order differencing.

The question now is whether a given time series is stationary. If not stationary, how many times do we need to difference the original data to obtain a stationary series? First, we emphasize that for a finite length time series, it is virtually impossible to determine whether the series is stationary or not from the data alone. Moreover, there is no clear demarcation between stationarity and nonstationarity. For a more detailed discussion, see BJR. With that in mind, let us see how to determine the degree of differencing necessary to produce a stationary time series in this case.

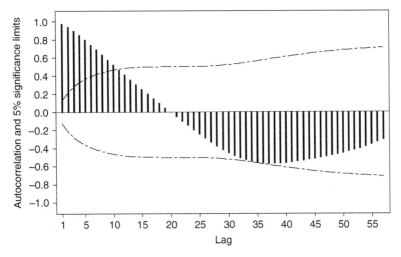

Figure 4.2 ACF for the temperature data.

The traditional tool for determining stationarity is the autocorrelation function (ACF) as discussed earlier. If a time series is nonstationary, the ACF will not die out quickly. In the present case, Figure 4.2 shows the ACF for the raw data. We see that the autocorrelation does not die out even for large lags.

We proceed to compute the autocorrelation for the first difference, $\nabla z_t = z_t - z_{t-1}$. The autocorrelation of the first difference, shown in Figure 4.3, is somewhat better behaved. However, to be on the safe side, we proceed to difference the data a second time. The autocorrelation for the twice-differenced data shown in Figure 4.4 is not much different from that of the first difference. To get a feel for the effect of differencing, Figure 4.5 shows, side by side, time series plots of the raw data as well as the first, second, and third differences. After the first

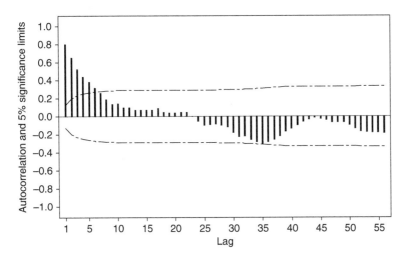

Figure 4.3 ACF of the first difference of series C.

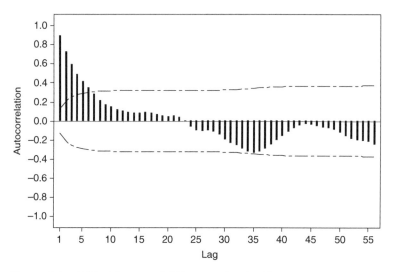

Figure 4.4 ACF of the second difference of series C.

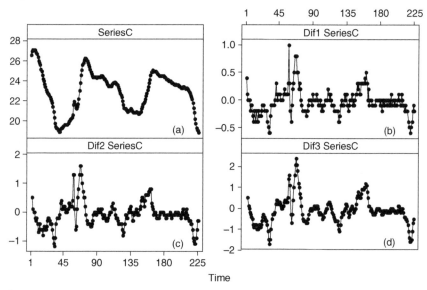

Figure 4.5 Time series plots of (a) the original data series C, (b) the first difference, (c) the second difference, and (d) the third difference.

difference, the data appears to be stationary, indicating that a single differencing may be enough. However, it could be argued that we may have to difference a second time. We will pursue this issue later.

4.2.2 Using the Variogram to Detect Nonstationarity

The traditional way of checking for stationarity is via the ACF, which has been illustrated in Section 4.2.1. An alternative tool is the variogram discussed in

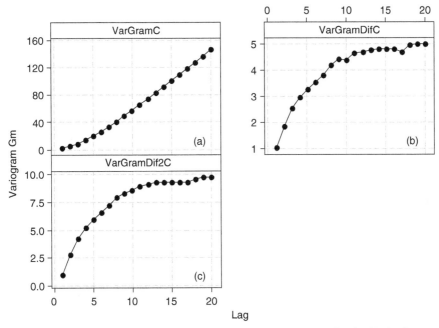

Figure 4.6 Sample variogram for (a) the raw temperature data, series C, (b) the first difference, and (c) the second difference.

Chapter 3. The appeal of the variogram is that, unlike the autocorrelation, it is defined (theoretically) for both stationary and (many) nonstationary processes, see for example, BJR. If a process is stationary, the variance of differences between observations far enough apart will stabilize. Hence, the variogram will flatten out and come to a stationary asymptote.

Figure 4.6 shows the estimated variogram from the original temperature data as well as those of the first and second differences. Panel a in Figure 4.6 shows that the variogram for the original temperature data continues to increase. This indicates that the process is nonstationary. However, both panels b and c show a leveling out of the variogram. This indicates that the first and the second differences may be stationary.

We (tentatively) conclude from both the autocorrelation and the variogram analyses that we need to difference the data once to achieve stationarity.

4.3 AUTOREGRESSIVE INTEGRATED MOVING AVERAGE (ARIMA) MODELS

In the case study presented above, the data is nonstationary. Differencing before we use the (stationary) ARMA model to fit the (differenced) data $w_t = \nabla z_t = z_t - z_{t-1}$ is therefore deemed appropriate. In general, we may need to difference d times. Thus, we may fit the dth difference, $w_t = \nabla^d z_t$, to a stationary ARMA model. Because the inverse operation of differencing is summing or integrating, an ARMA model applied to d differenced data is called an autoregressive *integrated* moving average process, ARIMA(p, d, q). In practice, the orders p, d, and q are seldom higher than 2.

Note that in the formulations of the AR(p), the MA(q), and the ARMA(p, q) models in Chapter 3, we have assumed that the "intercept" is zero. If the models are applied to stationary data, we subtract the average from the data before we model the process. This automatically makes the mean zero. If the model is applied to data that is nonstationary and hence needs to be differenced either once or twice, adding a nonzero "intercept" term to the model implies that there is an underlying deterministic first (linear) or second order polynomial trend in the data. A first or second order *deterministic* trend would require the process mean to evolve for future values according to a first or second order polynomial. In other words, the mean would never turn direction but continue to either increase or decrease. In our case, the physical context as well as the nonstationary behavior seen in Figure 4.1 does not suggest a deterministic trend. Indeed, Figure 4.1 indicates a *stochastic* trend. Thus, we assume the mean of the differenced data and the "intercept" term to be zero. Indeed, in most time series studies, it makes more sense to assume the more flexible stochastic trend model (no constant term), rather than a rigid deterministic trend. This is an important issue when selecting time series models using standard statistical software. The software may ask whether you wish to include a constant term. If you select a nonzero constant term, you are assuming a deterministic trend! We will come back to this issue in Chapter 7.

4.3.1 Model Identification and Estimation

In practice, we never know the type of model and specifically the orders p and q necessary to adequately model a given process. We need to determine the model

TABLE 4.1 Summary of Properties of Autoregressive (AR), Moving Average (MA), and Mixed Autoregressive Moving Average (ARMA) Processes

	AR(p)	MA(q)	ARMA(p, q)
Model	$w_t = \phi_1 w_{t-1} + \ldots$ $+ \phi_p w_{t-p} + a_t$	$w_t = a_t - \theta_1 a_{t-1} - \ldots$ $- \theta_q a_{t-q}$	$w_t = \phi_1 w_{t-1} + \cdots +$ $\phi_p w_{t-p} - \theta_1 a_{t-1}$ $+ \ldots - \theta_q a_{t-q} + a_t$
Autocorrelation function (ACF)	Infinite; damped exponentials and/or damped sine waves; Tails off	Finite; cuts off after q lags	Infinite; damped exponentials and/or damped sine waves; Tails off
Partial autocorrelation function (PACF)	Finite; cuts off after p lags	Infinite; damped exponentials and/or damped sine waves; Tails off	Infinite; damped exponentials and/or damped sine waves; Tails off

Source: Adapted from BJR.

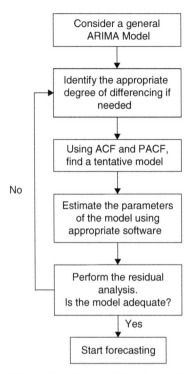

Figure 4.7 Stages of the time series model building process using ARIMA (Adapted from BJR, p. 18).

that best fits the data based on a look at the ACF and the partial autocorrelation function (PACF). Although we had discussed ACF and PACF in Chapter 3, we revisit these in Table 4.1, which summarizes how to interpret the ACF and the PACF. Further, since time series modeling requires judgment and experience, BJR suggested an iterative model building approach as shown in Figure 4.7.

We now demonstrate how the iterative model building process works. First of all, we can see from ACF plots of the first and second differences in Figures 4.3 and 4.4 that the time series needs to be differenced only once. With reference to Table 4.1, we can also see that the ACF of the first difference looks (somewhat) like a damped sine wave. The PACF for the first difference is shown in Figure 4.8. Again, with reference to Table 4.1, we see that the PACF cuts off after lag 1. This indicates that the differenced process $w_t = \nabla z_t = z_t - z_{t-1}$ can be modeled as an AR(1) process. Thus, the model we have tentatively identified is $w_t = \phi_1 w_{t-1} + a_t$, where $w_t = \nabla z_t = z_t - z_{t-1}$. Substituting the latter expression into the former, we get $z_t - z_{t-1} = \phi_1(z_{t-1} - z_{t-2}) + a_t$ or $z_t = (1 + \phi_1)z_{t-1} - \phi_1 z_{t-2} + a_t$.

4.3.2 Parameter Estimation

Now that we have "identified a model to be entertained," we proceed to estimate the parameters of that model. Depending on the software package and the method

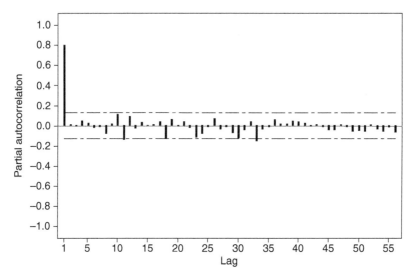

Figure 4.8 PACF for the differenced chemical process data.

TABLE 4.2 Estimation Summary from Fitting an AR(1) Model to the First Difference of the Chemical Process Data — Series C

Type	Coefficient	SE of coefficient	t	p
AR(1): $\hat{\phi}_1$	0.8239	0.0382	21.55	0.000

of estimation used, we should expect to obtain slightly different results when fitting a time series model. For further details on estimation methods, see BJR. Table 4.2 shows the summary statistics from fitting the AR(1) model to the differenced data. Furthermore, the residual sum of squares is 4.03183 with 224 degrees of freedom and therefore the mean square of the residuals is 0.01800. In other words, $\hat{\sigma}^2_{a_t} = 0.018$.

4.3.3 Model Diagnostic Checking

Once a model has been fitted to the data, we proceed to conduct a number of diagnostics checks. If the model fits well, the residuals should essentially behave like white noise. In other words, the residuals should be uncorrelated with constant variance. Moreover, in modeling, we often assume that the errors are normally distributed; hence, we expect the residuals to be more or less normally distributed. If not, it is a sign that something is amiss.

Standard checks are to compute the ACF and the PACF of the residuals as shown in Figure 4.9. Although there are a few autocorrelations and partial auto-correlations that are larger than the 5% significance limits, Figure 9a and b show no particularly alarming indications that the model does not fit well. To further check that the estimated autocorrelations $r_k(a), k = 1, 2, \ldots, K$, taken as a whole, do not indicate a model inadequacy, we compute the modified Ljung–Box–Pierce

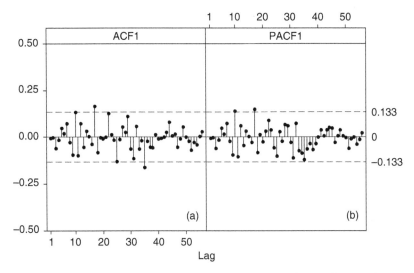

Figure 4.9   (a) ACF and (b) PACF of the residuals with 5% significance limits.

TABLE 4.3   Modified Ljung–Box–Pierce Chi-Square Statistic for the Residuals

Lag K	12	24	36	48
Chi square: $\tilde{Q}$	13.0	27.0	49.2	53.9
Degrees of freedom	11	23	35	47
p-values	0.292	0.254	0.056	0.229

chi-square statistics (Table 4.3),

$$\tilde{Q} = n(n+2) \sum_{k=1}^{K} (n-k)^{-1} r_k^2(\hat{a}) \tag{4.1}$$

This statistic is approximately distributed with a chi-square distribution with $(K - p - q)$ degrees of freedom, $\chi^2(K - p - q)$. Thus, if it is larger than expected, it indicates that the residuals are not independent. In the present case, there is a hint of a problem at lag 36 where the p-value is only 0.056. Otherwise, there does not seem to be a real problem.

Further diagnostics checking can be done by looking at the residuals in various ways. Figure 4.10a shows a normal plot of the residuals. If the residuals are normally distributed, they should all more or less lie on a straight upward sloping line. In the present case, most of the residuals look fine except for observations 58–60 that seem to be outliers. Figure 4.10b shows the residuals versus the fitted values. This plot should show a random pattern within a parallel horizontal band. In this case, most of the points behave that way except for observations 58–60. Figure 4.10c is a simple histogram of the residuals, which is not very useful but nonetheless a standard output. Figure 4.10d is a time series plot of the residuals.

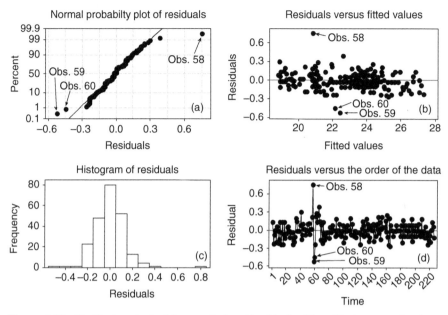

Figure 4.10 Residual analysis: (a) normal plot of residuals, (b) residuals versus fitted values, (c) a histogram of the residuals, and (d) time series plot of the residuals.

Again, we see that observations 58–60 look a bit suspicious. Of course, what we see around observations 58–60 as outliers is the peculiar episode in the data we already noticed in Figure 4.1 where the process seemed to have a bit of a hiccup.

4.3.4 Outliers

Except for the outliers around observations 58–60, we seem to have identified and estimated a good time series model for the chemical process temperature, series C. However, the outliers should worry us. For example, we may ask how much influence the outliers may have had on the final model estimates, or even more fundamentally, how much influence the outliers could have had on the selected model type.

The first thing one should do in practice, of course, is to talk to the process engineers and try to investigate the nature of this process abnormality. Was it a real process upset or a misreporting of the data? Could it be a problem with a temperature gauge? Did someone do something to the process around this time? If so, what did they do? We should always ask probing questions such as when, where, what, who, and how. One can imagine a number of scenarios. Unfortunately, in the present case, we are dealing with old data. Therefore, we cannot pursue this line of inquiry any further, but if this were your process, we would encourage you to do so.

There are a few general statistical approaches to investigate the effect of the outliers in time series that we briefly discuss now. First, there are two general types of outliers in time series. Each has different signatures. An additive outlier (AO) is an isolated outlier—a single data point—which does not seem to fit the normal pattern. This type of outlier is typically a recording error or gauge error at a single instance. Another type of outlier, called *innovation outlier* (IO), occurs as an outlier in the white noise process a_t at time t. You may recall that in Chapter 1 we discussed how an AR model can physically be thought of as similar to a pendulum being set in motion by random shots from a pea shooter. In this case, we can think of the outlier being produced by a particularly large pea hitting the pendulum at high speed. The effect of such an IO will tend to linger around and upset the general dynamics of the system for some time after the incident. Given the pattern in the data, we would be inclined to think our outlier is of the latter type. One useful way to look at the data to identify the type of outliers is to produce a scatter plot as in Figure 4.11 of the differenced data, w_t versus the lagged differenced data one time unit earlier w_{t-1}. This plot shows the lag 1 autocorrelation. We see that the consecutive observations 58, 59, and 60 all deviate from the general elliptical pattern one would otherwise expect to find.

There are a number of approaches to dealing with outliers in time series—some sophisticated and others simple commonsense type. Here, we will discuss a few of the latter type; see BJR Chapter 12 for a more detailed analysis. In the present case, we have 226 observations and the upset happened at observation 58. Thus, we have enough data to identify and fit separate time series models to the data both before and after the upset. We will not show all the details but just report that the appropriate model seems to be the same as

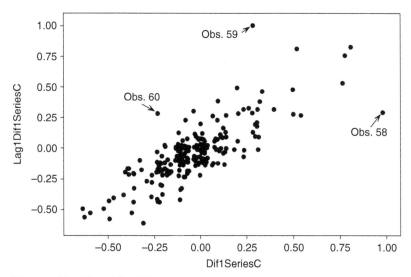

Figure 4.11 Plot of the differenced process data versus the same differenced data lagged by one time unit (jitters added).

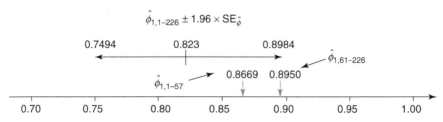

Figure 4.12 Estimates of the autoregressive parameter $\hat{\phi}_{1,1-226}$ and a 95% confidence interval based on the entire data set as well as parameter estimates $\hat{\phi}_{1,1-57}$ for the first 57 observations before the process upset and the parameter estimates $\hat{\phi}_{1,61-226}$ based on the observations 61–226 after the process upset.

we identified for the entire data and that the estimated coefficients differ only slightly. Indeed, in Figure 4.12, we show graphically the estimated coefficient $\hat{\phi}_{1,1-226} = 0.8239$ for the entire dataset and a 95% confidence interval. Superimposed are $\hat{\phi}_{1,1-57} = 0.8950$, the estimate from the first 57 observations, and $\hat{\phi}_{1,61-226} = 0.8669$, the estimate based on observations 61–226. These two estimates do deviate from the original estimate. Thus, the outliers did have an effect on the estimated coefficient, but they fall within a 95% confidence interval. A more significant difference was in the estimated residual variance. For the entire data set, $\hat{\sigma}_{a,1-226}^2 = 0.018$. However, for the first 57 observations, $\hat{\sigma}_{a,1-57}^2 = 0.0148$ and for observations 61–226 it was $\hat{\sigma}_{a,61-226}^2 = 0.0121$. In the latter case, this is an approximately 32% reduction in variance.

We also replaced observations 58–60 by linear interpolation between observations 57 and 61. This yielded a new estimate of ϕ_1 that also fell within the 95% confidence interval. However, linear interpolation is not the best approach. Better methods are discussed in Chapter 7.

We conclude this section feeling reasonably safe that a good model for the temperature of the chemical process seems to be the nonstationary ARIMA(1, 1, 0) model

$$z_t = 1.824 z_{t-1} - 0.824 z_{t-2} + a_t \tag{4.2}$$

4.3.5 Alternative Model

We indicated above that time series analysis typically does not yield a single definitive model. In the present case, an alternative may be to difference the data twice. There are several indications that second order differencing may be appropriate. If the data only needed to be differenced once, we would expect the variogram of the original data shown in Figure 4.6a to increase linearly. However, it shows some slight curvature, like a second order polynomial, indicating that a second differencing might be appropriate. Further, in the final model fitted to the first difference discussed above, the AR parameter $\hat{\phi}_1 = 0.824$ was close to 1. Now consider our final model $z_t = (1 + \phi_1) z_{t-1} - \phi_1 z_{t-2} + a_t$. We can rewrite this as $(z_t - z_{t-1}) = \phi_1 (z_{t-1} - z_{t-2}) + a_t$ or as $(z_t - z_{t-1}) - \phi_1 (z_{t-1} - z_{t-2}) =$

a_t. Now, suppose $\phi_1 = 1$. Then we would have $(z_t - z_{t-1}) - (z_{t-1} - z_{t-2}) = a_t$. Using the notation introduced above, this can be written as $\nabla^2 z_t = a_t$. In other words, the second order difference is white noise or the temperature could be modeled as an ARIMA(0, 2, 0) process.

4.4 FORECASTING USING ARIMA MODELS

One of the main goals in time series modeling is to gain the ability to make forecasts about the future on the basis of the model that has been entertained. In this section, we will see how we can make forecasts based on an ARIMA(p, d, q) model given as:

$$z_t = \sum_{i=1}^{p+d} \varphi_i z_{t-i} + a_t - \sum_{i=1}^{q} \theta_i a_{t-i} \tag{4.3}$$

Note that we switched from using ϕ's to φ's in Equation (4.3) because that model also includes the differencing. To see the relationship between ϕ's and φ's, consider the ARIMA(2, 1, 1) process:

$$w_t = \phi_1 w_{t-1} + \phi_2 w_{t-2} + a_t - \theta_1 a_{t-1} \tag{4.4}$$

where $w_t = (1 - B)z_t$. Replacing w_t with $(1 - B)z_t$, we get the following equation:

$$(1 - B)z_t = \phi_1(1 - B)z_{t-1} + \phi_2(1 - B)z_{t-2} + a_t - \theta_1 a_{t-1} \tag{4.5}$$

or

$$\begin{aligned} z_t &= z_{t-1} + \phi_1 z_{t-1} - \phi_1 z_{t-2} + \phi_2 z_{t-2} - \phi_2 z_{t-3} + a_t - \theta_1 a_{t-1} \\ &= (1 + \phi_1)z_{t-1} - (\phi_1 - \phi_2)z_{t-2} - \phi_2 z_{t-3} + a_t - \theta_1 a_{t-1} \end{aligned} \tag{4.6}$$

Hence in the notation used in Equation (4.3), we have

$$\begin{aligned} \varphi_1 &= (1 + \phi_1) \\ \varphi_2 &= -(\phi_1 - \phi_2) \\ \varphi_3 &= -\phi_2 \end{aligned} \tag{4.7}$$

Now, our goal is to make a forecast at time t about an observation l time units ahead, z_{t+l}. We will denote the forecast $\hat{z}_t(l)$. In order to judge how good our forecast is, we will consider the mean square error of the forecast defined as $E[(z_{t+l} - \hat{z}_t(l))^2]$. It turns out that the forecast that minimizes the mean square error is the conditional expectation of z_{t+l} at time t, that is, $E_t[z_{t+l}|z_t, z_{t-1}, \ldots]$. To arrive at this conclusion, we have to assume that $E_t[a_{t+j}|z_t, z_{t-1}, \ldots] = 0$ for $j > 0$ and for that we assume that $\{a_t\}$s are independent. Note that this is different than our usual assumption of uncorrelated $\{a_t\}$s that we have been making so far. For most cases, it will be safe to make the independence assumption. However, as indicated in BJR, the independence assumption may not hold for "certain types of intrinsically nonlinear processes."

Now we figured out that the best forecast of a future observation in the mean square error sense is what we would expect it to be at time t, given the data up to and including time t. While this makes intuitive sense, our next task is to actually find out what this expectation is. There are several approaches to this problem. We will pick what we perceive as the simplest approach where we use the difference equation as in Equation (4.3). In the following equation, we will adopt the notation used in BJR and define $[z_{t+l}]$ as the conditional expectation of z_{t+l}, that is, $E_t[z_{t+l}|z_t, z_{t-1}, \ldots]$ and $[a_{t+l}]$ as the conditional expectation of $[a_{t+l}]$, that is, $E_t[a_{t+l}|z_t, z_{t-1}, \ldots]$. Hence, our forecast for z_{t+l} is as follows:

$$\hat{z}_t(l) = [z_{t+l}] = \sum_{i=1}^{p+d} \varphi_i [z_{t+l-i}] + [a_{t+l}] - \sum_{i=1}^{q} \theta_i [a_{t+l-i}] \tag{4.8}$$

The actual calculation of the forecasts is done based on the following rules:

1. The conditional expectations of the present and past observations are actually themselves. That is, $[z_{t-j}] = z_{t-j}$ for $j = 0, 1, 2, \ldots$
2. The conditional expectations of the future observations are replaced by their forecasts. That is, $[z_{t+j}] = \hat{z}_t(j)$ for $j = 1, 2, \ldots$
3. The conditional expectations of the present and past shocks are actually themselves. But since these random shocks are not observed, we replace them by one-step-ahead forecast errors. That is, $[a_{t-j}] = e_{t-j}(1) = z_{t-j} - \hat{z}_{t-j-1}(1)$ for $j = 0, 1, 2, \ldots$
4. The conditional expectations of the future shocks are replaced by zero as we assumed earlier. That is, $[a_{t+j}] = 0$ for $j = 1, 2, \ldots$

Now as an example, consider the model in Equation (4.2) that we had for the temperature of the chemical process.

$$z_t = 1.824 z_{t-1} - 0.824 z_{t-2} + a_t$$

Now following the rules we stated above, the one-step-ahead forecast, $\hat{z}_t(1)$, can be written as:

$$\hat{z}_t(1) = 1.824[z_t] - 0.824[z_{t-1}] + [a_{t+1}]$$
$$= 1.824 z_t - 0.824 z_{t-1} \tag{4.9}$$

Similarly, two-step-ahead forecast $\hat{z}_t(2)$ is

$$\hat{z}_t(2) = 1.824[z_{t+1}] - 0.824[z_t] + [a_{t+2}]$$
$$= 1.824 \hat{z}_t(1) - 0.824 z_t \tag{4.10}$$

Note that we replaced $[z_{t+1}]$ with its forecast. Then in general, the l-step-ahead forecast for the temperature of the chemical process for $l = 3, 4, \ldots$ is given by

$$\hat{z}_t(l) = 1.824 \hat{z}_t(l-1) - 0.824 \hat{z}_t(l-2) \quad l = 3, 4, \ldots \tag{4.11}$$

The next question is of course whether we can somehow quantify the uncertainty in our predictions, for example, in the form of prediction intervals.

If we assume that a_t is normally distributed, it can be shown that the l-step-ahead prediction errors, $e_t(l) = z_{t+l} - \hat{z}_t(l)$, are also normally distributed with mean zero and variance $\sigma_e^2(l)$. Therefore, the 95% prediction interval for z_{t+l}, for example, is given as

$$\hat{z}_t(l) \pm 1.96\sigma_e(l) \tag{4.12}$$

To calculate $\sigma_e(l)$, we use the infinite weighted sum of current and past shocks representation of z_{t+l}

$$z_{t+l} = a_{t+l} + \psi_1 a_{t+l-1} + \psi_2 a_{t+l-2} + \dots \tag{4.13}$$

We can show that the weights $\{\psi_t\}$ can be calculated from the following relationship

$$(1 - \varphi_1 B - \dots - \varphi_{p+d} B^{p+d})(1 + \psi_1 B + \psi_2 B^2 + \dots)$$
$$= (1 - \theta_1 B - \dots - \theta_q B^q) \tag{4.14}$$

with

$$\psi_1 - \varphi_1 = -\theta_1$$
$$\psi_2 - \varphi_1 \psi_1 = -\theta_2$$
$$\vdots$$

Furthermore, we can also show that

$$e_t(l) = z_{t+l} - \hat{z}_t(l) = a_{t+l} + \psi_1 a_{t+l-1} + \psi_2 a_{t+l-2} + \dots + \psi_{l-1} a_{t+1} \tag{4.15}$$

and

$$\text{Var}(e_t(l)) = \sigma_e^2(l) = (1 + \psi_1^2 + \psi_2^2 + \dots + \psi_{l-1}^2)\sigma_a^2 \tag{4.16}$$

where $\hat{z}_t(l)$ is the minimum mean square error forecast. Most statistical software packages with time series analysis capabilities will automatically generate prediction intervals when doing forecasting. In the next section, we go through another example from BJR and consider modeling and forecasting issues using ARIMA models.

4.5   EXAMPLE 2: CONCENTRATION MEASUREMENTS FROM A CHEMICAL PROCESS

In the following sections, the time series we will consider consists of 197 consecutive concentration observations from a full-scale chemical process observed every 2 h for $2 \times 197 = 394$ h. The data set is series A from BJR provided in the Appendix A. The purpose of the study was to develop a statistical control strategy to help the process operators keep the concentration as close to a fixed target as possible.

While the data was sampled, the process operators manipulated another variable to approximately cancel out the variation in the output. From a process modeling standpoint, such control actions are usually counterproductive. To develop a control strategy, we need to observe the process dynamics when the process is left free to change on its own. To appreciate why this is important, consider for a moment the temperature in a room carefully controlled by a thermostat. If the thermostat is doing its job, the temperature will essentially be constant with only minor random fluctuations around the mean. Studying such a carefully controlled process will provide no information about how to keep the process in control.

Fortunately, in the present case, the history of each of the individual control manipulations was carefully recorded. Thus, it was possible afterward to reconstruct the record of the concentration readings as if the manipulations had not been imposed. This essentially allows us to see what the process would have looked like, had it been left alone.

Figure 4.13 shows the time series plot of the 197 concentrations from the chemical process. To a trained eye, it is immediately seen that the data is positively autocorrelated and possibly nonstationary. Indeed, one of our objectives in this book is to show how to look at plots of time series data to get a feel for what it means to be autocorrelated. Thus for comparison, Figure 4.14 shows 197 simulated normally distributed random and independent observations with the same average and standard deviation as the data in Figure 4.13. It can be seen that the independent data in Figure 4.14 fluctuates more randomly, have a more ragged appearance, and show no particular pattern. For comparison, the data in Figure 4.13 exhibits a tendency for a high observation to be followed by a high observation and a low observation to be followed by a low observation. Indeed, the data in Figure 4.13 has a somewhat more smooth or wavy appearance compared to the data in Figure 4.14, which is typical for positively autocorrelated data. Further, if you look at Figure 4.13, imagine a freehand line tracing the center

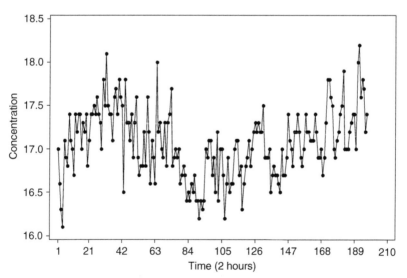

Figure 4.13 Time series plot of chemical process concentration readings sampled every 2 h: series A from BJR.

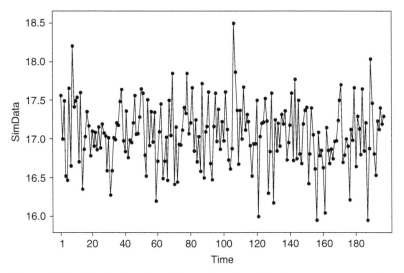

Figure 4.14   Simulated random independent data with the same average and standard deviation as in Figure 4.13.

of the data as you move across the graph, it seems as if the (short-term) mean changes, first moving up, then down for a while, and then back up again. This indicates that the process may be slightly nonstationary.

4.5.1   Preliminary Time Series Analysis

Now that we have conjectured that the data is autocorrelated and possibly non-stationary, we proceed to compute the sample ACF shown in Figure 4.15. The

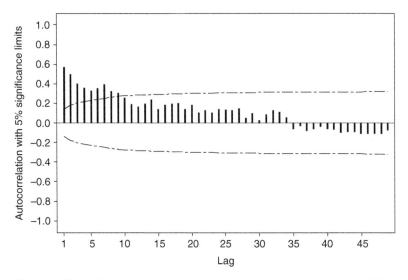

Figure 4.15   ACF of the chemical process concentration: series A from BJR.

most noticeable feature in Figure 4.15, beside the positive autocorrelation, is that the autocorrelation does not die out even for large lags. As we mentioned earlier, this is typical for a process that may be nonstationary. The PACF shown in Figure 4.16 has two large and significant spikes for lags 1 and 2 indicating that perhaps we need an AR model of order 2. However, the indication of the lingering autocorrelation in Figure 4.15 remains to be the most important feature of the data.

To further investigate the possible nonstationarity, we have in Figure 4.17a plotted the sample variogram of the original data and in Figure 4.17b the sample variogram of the first difference. Figure 4.17a shows an increasing trend in the

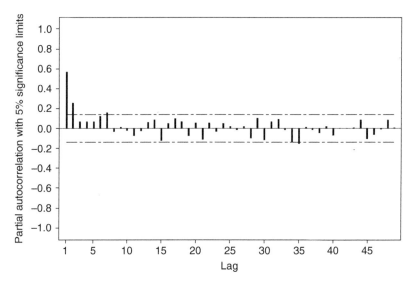

Figure 4.16 PACF of the chemical process concentration: series A from BJR.

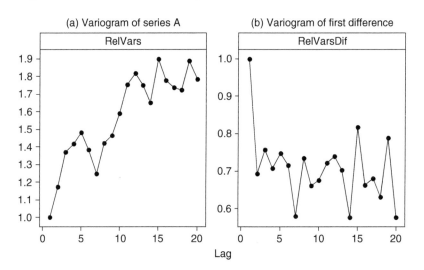

Figure 4.17 Variogram of (a) series A and (b) the first difference of series A from BJR.

sample variogram. This provides further evidence that the process is indeed non-stationary. However, the plot of the sample variogram of the first difference in Figure 4.17b, which quickly becomes stable, indicates that the first difference is stationary. This impression is further corroborated by studying Figure 4.18, the time series plot of the first differences of the process data.

Figure 4.19 shows the autocorrelation of the first difference of the data. We see that not only is there a large negative autocorrelation at lag 1 but also the sample autocorrelation is negligible for larger lags. PACF of the first difference

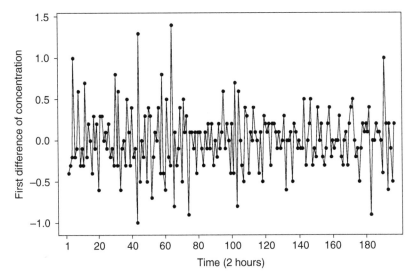

Figure 4.18   First difference of chemical process concentration: series A from BJR.

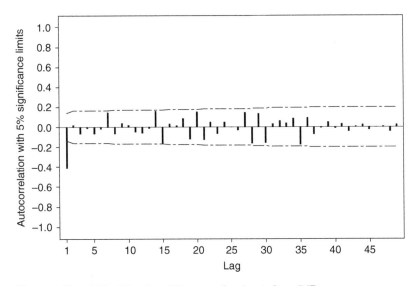

Figure 4.19   ACF of the first difference of series A from BJR.

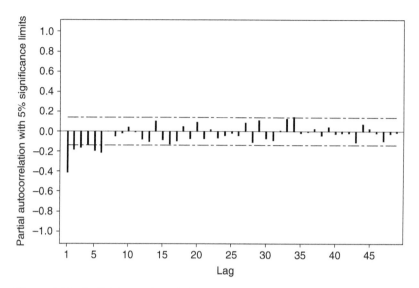

Figure 4.20   PACF for the first difference of series A from BJR.

TABLE 4.4   Summary of Information from Fitting an IMA(1, 1) Model to the Concentration Data

Term	Estimate	Standard error of coefficient	t	p
θ	0.7050	0.0507	13.90	0.000
Residual Sum of Squares (SS)	19.6707	$df = 195$		
Mean Squares (MS)	0.1009			

in Figure 4.20 tails out slowly. Comparing these findings with the summary of properties of time series models in Table 4.1 suggests that the first difference of the data may be modeled as a first order moving average process. In other words, the time series itself can be modeled as a first order integrated moving average (IMA) process, IMA(1, 1) (or ARIMA(0, 1, 1)).

$$\nabla z_t = z_t - z_{t-1} = w_t = a_t - \theta a_{t-1} \qquad (4.17)$$

To check this tentatively identified IMA(1, 1) model, we proceed to fit the model. The summary information for the fitted model is provided in Table 4.4 where we see that the MA coefficient θ is significant with a small p-value.

4.5.2   Diagnostic Checking

To further check how well the IMA(1, 1) model fits the data, we proceed to conduct the usual residual diagnostic checks. Figure 4.21 shows the sample ACF and Figure 4.22 the sample PACF of the residuals. None of these graphs show any significant autocorrelations. Figure 4.23 provides a summary of residual plots. Again, there does not appear to be any worrisome deviations from normality and

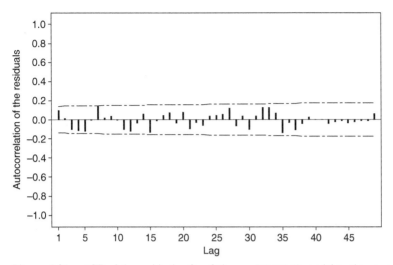

Figure 4.21 ACF of the residuals after fitting an IMA(1,1) model to the concentration data: series A from BJR.

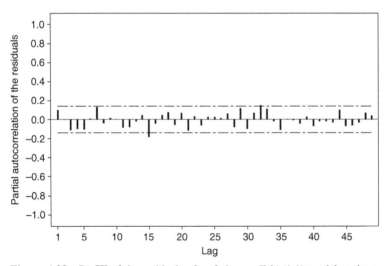

Figure 4.22 PACF of the residuals after fitting an IMA(1,1) model to the concentration data: series A from BJR.

independence except for a few not too large outliers. Overall, the appearance of the residuals in Figures 4.21–4.23 indicates that the IMA(1, 1) model fits the data reasonably well.

4.5.3   Implications and Interpretations of the IMA(1, 1) Model

We tentatively identified and fitted an IMA process (1, 1) to the chemical process data. Such a model is nonstationary. It has been argued that the natural state of any industrial process is in fact nonstationary. Once again the argument is simply

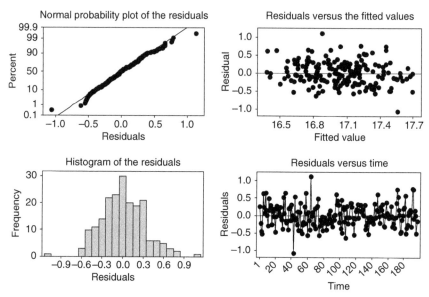

Figure 4.23 Summary of residual plots after fitting an IMA(1,1) model to the concentration data: series A from BJR.

that according to the second law of thermodynamics, any process will eventually deteriorate and drift off (unless there is added energy to the system, e.g., periodic maintenance). Thus, we should expect processes to be nonstationary. Processes may temporarily (i.e., short term) behave as if they are stable. However, if run long enough and deprived from some form of control, the natural state of any process is that it is nonstationary.

The IMA(1, 1) process is perhaps the most commonly encountered nonstationary model in practical applications. It implies that the changes, $\nabla z_t = z_t - z_{t-1}$, to the process are stationary. Further, these changes are only short-term autocorrelated. This further implies that the process itself does not have a stable mean. Rather, the mean keeps changing.

To further appreciate its properties, we may rewrite the IMA(1, 1) process as $z_t = z_{t-1} + a_t - \theta a_{t-1}$. By changing the time parameter by one unit, this implies that $z_{t-1} = z_{t-2} + a_{t-1} - \theta a_{t-2}$. If we substitute the latter expression into the former, we get $z_t = (z_{t-2} + a_{t-1} - \theta a_{t-2}) + a_t - \theta a_{t-1} = z_{t-2} + (1 - \theta)a_{t-1} - \theta a_{t-2} + a_t$. Recursively continuing this substitution of previous z_t's, the IMA (1, 1) model can eventually be written as

$$z_t = (1 - \theta) \sum_{j=1}^{\infty} a_{t-j} + a_t \tag{4.18}$$

This expression in terms of the random shocks, a_t's, shows that the IMA (1, 1) process is a noisy random walk. Indeed, if $\theta = 0$, the process is a pure random walk $z_t = \sum_{j=0}^{\infty} a_{t-j}$ and if $\theta = 1$, the IMA(1, 1) model becomes white

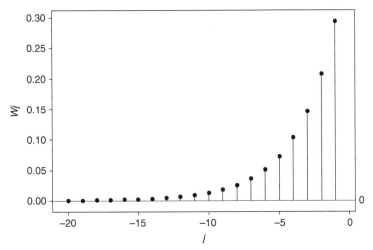

Figure 4.24 Weights assigned to past observations in an exponentially weighted moving average when $\theta = 0.705$.

noise, $z_t = a_t$. For $0 < \theta < 1$, the IMA(1, 1) model is a noisy random walk with properties ranging between the two extremes of a stationary to a nonstationary process. Therefore, the IMA(1, 1) model is quite flexible and can be used to model a wide range of practical situations.

Yet another way to write the IMA(1, 1) model is to substitute expressions for previous a_t's. This produces an infinite AR process in terms of previous z_t's.

$$z_t = (1 - \theta)(z_{t-1} + \theta z_{t-2} + \theta^2 z_{t-3} + \ldots) + a_t \qquad (4.19)$$

We recognize the expression in the second set of parenthesis as an exponentially weighted moving average (EWMA) of past observations. (The term $(1 - \theta)$ is just a scaling factor.) In other words, unlike a regular average that assigns equal weight to all observations, an EWMA has a relatively short memory that assigns decreasing weights to past observations since $|\theta| < 1$. The weights assigned to previous observations $z_{t-j}, j = 1, 2, 3, \ldots$ for $\theta = 0.705$, for example, is shown in Figure 4.24. We see that for this value of θ, the memory is relatively short and that observations that are more than 15 time units old are essentially discounted. Note further that Muth (1960) showed that the EWMA is the optimal forecast model for an IMA(1, 1) process; see Box and Luceno (1997, Chapter 5).

To get an intuitive feel for what the IMA(1, 1) model means, we have superimposed the fitted model values upon the original observations in Figure 4.25. The fitted model values are in this case the same as the EWMA forecasts with $\hat{\theta} = 0.705$. We see that the EWMA or the fitted values provide a smooth curve that tracks the local mean of the concentration process. Therefore, we can interpret the EWMA as a local estimator of the current mean as we move across the nonstationary time series.

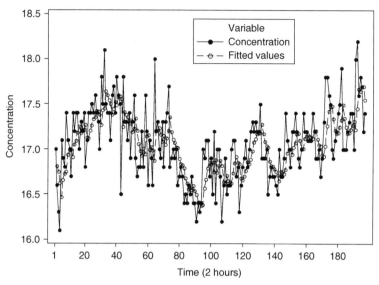

Figure 4.25 Time series plot of the concentration data (solid dots) with the fitted values from the IMA(1, 1) model superimposed (open dots).

4.5.4 To Be or Not To Be Stationary

We have previously indicated that the choice between a stationary and a nonstationary model is not easily made based on a relatively short series of data alone. The physical circumstances of the process should also be taken into consideration. In the present case, it could be argued that a stationary model could also fit the data. Indeed, a stationary ARMA model ARIMA(1, 0, 1) would fit the data well. The parameter estimates are given in Table 4.5.

In the ARIMA(1, 0, 1) model, we have $\hat{\phi} = 0.9151$. With an estimated standard error of 0.0433, we see that $\hat{\phi}_1$ is relatively close to 1. Indeed, if we let $\phi_1 \to 1$ from below we essentially end up with the IMA model $z_t - z_{t-1} = a_t - \theta a_{t-1}$. Therefore, the selection of a stationary or nonstationary model must often be made on the basis of not only the data but also a physical understanding of the process. If the AR parameter as in this case is close to 1, this may be an indication that it might be better to consider a nonstationary model.

TABLE 4.5 Estimated Parameters and Summary Statistics from Fitting a Stationary ARIMA (1, 0, 1) Model to the Concentration Data, BJR Series A

Model term	Coefficient	Standard error	t	p
AR: ϕ_1	0.9151	0.0433	21.11	0.000
MA: θ_1	0.5828	0.0849	6.87	0.000
Constant	1.44897	0.0094	154.38	0.000

[a] Residual SS = 19.188; MS = 0.0989; df = 194.

In the present case, the deciding point is this: is it reasonable to assume that this chemical process is stationary if left alone? We think not! A stationary model would imply that the process, if left alone, eventually would return to its (fixed and stable) mean value. This does not seem realistic. Rather, it is much more reasonable to assume that if the process is left alone, it will eventually drift off and behave as if out of control. Therefore, we are inclined to favor the nonstationary model with the interpretation that the exponentially moving average represents the best estimate of the current level of the process.

4.6   THE EWMA FORECAST

One of the simplest and most popular forecast equations is the exponentially weighted moving average (EWMA), the same tool used for quality control and process monitoring; see Box (1991) and Montgomery *et al.* (2008). The EWMA, also called exponential smoothing, is a simple prototype for all time-series-based forecasting equations. Therefore, let us digress to explain how the EWMA works. Suppose z_t is the observed sales in month t. Using the hat symbol to indicate that it is an estimate, we denote by $\hat{z}_t(1)$, the 1-step-ahead forecast of the sales in month $t + 1$ made in month t. The EWMA forecast of the next month sales $\hat{z}_t(1)$ is given by

$$\hat{z}_t(1) = (1 - \theta)z_t + \theta\hat{z}_{t-1}(1) \tag{4.20}$$

where θ is a constant between 0 and 1. In other words, the forecast $\hat{z}_t(1)$ is a linear combination of the old forecast, $\hat{z}_{t-1}(1)$, and the new observation, z_t. This is a convenient expression because all we need to keep record of to make a forecast is the last observation z_t and the last forecast $\hat{z}_{t-1}(1)$.

Another way to write the EWMA forecast Equation (4.20) is $\hat{z}_t(1) = \hat{z}_{t-1}(1) + (1 - \theta)[z_t - \hat{z}_{t-1}(1)]$. In other words, the new forecast is the last forecast plus $1 - \theta$ times a correction for the forecast error $[z_t - \hat{z}_{t-1}(1)]$ we made in the previous period.

Now, if we change the time index in Equation (4.20), we get $\hat{z}_{t-1}(1) = (1 - \theta)z_{t-1} + \theta\hat{z}_{t-2}(1)$. We may then substitute this into Equation (4.20) to get

$$\begin{aligned}\hat{z}_t(1) &= (1 - \theta)z_t + \theta[(1 - \theta)z_{t-1} + \theta\hat{z}_{t-2}(1)] \\ &= (1 - \theta)[z_t + \theta z_{t-1}] + \theta^2\hat{z}_{t-2}(1) \end{aligned} \tag{4.21}$$

Doing this one more time, we get $\hat{z}_{t-2}(1) = (1 - \theta)z_{t-2} + \theta\hat{z}_{t-3}(1)$ and

$$\begin{aligned}\hat{z}_t(1) &= (1 - \theta)[z_t + \theta z_{t-1}] + \theta^2[(1 - \theta)z_{t-2} + \theta\hat{z}_{t-3}(1)] \\ &= (1 - \theta)[z_t + \theta z_{t-1} + \theta^2 z_{t-2}] + \theta^3\hat{z}_{t-3}(1) \end{aligned} \tag{4.22}$$

If we continue in this manner, we eventually get

$$\hat{z}_t(1) = (1 - \theta)[z_t + \theta z_{t-1} + \theta^2 z_{t-2} + \theta^3 z_{t-3} + \theta^4 z_{t-4} + \cdots] \tag{4.23}$$

which is an infinite AR representation of the EWMA. It is from this expression that the name exponential moving average derives; it is a weighted average of past observations with weights that are exponentially (or geometrically) decreasing if $0 < \theta < 1$. For later reference, we denote the exponentially declining weights $(1 - \theta)\theta^j$ by π_j. In other words, $\pi_j = (1 - \theta)\theta^j, j = 1, 2, \ldots$ are the weights we assign to past observations z_{t+1-j} in the forecast equation

$$\hat{z}_t(1) = \pi_1 z_t + \pi_2 z_{t-1} + \pi_3 z_{t-2} + \pi_4 z_{t-3} + \pi_4 z_{t-4} + \cdots$$

$$= \sum_{j=1}^{\infty} \pi_j z_{t+1-j} \tag{4.24}$$

The EWMA forecast was originally proposed by Holt (1957) as an ad hoc forecast equation; it made practical sense that a forecast should be a weighted average that assigns most weight to the most immediate past observation, somewhat less weight to the second to the last observation, and so on. Holt did not postulate any underlying time series model or invoke any theory to justify his forecast equation. It just made good practical sense. However, Muth (1960) asked the seminal question: for which time series model is the EWMA forecast optimal? He then went on to prove that if the underlying time series is the IMA(1, 1) process $z_t = z_{t-1} + a_t - \theta a_{t-1}$, a nonstationary time series, then the EWMA forecast is optimal.

EXERCISES

4.1 Time series plot of a data set and its first difference together with corresponding ACF and PACF plots are given below. Determine the orders p, d, and q of a tentative ARIMA(p, d, q) model that can be used for this data.

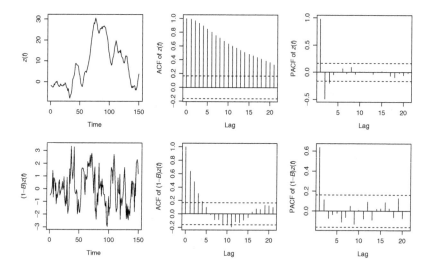

4.2 For the data in Exercise 4.1 and its first difference, the variograms are given below. Comment on the plots.

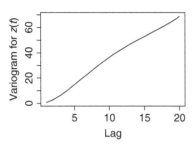 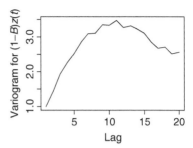

4.3 Repeat Exercise 4.1 for the plots below of a new data set.

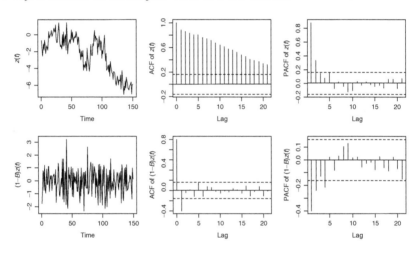

4.4 For the data in Exercise 4.3 and its first difference, the variograms are given below. Comment on the plots.

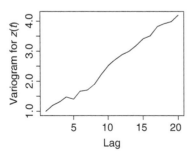 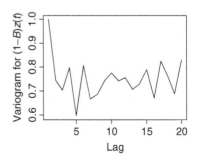

4.5 Repeat Exercise 4.1 for the plots below of a new data set.

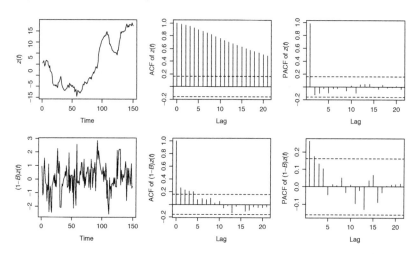

4.6 For the data in Exercise 4.5 and its first difference, the variograms are given below. Comment on the plots.

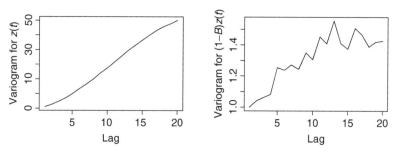

4.7 Repeat Exercise 4.1 for the plots below of a new data set.

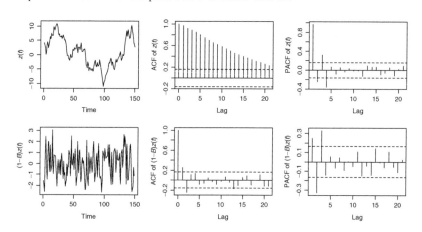

4.8 For the data in Exercise 4.7 and its first difference, the variograms are given below. Comment on the plots.

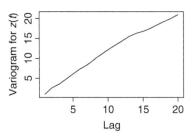 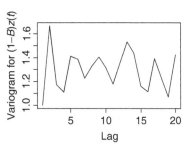

4.9 Repeat Exercise 4.1 for the plots below of a new data set.

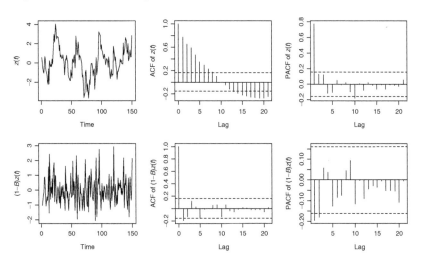

4.10 For the data in Exercise 4.9 and its first difference, the variograms are given below. Comment on the plots.

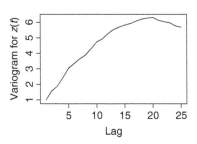 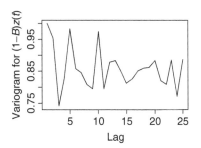

4.11 Table B.6 consists of the daily average exchange rates between US dollar and the euro from January 1, 2008 to June 1, 2008. Fit an appropriate ARIMA model for this data.

4.12 For the data in Exercise 4.11

a. Forecast the exchange rate for the next 7 days following June 1, 2008, using the model you developed in Exercise 4.11.

b. The exchange rates from June 2 to June 8 of that year turn out to be 0.6431, 0.6435, 0.644, 0.6473, 0.6469, 0.6386, and 0.6356. Using the model you developed in Exercise 4.11, make a 1-day-ahead forecast on June 1, 2008. Then assume that the current date is June 2 and the true exchange rate, 0.6431, is available. Fit a new model to the original data set extended with the observation for June 2. Proceed with making a 1-day-ahead forecast for June 3. Repeat this procedure until June 8.

c. Compare the sum of squared prediction errors $\left(\sum (z_t - \hat{z}_{t-1}(1))^2\right)$ for the forecasts in a and b.

4.13 For the data in Exercise 4.11, develop EWMA 1-day-ahead forecasts as in Exercise 4.12b. For the EWMA, try $\theta = 0.1, 0.3, 0.5, 0.7, 0.9$. Which θ value gives the smallest sum of squared prediction errors?

4.14 Find the "optimal" θ for the EWMA using the following procedure:

(i) For the data in Exercise 4.11, consider the EWMA $\tilde{y}_1$ as the forecast for $y_2, \tilde{y}_2$ as the forecast for y_3, and so on.

(ii) Using $\theta = 0.1, 0.2, 0.3, 0.4, 0.5, 0.6, 0.7, 0.8, 0.9$ calculate the EWMA.

(iii) Plot the sum of squared predictions errors for various θ values versus θ for the data in Exercise 4.11.

(iv) Find the θ value that gives the minimum sum of squared prediction errors. This is the optimal θ.

4.15 Repeat the forecasting procedure described in Exercise 4.12b using EWMA with the optimal θ obtained in Exercise 4.14 and compare the results.

4.16 Table B.7 shows the monthly US unemployment rates from January 1980 to September 2010. Fit an appropriate ARIMA model using only the data from January 1980 to December 2009.

4.17 For the data in Exercise 4.16

a. Forecast the unemployment rate for the first 9 months of 2010 based on the ARIMA model you developed in Exercise 4.16, that is, assuming that you are currently in December 2009.

b. Using the model you developed in Exercise 4.16, make a 1-month-ahead forecast as in Exercise 4.12b.

c. Compare the sum of squared prediction errors $\left(\sum (z_t - \hat{z}_{t-1}(1))^2\right)$ for the forecasts in a and b.

4.18 For the data in Exercise 4.16, develop EWMA 1-month-ahead forecast as in Exercise 4.17b. For the EWMA, try $\theta = 0.1, 0.3, 0.5, 0.7, 0.9$. Which θ value gives the smallest sum of squared prediction errors?

4.19 Find the "optimal" θ for the EWMA for the data in Exercise 4.16 as in Exercise 4.14.

4.20 Repeat the forecasting procedure described in Exercise 4.17b using EWMA with the optimal θ obtained in Exercise 4.19 and compare the results.

CHAPTER 5

SEASONAL MODELS

5.1 SEASONAL DATA

For the models we discussed so far, the serial dependence of the current observation to the previous observations was often strongest for the immediate past and followed a decaying pattern as we move further back in time. For some systems, however, this dependence shows a repeating, cyclic behavior. This is particularly true for business and financial data. The inventory level on a Monday will be somewhat similar to the levels on previous Mondays or the sales of ice cream in June of one year will show a strong correlation not only with last month's sales but also with the sales in June of the previous years. This cyclic pattern or as more commonly called *seasonal pattern* can be very effectively used to further improve the forecasting performance. The ARIMA models we have discussed are flexible enough to allow for modeling both seasonal and nonseasonal dependence. In this chapter, we will extend the use of ARIMA models to seasonal data and go through the modeling process with the help of examples.

5.1.1 Example 1: The International Airline Passenger Data

We use the international airline data originally analyzed by Brown (1962) and modeled with a seasonal ARIMA time series model by Box and Jenkins (1970) to demonstrate how seasonal ARIMA models can be used to model cyclic data and how the model can be used for short-term forecasting. The data is available as series G in BJR and is reproduced in Table 5.1.

The monthly totals of international airline passengers from January 1949 to December 1960 are plotted in Figure 5.1. It is immediately obvious that the data exhibits an upward trend as well as a yearly seasonal pattern. The presentation of the data in Table 5.1 as a two-way table with years in the vertical direction and month in the horizontal direction inspires a two-way analysis of variance (ANOVA) way of thinking. Therefore, to get a closer look at the seasonal pattern, we have used an interaction plot in Figure 5.2, usually used to display two-way ANOVA type data and to show the trend and seasonality in more close-up format. Showing seasonal time series data with an interaction plot is perhaps unusual, but provides a useful way of studying the trend and seasonality patterns. From Figures 5.1 and 5.2, a number of important aspects of the data emerge. Clearly, there is an overall increasing trend in the number of passengers over the

---

Time Series Analysis and Forecasting by Example, First Edition. Søren Bisgaard and Murat Kulahci.
© 2011 John Wiley & Sons, Inc. Published 2011 by John Wiley & Sons, Inc.

TABLE 5.1 Monthly Passenger Totals (Measured in Thousands) in International Air Travel – BJR Series G

Year	January	February	March	April	May	June	July	August	September	October	November	December
1949	112	118	132	129	121	135	148	148	136	119	104	118
1950	115	126	141	135	125	149	170	170	158	133	114	140
1951	145	150	178	163	172	178	199	199	184	162	146	166
1952	171	180	193	181	183	218	230	242	209	191	172	194
1953	196	196	236	235	229	243	264	272	237	211	180	201
1954	204	188	235	227	234	264	302	293	259	229	203	229
1955	242	233	267	269	270	315	364	347	312	274	237	278
1956	284	277	317	313	318	374	413	405	355	306	271	306
1957	315	301	356	348	355	422	465	467	404	347	305	336
1958	340	318	362	348	363	435	491	505	404	359	310	337
1959	360	342	406	396	420	472	548	559	463	407	362	405
1960	417	391	419	461	472	535	622	606	508	461	390	432

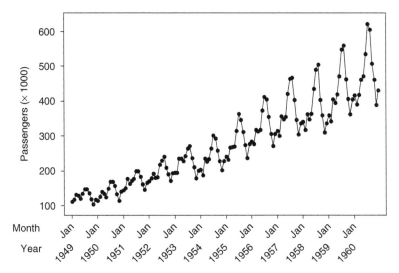

Figure 5.1   Time series plot of the monthly international airline passenger data—BJR series G.

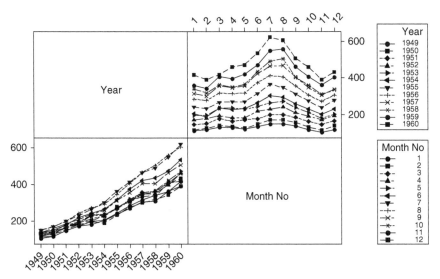

Figure 5.2   Interaction plot of the monthly international airline passenger data—BJR series G.

entire 12-year period. This trend may have a slightly increasing slope over time, possibly being exponential. Furthermore, the number of passengers is seasonal with a consistent peak in July and the lowest numbers in January and November. Most importantly, for forecasting purposes, the pattern seems to be consistent from year to year. This is good news because consistent patterns of trends and seasonality are what allow us to make good forecasts.

5.1.2 Data Transformation for the Airline Passenger Data

The keen data analyst will notice, especially from Figure 5.1, that the differences between peaks and valleys increase over time. In other words, the variability is not constant, but increases over time and gets larger. In fact, the variability seems to get larger with increasing (average) number of passengers over the years. Such a pattern is common in economic data and suggests the need for a transformation as discussed in Chapter 3. An important preliminary step in any data analysis is to consider the possibility of a data transformation. The goal of the transformation is to identify a scale where the residuals, after fitting a model, will have homogeneous variability.

To consider a possible transformation, let us for the moment ignore the seasonality issue. In Chapter 3, we discussed the log transformation, and related it to the percentage growth from time t to time $t + 1$, which is computed as

$$p_t = \frac{y_{t+1} - y_t}{y_t} \times 100$$

We then defined $p_t^* = p_t/100$ as the growth rate and we showed that using the relationship $y_{t+1} = (1 + p_t^*)y_t$, the difference on the log scale between successive observations can be written as

$$z_{t+1} - z_t = \log(y_{t+1}) - \log(y_t)$$
$$= \log\left(\frac{y_{t+1}}{y_t}\right)$$
$$= \log\left(\frac{(1 + p_t^*)y_t}{y_t}\right)$$
$$= \log(1 + p_t^*)$$

However, for small values of p_t^*, the logarithmic function is approximately a linear function. In other words, $\log(1 + p_t^*) \approx p_t^*$ and hence $z_{t+1} - z_t \approx p_t^*$. This means that on the log scale, the difference between successive observations therefore remains stable in time. This in turn makes the variance more stable. Therefore, for the airline passenger data we will also use the log transformation.

Figures 5.3 and 5.4 show the log transformed data. We now see that the trend is more nearly linear, and that the variability is more or less constant (homogeneous) when comparing over the months and over the years. In fact, the lines connecting the months on the right-hand side of Figure 5.4 are now almost "parallel." In the ANOVA context, we would say that there does not seem to be any interactions when the lines are "parallel." That means the month and year effects are more or less additive or linear. Therefore, the logarithmic transformation seems to have removed the nonadditivity or nonlinearity in the data.

Another rough graphical check for the "right" transformation is the range–mean plot (Jenkins, 1979, pp. 95–97). A range–mean plot is produced by dividing the time series into smaller segments and plotting the range versus

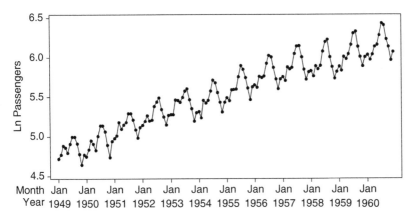

Figure 5.3   Time series plot of the natural logarithm of the number of airline passengers.

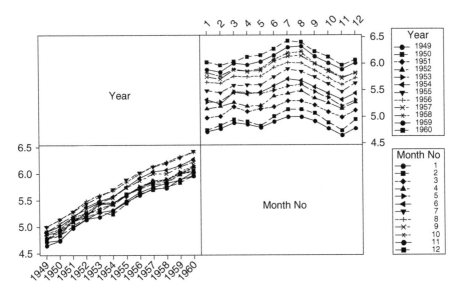

Figure 5.4   Interaction plot of the natural logarithm of the number of airline passengers.

the average of each segment on a scatter plot. In the present case, we divided the time series into years, computed the yearly range and average, and plotted those as a scatter plot in Figure 5.5. We see that there seems to be a linear relationship between the average and the range, indicating that a log transformation is appropriate.

Data transformations should ultimately be checked on the residuals after fitting an appropriate time series model. Therefore, we tentatively proceed with the log transformation of the data and return to the issue of checking the residuals later.

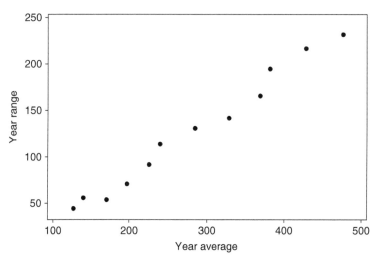

Figure 5.5   Range–mean plot of the airline passenger data.

5.2   SEASONAL ARIMA MODELS

The most important structural issue to recognize about seasonal time series is that if the season is s periods long, then observations that are s time intervals apart are alike. In our case, $s = 12$ months. Thus, what happened in, say, the holiday month of December of one year will be similar to and correlated with what we can expect to happen in December of the following year. However, we should also expect that an observation in December is correlated with what happened in the immediately preceding month, November. With reference to Table 5.1, this is somewhat similar to a two-way ANOVA model where we often find similarities in the same columns (representing months) and same rows (representing years). We therefore have two relationships going on simultaneously (i) between observations for successive months within the same year, and (ii) between observations for the same month in successive years. Therefore, we essentially need to build two time series models—one for the relationship between successive months within the years and one that links the same months in successive years and then combine the two.

With this in mind, the model building process for seasonal models is otherwise similar to that used for nonseasonal models. First, if the data is nonstationary, we use difference operations to make the data stationary so the autocorrelation dies out relatively quickly. However, for seasonal data, we may need to use not only the regular difference $\nabla z_t = z_t - z_{t-1}$ but also a seasonal difference $\nabla_s z_t = z_t - z_{t-s}$. Sometimes, we may even need both to obtain an autocorrelation function (ACF) that dies out sufficiently quickly. To find out what works requires trial and error and practical experience with time series analysis.

For the log transformed airline data, with an $s = 12$ month seasonality, we proceed iteratively and try a regular first difference $\nabla z_t = z_t - z_{t-1}$, a seasonal difference $\nabla_{12}z_t = z_t - z_{t-12}$, and the two combined $\nabla\nabla_{12}z_t = \nabla(z_t - z_{t-12}) = z_t - z_{t-1} - z_{t-12} + z_{t-13}$. The latter difference transformation is most easily done by differencing twice, one time with lag 12 and one subsequent difference with lag 1 or in the opposite order. The log transformed airline passenger numbers z_t, the first difference ∇z_t, the seasonal difference $\nabla_{12}z_t$, and the combined difference $\nabla\nabla_{12}z_t$ are all shown as time series plots in Figure 5.6 and the corresponding ACFs in Figure 5.7. We see from the first difference ∇z_t shown in Figures 5.6b and 5.7b that the seasonal component is quite persistent over time with large spikes in the ACF at lag 12, 24, and 36. For the seasonally differenced data

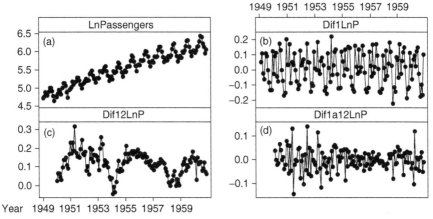

Figure 5.6 Time series plots of (a) the log transformed series z_t, (b) the first difference ∇z_t, (c) the seasonal difference $\nabla_{12}z_t$, and (d) the combined first difference and seasonal difference $\nabla\nabla_{12}z_t$.

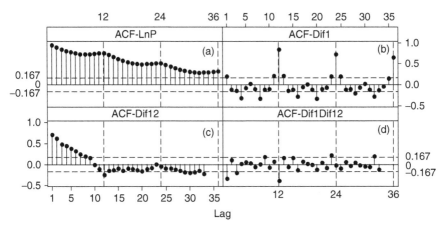

Figure 5.7 ACFs for (a) the log transformed series z_t, (b) the first difference ∇z_t, (c) the seasonal difference $\nabla_{12}z_t$, and (d) the combined first difference and seasonal difference $\nabla\nabla_{12}z_t$. The horizontal lines at ± 0.167 are approximate $2\times$ standard error limits for the sample autocorrelations $\pm 2\mathrm{SE} \approx \pm 2/\sqrt{n}$ where n is the sample size.

$\nabla_{12}z_t$ in Figures 5.6c and 5.7c, we see that the data remains nonstationary with the autocorrelation lingering on for large lags. Finally, when the data is differenced twice, the time series plot of $\nabla\nabla_{12}z_t$ seems to indicate that the data is stationary and that the seasonal autocorrelation spikes at 12, 24, and 36 now fade out. Thus, we tentatively conclude that we need to use one regular and one seasonal difference, that is, we need to use $\nabla\nabla_{12}z_t$ for further time series modeling.

5.2.1 Model Identification

Having transformed the data to stationary form using the $\nabla\nabla_{12}z_t$ difference operation, we are now ready to identify the seasonal time series model. The methodology for identifying stationary seasonal models is a modification of the one used for regular ARMA time series models where the patterns of the sample ACF and the sample partial autocorrelation function (PACF) provide guidance. The main patterns in ACF and PACF we are looking for in nonseasonal ARIMA models have been summarized in Table 4.1. We will make use of the same table for seasonal ARIMA models as well. The main difference between regular and seasonal time series model identification is that we are looking for telltale signs of two models—one seasonal and one regular month-to-month model. In addition, note that in nonseasonal data, it is usually sufficient to consider the ACF and the PACF for up to 20–25 lags. But for seasonal data, we recommend increasing the number of lags to at least 3 or 4 multiples of the seasonality, say, 36 to 48 in our example.

Starting with the seasonal pattern, we look for similarities that are 12 lags apart. From Figure 5.7d showing the ACF for the time series $w_t = \nabla\nabla_{12}z_t$, we notice an isolated significant negative spike at lag 12, after which the seasonal autocorrelation pattern cuts off. In other words, the autocorrelation appears to be insignificant at lags 24 and 36. With reference to Table 4.1, this is the telltale sign of a moving average model applied to the 12-month seasonal pattern.

Now, turning to the regular time series model, we are looking for patterns between successive months. From Figure 5.7d, we see a significant autocorrelation at lag 1, after which the ACF seems to cut off. This would indicate a first order moving average term in the regular model. From Figure 5.8 showing the PACF of $w_t = \nabla\nabla_{12}z_t$, we see two major spikes at lag 1 and 12 followed by what looks like tail-off pattern between lag 1 and 11 and repeated after lag 12. Again referring to Table 4.1, the tailing-off pattern in the PACF confirms the need for a first order moving average model in both the regular and the seasonal models.

An alternative and somewhat safer way to identify a seasonal model is to proceed iteratively, first identifying and fitting a seasonal model, and then looking at the residuals after fitting such a model to identify the regular model. We now briefly show how that works.

As before, from Figure 5.7d we see that the ACF shows the pattern of a first order seasonal moving average (SMA) term, one spike at 12 and then a cutoff. We therefore fit a first order SMA model and compute the residuals. From those residuals, we compute the ACF shown in Figure 5.9. From Figure 5.9, we see that the ACF shows a significant negative spike at lag 1, indicating that we need an additional regular moving average term.

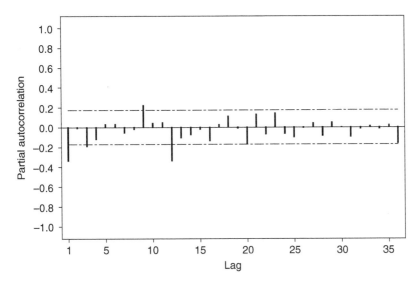

Figure 5.8   The sample PACF for $w_t = \nabla \nabla_{12} z_t$.

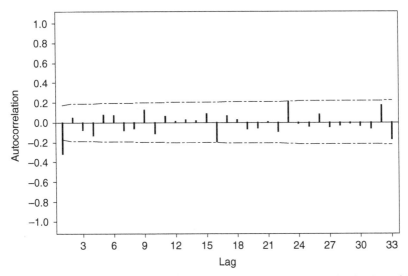

Figure 5.9   The sample ACF of the residuals after fitting a first order SMA model to $w_t = \nabla \nabla_{12} z_t$.

5.2.2   The Anatomy of Seasonal Models

We will now show more specifically what it means to combine two models, a seasonal and a regular. In general, we see from Table 4.1 that a first order moving average model is of the form $w_t = a_t - \theta_1 a_{t-1}$. As for the 12-month seasonal model, the first order moving average model is modified as $w_t = a_t - \theta_1 a_{t-12}$.

However, to avoid confusion, we need to change the symbols and so we will write our SMA model as $w_t = b_t - \Theta_1 b_{t-12}$ where b_t is the "error" term and Θ_1 the SMA parameter. Note that in the seasonal model, we do not necessarily assume that the "error" b_t is uncorrelated over time. This SMA model will be used to model the year-to-year seasonal dynamics indicated by the significant autocorrelation at lag 12.

Now, turning to the regular time series model the pattern in Figure 5.7d of the sample autocorrelation, or the even clearer pattern seen in Figure 5.9, indicates that there is some significant autocorrelation at lag 1. Therefore to model the month-to-month dynamic, we need an additional regular first order moving average model stated in generic terms as $w_t = a_t - \theta_1 a_{t-1}$. Now w_t is the symbol we use for the basic sequence of stationary time observations. However, we need this generic first order moving average model to model the "error" term b_t in the seasonal model above. In other words, we need to substitute b_t for w_t in the generic model. We therefore get $b_t = a_t - \theta_1 a_{t-1}$, where a_t is the actual error term assumed to be uncorrelated white noise. Combining the two models, $w_t = b_t - \Theta_1 b_{t-12}$ and $b_t = a_t - \theta_1 a_{t-1}$, we get

$$
\begin{aligned}
w_t &= b_t - \Theta_1 b_{t-12} \\
&= (a_t - \theta_1 a_{t-1}) - \Theta_1(a_{t-12} - \theta_1 a_{t-13}) \\
&= a_t - \theta_1 a_{t-1} - \Theta_1 a_{t-12} + \Theta_1 \theta_1 a_{t-13}
\end{aligned}
\tag{5.1}
$$

The model for w_t is for the data after the log transformation and after achieving a stationary time series by differencing with a regular and a seasonal difference. Thus, $w_t = \nabla \nabla_{12} z_t = z_t - z_{t-1} - z_{t-12} + z_{t-13}$. Putting all this together, the model for the log of the airline passenger number is

$$
z_t - z_{t-1} - z_{t-12} + z_{t-13} = a_t - \theta_1 a_{t-1} - \Theta_1 a_{t-12} + \Theta_1 \theta_1 a_{t-13}
\tag{5.2}
$$

Such a model is often called a *seasonal ARIMA model* and referred to as $\text{ARIMA}(p, d, q) \times (P, D, Q)_{12}$ model with $p = 0, d = 1, q = 1, P = 0, D = 1$, and $Q = 1$.

5.2.3 Model Fitting and Model Checking

The $\text{ARIMA}(0, 1, 1) \times (0, 1, 1)_{12}$ model can be fitted directly to the logged time series data z_t with standard statistical software such as MINITAB or SAS JMP by specifying one seasonal difference, one regular difference, one SMA term, one regular moving average term, and no constant.

In the present case, we have used the entire dataset of 144 observations to identify the model. However, we like to demonstrate how to use the model to make short-term forecasts. We therefore set aside the last year's observations when we fit the model and so we use the last year to compare the forecast with the actual data. Thus, we use observations 1 through 132 for model fitting. Edited output from fitting a seasonal ARIMA $(0, 1, 1) \times (0, 1, 1)_{12}$ model to the data from January 1949 to December 1959 is given in Table 5.2.

TABLE 5.2   ARIMA $(0, 1, 1) \times (0, 1, 1)_{12}$ Model Summary for the Airline Passenger Data

Model term	Coefficient	Standard error	t	p
MA 1	0.3407	0.0868	3.93	0.000
SMA 12	0.6299	0.0766	8.23	0.000

Differencing: 1 regular, 1 seasonal of order 12.
Number of observations: Original series 132, after differencing 119.
Residuals: SS $= 0.151421$; MS $= 0.001294$; $df = 117$.
Modified Box–Pierce (Ljung–Box) chi-square statistic:

Lag	12	24	36	48
Chi square	7.5	19.6	30.5	38.7
df	10	22	34	46
p-value	0.679	0.607	0.638	0.770

We see from Table 5.2 that both the regular and SMA terms are significant. Furthermore, the Ljung–Box chi-square statistics discussed in Chapter 4 indicates that for all lags there are no signs of significant remaining autocorrelation in the residuals. Figures 5.10 and 5.11 provide a summary plots of the standard residual checks. Neither the ACF nor the PACF in Figures 5.10a and 5.10b of the residuals and the residual plots in Figure 5.11 provide indications of serious lack of fit except for an outlier in February 1954 (observation number 62). Furthermore, the normal probability plot of the residuals and the plot of residuals versus the fitted values in Figure 5.11 indicate that the log transformation made the residuals look more or less normally distributed and the variance relatively stable and independent of the mean.

TABLE 5.3   Forecasts with 95% Prediction Intervals for 1960 for the Airline Passenger Data on Log Scale After Fitting a Seasonal ARIMA $(0, 1, 1) \times (0, 1, 1)_{12}$ Model to the Log Airline Data from January 1949 to December 1959

Time	Number	Forecast	Lower	Upper	Actual	Difference
1960–1	133	6.03771	5.96718	6.10823	6.03309	−0.00462
1960–2	134	5.99099	5.90652	6.07546	5.96871	−0.02228
1960–3	135	6.14666	6.05023	6.24308	6.03787	−0.10879
1960–4	136	6.12046	6.01341	6.22751	6.13340	0.01293
1960–5	137	6.15698	6.04026	6.27369	6.15698	−0.00000
1960–6	138	6.30256	6.17692	6.42819	6.28227	−0.02029
1960–7	139	6.42828	6.29432	6.56224	6.43294	0.00466
1960–8	140	6.43857	6.29677	6.58037	6.40688	−0.03169
1960–9	141	6.26527	6.11604	6.41450	6.23048	−0.03479
1960–10	142	6.13438	5.97807	6.29069	6.13340	−0.00098
1960–11	143	6.00539	5.84231	6.16846	5.96615	−0.03924
1960–12	144	6.11358	5.94401	6.28316	6.06843	−0.04515

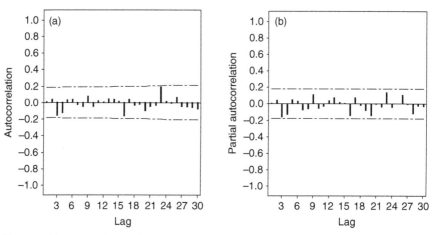

Figure 5.10 Plots of (a) ACF and (b) PACF of the residuals after fitting a seasonal ARIMA $(0, 1, 1) \times (0, 1, 1)_{12}$ model to the log airline data from January 1949 to December 1959.

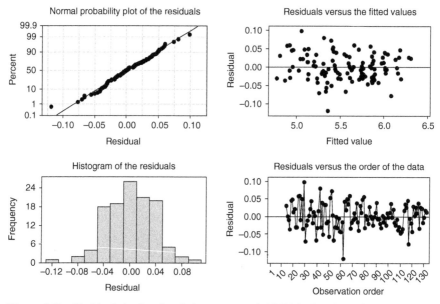

Figure 5.11 Residual checks after fitting a seasonal ARIMA $(0, 1, 1) \times (0, 1, 1)_{12}$ model to the log airline data from January 1949 to December 1959.

Seasonal data allows for additional ways of checking residuals besides those provided by standard time series software that we now demonstrate. Figures 5.12 and 5.13 show the residuals after fitting the seasonal ARIMA model to all the observations from January 1949 to December 1960. Figure 5.12 is a plot of the residuals versus the months and Figure 5.13 is the residuals versus the years. We

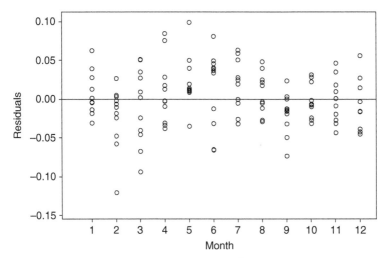

Figure 5.12 Residuals versus month. Residuals are computed after fitting a seasonal ARIMA $(0,1,1) \times (0,1,1)_{12}$ model to the log airline data from January 1949 to December 1960.

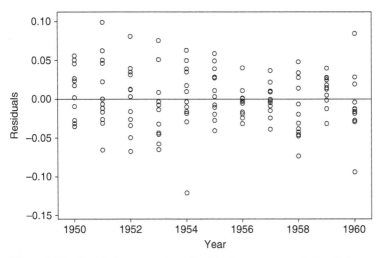

Figure 5.13 Residuals versus years. Residuals are computed after fitting a seasonal ARIMA $(0,1,1) \times (0,1,1)_{12}$ model to the log airline data from January 1949 to December 1960.

see that although log transformation did a fairly good job in homogenizing the residual variation, it did not do a "perfect" job as far as visual inspection goes. For example, it is now apparent that something unusual happened in February 1954 where we see a significant outlier. From Figure 5.12, we see that the month of October consistently seems to have much less variability than the rest, and Figure 5.13 seems to indicate that the residuals for years 1956, 1957, and 1959 also were less variable. We also see from Figure 5.12 that some seasonality remains as they follow a slight trend from month to month. This prompted us to try a variety of alternative transformations and models but none seemed to do a better job. Perfection is an unrealistic goal and more formal residual tests lead us to believe that the log transformation and the ARIMA $(0, 1, 1) \times (0, 1, 1)_{12}$ model provide a fairly useful description of the dynamics in this data. Thus, we conclude that a reasonably good model for the log of the number of international airline passengers, z_t, is given by

$$z_t = z_{t-1} + z_{t-12} - z_{t-13} + a_t - 0.34a_{t-1} - 0.63a_{t-12} + 0.214a_{t-13} \quad (5.3)$$

5.3   FORECASTING USING SEASONAL ARIMA MODELS

To develop a forecast equation for a seasonal time series model, we follow a procedure similar to that we discussed for regular time series models in Chapter 4. The easiest way to proceed is to start with the model for the fitted time series

$$z_t = z_{t-1} + z_{t-12} - z_{t-13} + a_t - \theta_1 a_{t-1} - \Theta_1 a_{t-12} + \Theta_1 \theta_1 a_{t-13} \quad (5.4)$$

If we want a forecast of z_t, say ℓ time units ahead of time t, we start with the time series model rewritten as

$$z_{t+\ell} = z_{t+\ell-1} + z_{t+\ell-12} - z_{t+\ell-13} + a_{t+\ell}$$
$$-\theta_1 a_{t+\ell-1} - \Theta_1 a_{t+\ell-12} + \Theta_1 \theta_1 a_{t+\ell-13} \quad (5.5)$$

Next, we denote by $\hat{z}_t(\ell)$ the forecast of $z_{t+\ell}$ made at time t using all information available up to and including time t. The forecast equation is then obtained from the time series model as

$$\hat{z}_t(\ell) = \left[z_{t+\ell-1} + z_{t+\ell-12} - z_{t+\ell-13} + a_{t+\ell} \right.$$
$$\left. -\theta_1 a_{t+\ell-1} - \Theta_1 a_{t+\ell-12} + \Theta_1 \theta_1 a_{t+\ell-13} \right]$$
$$= [z_{t+\ell-1}] + [z_{t+\ell-12}] - [z_{t+\ell-13}] + [a_{t+\ell}]$$
$$-\theta_1 [a_{t+\ell-1}] - \Theta_1 [a_{t+\ell-12}] + \Theta_1 \theta_1 [a_{t+\ell-13}] \quad (5.6)$$

where the square bracket notation in $[z_{t+l}]$ and $[a_{t+l}]$ represents the conditional expectations (mean) of $z_{t+\ell}$ or $a_{t+\ell}$ taken at time t, given all information available up until and including time t. This implies that when forecasting any a_{t+j} term with $j > 0$, a_{t+j} has not yet occurred. Our best "guess" for it then will be its expected value, which is assumed to be zero. Symbolically, we write that as $[a_{t+j}] = 0$, for $j > 0$. However, if an error term a_{t+j} appears in the forecasting equation with an index $t + j$ where $j \leq 0$ so that the time index is before or at the current time t, then a_{t+j} has occurred and its actual value should be used. However, the random error a_{t+j} itself cannot be observed. Instead, we estimate it by using the 1-step-ahead forecast error. Hence, we have $[a_{t+j}] = z_{t+j} - \hat{z}_{t+j-1}(1)$ for $j \leq 0$. The same is applicable for any values of $[z_{t+l}]$. If the time index refers to an already observed value, we have $[z_{t+j}] = z_{t+j}, j \leq 0$. However, if the time index refers to a future value, then we set $[z_{t+j}]$ equal to the forecasted value, that is, $[z_{t+j}] = \hat{z}_t(j)$ for $j > 0$. For example, to get the 3-month-ahead forecast at time t, we have

$$\hat{z}_t(3) = [z_{t+3-1}] + [z_{t+3-12}] - [z_{t+3-13}] + [a_{t+3}]$$
$$-\theta_1[a_{t+3-1}] - \Theta_1[a_{t+3-12}] + \Theta_1\theta_1[a_{t+3-13}]$$
$$= [z_{t+2}] + [z_{t-9}] - [z_{t-10}] + [a_{t+3}] - \theta_1[a_{t+2}]$$
$$-\Theta_1[a_{t-9}] + \Theta_1\theta_1[a_{t-10}] \tag{5.7}$$

Using the rules we just discussed, we get

$$\hat{z}_t(3) = [z_{t+2}] + [z_{t-9}] - [z_{t-10}] + [a_{t+3}] - \theta_1[a_{t+2}] - \Theta_1[a_{t-9}] + \Theta_1\theta_1[a_{t-10}]$$
$$= \hat{z}_t(2) + z_{t-9} - z_{t-10} - \Theta_1\hat{a}_{t-9} + \Theta_1\theta_1\hat{a}_{t-10} \tag{5.8}$$

However, $\hat{a}_{t-9} = z_{t-9} - \hat{z}_{t-10}(1)$ and $\hat{a}_{t-10} = z_{t-10} - \hat{z}_{t-11}(1)$. Thus, the recursive forecasting equation is

$$\hat{z}_t(3) = \hat{z}_t(2) + z_{t-9} - z_{t-10} - \Theta_1(z_{t-9} - \hat{z}_{t-10}(1)) + \Theta_1\theta_1(z_{t-10} - \hat{z}_{t-11}(1))$$
$$= z_t(2) + (1 - \Theta_1)z_{t-9} - (1 - \Theta_1\theta_1)z_{t-10} + \Theta_1\hat{z}_{t-10}(1)$$
$$-\Theta_1\theta_1\hat{z}_{t-11}(1) \tag{5.9}$$

where we substitute in our parameter estimates $\hat{\Theta}_1 \approx 0.63$ and $\hat{\theta}_1 \approx 0.34$ to obtain our forecast.

Fortunately, this forecast equation is programmed as an option in standard statistical software such as MINITAB and JMP. We therefore need not get involved in the tedium of producing them. To demonstrate how this works, suppose we, at the end of December 1959, want to forecast the number of airline

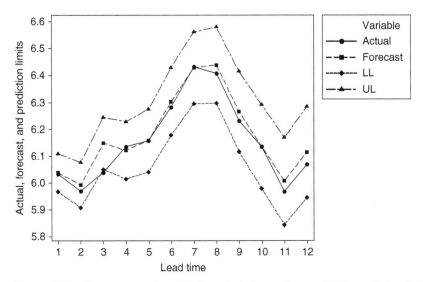

Figure 5.14 Forecasts together with the actual observations and 95% prediction intervals for 1960 for the airline passenger data on the log scale after fitting a seasonal ARIMA $(0, 1, 1) \times (0, 1, 1)_{12}$ model to the log airline data from January 1949 to December 1959.

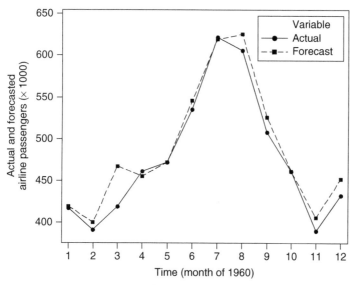

Figure 5.15 Forecasts together with the actual observations for 1960 for the airline passenger data in actual units.

passengers for the coming year, 1960. We then simply specify the forecast origin and the lead time, and the forecasts are automatically produced including typically 95% prediction intervals. We refer the readers to BJR for a detailed discussion on how the prediction intervals are computed.

The forecasts, the prediction intervals, and the actual observations are shown in Table 5.3 and plotted in Figure 5.14. Now, of course, the forecasts are on the log scale. In the original scale, the forecasts can be approximated by the antilog, $y_t = \exp(z_t)$. The forecasts and the actuals on the original $1000\times$ passenger scale are shown in Figure 5.15. It is striking how well the forecasts in this case match the actual observations.

5.4 EXAMPLE 2: COMPANY X'S SALES DATA

In the previous section, we discussed the use of the logarithmic transformation. In this section, we provide a more general discussion of the use of the class of Box–Cox transformations that include the log transformation as a special case. Specifically, we demonstrate how to determine a Box–Cox transformation in the context of seasonal time series modeling. For our demonstration, we use a much discussed, somewhat notorious, but educational example from the literature—the sales data of an engineering product provided by Chatfield and Prothero (1973). Their article generated a heated but interesting discussion about the appropriate use of transformations for time series data. The consensus at the end of the debate was that, as for any other model fitting effort, the "right" transformation often leads to a better fit and subsequently better forecasts for time series models.

By working through this example, we demonstrate in a step-by-step manner of how to use the general family of Box–Cox transformations to time series data. Our analysis also serves as a vehicle for discussing a number of other issues including the effect of outliers.

In their article entitled "Box–Jenkins seasonal forecasting: Problems in a case study," Chatfield and Prothero (1973) reported on problems they encountered when they were engaged by Company X, an engineering firm, to provide sales forecasts for an engineered product for lead times up to 12 months. Chatfield and Prothero (CP) provided monthly sales data for almost 6.5 years, from January 1965 to May 1971, as shown in Table 5.4. The 77 observations are plotted as a time series in Figure 5.16.

Three striking features of the sales data immediately come to light in Figure 5.16. First, the data clearly exhibits a seasonal pattern. Second, there is a consistent upward "linear" trend in the sales from 1965 to 1971. Third, a keen observer will notice that the amplitude of the seasonal cycle increases over time. In the previous section, where we analyzed the international airline data, we showed how the problem of an increasing amplitude can often be overcome by applying a log transformation. This was also the transformation CP used for Company X's sales data. However, as pointed out by Wilson (1973) and Box and Jenkins (1973) in the subsequent debate, the log transformation may not always be appropriate. Sometimes we need to resort to a more general class of transformations. Specifically, we may try the class of power transformations where in the simplest version, the response y_t is raised to some power y_t^λ where λ is a coefficient to be determined. For example, if $\lambda = 0.5$, then $y_t^{0.5} = \sqrt{y_t}$, and in the special case where $\lambda = 0$, we define the transformation to be $\ln(y_t)$.

The purposes of using a transformation are primarily (i) to decouple the mean and the variance so that the variability is more or less constant and does not depend on the mean, (ii) that the model is additive and relatively simple, and (iii) that the residuals, after fitting the model, are more or less normally distributed with zero mean and constant variance.

To find the "right" transformation is a bit of a chicken-and-egg dilemma— to judge how well a given transformation transforms the response to achieve constant variance and normally distributed residuals, we need to know the "right" statistical model. However, to determine the "right model," we need to know the "right" transformation. The way out of this dilemma is to proceed iteratively by tentatively entertaining a transformation, then tentatively identifying and fitting a model, then checking the transformation, and so on. To start this iterative process given in Figure 5.17, which we have already discussed in Chapter 4, we typically apply a number of graphical methods.

5.4.1 Preliminary Graphical Analysis

For our preliminary graphical analysis of the sales data, we first use the interaction plot demonstrated earlier in the analysis of the airline passenger data. This plot provides us with an understanding of the structure of the variation in the data over the years, over the month, and over the entire period of time. Figure 5.18a

TABLE 5.4 Monthly Sales Data for Company X from January 1965 to May 1971

Year	January	February	March	April	May	June	July	August	September	October	November	December
1965	154	96	73	49	36	59	95	169	210	278	298	245
1966	200	118	90	79	78	91	167	169	289	347	375	203
1967	223	104	107	85	75	99	135	211	335	460	488	326
1968	346	261	224	141	148	145	223	272	445	560	612	467
1969	518	404	300	210	196	186	247	343	464	680	711	610
1970	613	392	273	322	189	257	324	404	677	858	895	664
1971	628	308	324	248	272	—	—	—	—	—	—	—

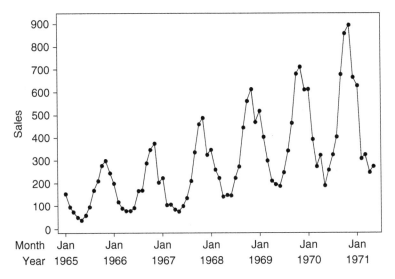

Figure 5.16 Time series plot of Company X's monthly sales of a single product.

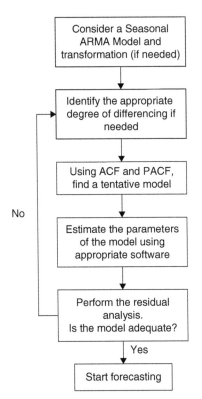

Figure 5.17 Stages of the time series model building process using Seasonal ARIMA models (Adapted from BJR)

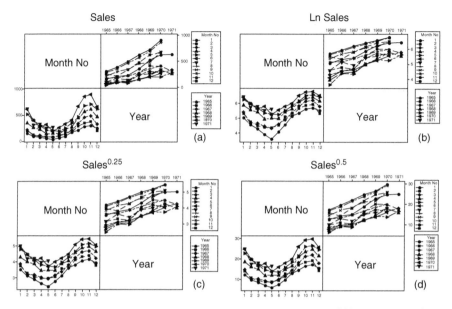

Figure 5.18   Interaction graphs for (a) sales, (b) ln(sales), (c) sales$^{0.25}$, and (d) sales$^{0.5}$.

is an interaction plot of the untransformed sales data. In the panel to the right in Figure 5.18a showing the monthly data over the years, we see that the trend lines are not parallel but "fan out." In other words, the variability increases as the average sales increases. Furthermore, in the panel to the left in Figure 5.18a showing the within-year monthly pattern, we see that there is not only a seasonal pattern in the sales but also a variability change from month to month. For example, the variability between years is much higher in October and November than in May and June. Therefore, we need to explore the possibility of a variance stabilizing transformation.

Before we engage in the search for an appropriate transformation, it should be pointed out that if the data spans only a narrow range, then a power transformation may not be of much help. The standard rule of thumb is that $\max(y_t)/\min(y_t) > 3$ for a power transformation to have much effect. In the present case, $\max(y_t)/\min(y_t) = 895/36 \approx 24$ so the sales data is definitely a candidate for transformation.

The most natural choice of transformation to try first is the log transformation. Figure 5.18b shows an interaction plot of the log transformed sales data, $\ln(y_t)$. The trend lines for the month over the years in the right-hand panel of Figure 5.18b now "fan in," showing almost the reverse pattern of Figure 5.18a. The left-hand panel of Figure 5.18b shows that the variability in May and June is the largest, exhibiting a reversed pattern from Figure 5.18a. This indicates that the log "overtransforms" the data. Therefore, we need something "milder" than the log. Examples of power transformations that "bend" the data less severely than the log are the square root $y_t^{0.5} = \sqrt{y_t}$ and the square root of the square root

$y_t^{0.25} = \sqrt{\sqrt{y_t}}$, the transformation indexed by $\lambda = 0.25$ being somewhat stronger and closer to the log, which corresponds to $\lambda = 0$.

We have applied the transformations, $y_t^{0.25}$ and $y_t^{0.5}$ to the sales data in Figure 5.18c and d, respectively. Figure 5.18c shows that for $y_t^{0.25}$, the trend lines, year by year and month by month, are almost parallel and the variability appears homogenous. In other words, the $y_t^{0.25}$ transformation seems to eliminate the dependency of the variability on the mean. Figure 5.18d shows the trend lines after the $y_t^{0.5}$ transformation. We see that with this transformation, the data still exhibits a bit of the "fanning out" of the trend lines. Thus, it appears that $y_t^{0.25}$ may be the more appropriate transformation. Indeed, the main issue in the heated debate following the original publication of the Chatfield and Prothero (1973) article referred to above was that for this particular data, applying the log transformation "overtransforms" the data, which in turn leads to a poor model fit and forecasts.

As discussed in the previous section, another useful way to graphically determine an appropriate transformation is to use the range–mean chart discussed by Jenkins (1979). The range–mean chart is a plot of the ranges versus the averages of subsections of a time series. For the sales data, one way to subsection the data is by years, using each year from 1965 to 1970, leaving out the last five observations since we do not have the data for the whole year 1971. Figure 5.19a shows the ranges of sales for each year between 1965 and 1970 versus the averages for each year during the same period. We see from

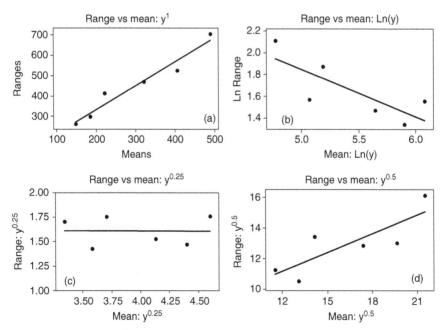

Figure 5.19 Range–mean charts for (a) no transformation, (b) $\ln(y_t)$, (c) $y_t^{0.25}$, and (d) $y_t^{0.5}$.

a somewhat different perspective that with no transformation, the within-year ranges increase with the yearly averages. The plot in Figure 5.19b of the ranges versus the averages after log transforming the sales data shows a negative trend, implying that as the averages increase the ranges decrease. Thus, we confirm that the log "overtransforms" the data. Figure 5.19c shows the $y_t^{0.25}$ transformation of the sales data. Here, we see an almost complete independence between the range and the average. Figure 5.19d shows the $y_t^{0.5}$ transformation of the sales data. In terms of strength, $y_t^{0.5}$ is in between y_t^1, no transformation, and $y_t^{0.25}$. We see that $y_t^{0.5}$ "undertransforms" the data. Thus, the $y_t^{0.25}$ transformation seems to be the most appropriate for this dataset. To further confirm this, we have shown the time series plots of (a) the sales data y_t, (b) $\ln(y_t)$, (c) $y_t^{0.25}$, and (d) $y_t^{0.5}$ in Figure 5.20. Even from this less detailed plot, it is clearly visible that the $y_t^{0.25}$ transformation seems to do the best job at keeping the seasonal amplitude stable.

5.4.2 Model Identification

Following the procedure outlined in Figure 5.17, now that we have tentatively identified a transformation, we proceed to identify an appropriate ARIMA time series model. The first step is to determine the degree of differencing necessary to make this clearly nonstationary seasonal data look stationary. Figure 5.21a shows the ACF for $z_t = y_t^{0.25}$. It shows a cyclic pattern that does not seem to die out even for large lags. This suggests, as a minimum, a first difference $\nabla^1 z_t = z_t - z_{t-1}$. The ACF of $\nabla^1 z_t$ at lags 12 and 24 shown in Figure 5.21b exhibits a persistent autocorrelation pattern, indicating the need for a seasonal differencing. Figure 5.21c shows the ACF for $\nabla^{12}\nabla^1 z_t = w_t$, which appears to

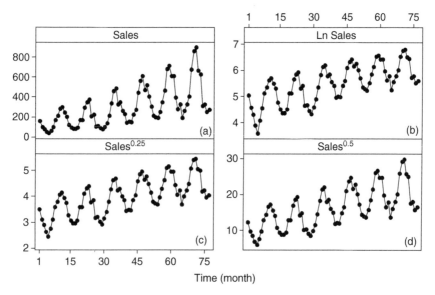

Figure 5.20 Time series plots of (a) sales, (b) ln(sales), (c) sales$^{0.25}$, and (d) sales$^{0.5}$.

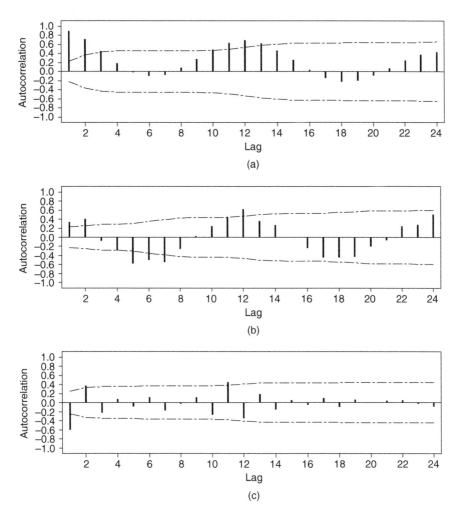

Figure 5.21 Autocorrelations for (a) $y_t^{0.25}$, (b) $\nabla^1 y_t^{0.25}$, and (c) $\nabla^{12}\nabla^1 y_t^{0.25}$.

be the one associated with a stationary time series. Further, we see a significant negative lag 1 autocorrelation followed by a tailing-off pattern. However, around lag 12, we see some borderline significant autocorrelations. Figure 5.22 shows the PACF of $\nabla^{12}\nabla^1 z_t = w_t$ exhibiting a large significant lag 1 spike followed by what looks like a cut off/zero partial autocorrelations for larger lags except around the seasonal length of 12.

With reference to the properties of ARMA models shown in Table 4.1, the strong lag 1 partial autocorrelation suggests a regular first order autoregressive model, AR(1). The ACF and PACF pattern around lag 12 further suggests a seasonal moving average term, SMA(1). In other words, a tentative model would be a seasonal ARIMA$(1, 1, 0) \times (0, 1, 1)_{12}$ model for $\nabla^{12}\nabla^1 z_t = w_t$. This is the model Chatfield and Prothero (1973) suggested for the log transformed data. However, the situation is not entirely unambiguous. A careful look at the

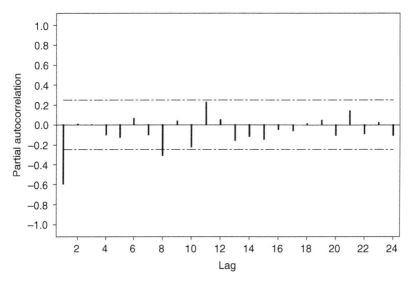

Figure 5.22 Partial autocorrelation for $\nabla^{12}\nabla^{1}y_t^{0.25}$.

ACF of $\nabla^{12}\nabla^{1}z_t = w_t$ indicates an alterative interpretation of a significant lag 1 autocorrelation followed by a cutoff and zero autocorrelations up to about lag 11 and 12. Thus, a seasonal ARIMA$(0, 1, 1) \times (0, 1, 1)_{12}$ model may also be appropriate. However, because of the historical background for this example, we proceed with the ARIMA$(1, 1, 0) \times (0, 1, 1)_{12}$ model advocated by Chatfield and Prothero (1973) and used by Box and Jenkins (1973) in their rebuttal.

5.4.3 Model Estimation and Diagnostic Checks

Now that we have decided tentatively to entertain the ARIMA$(1, 1, 0) \times (0, 1, 1)_{12}$ model for $z_t = y_t^{0.25}$, we proceed to fit this model to all 77 observations given in Table 5.5. The estimation summary is given in Table 5.5. Both the regular autoregressive parameter and the SMA parameter are significant. However, the significant Box–Pierce/Ljung–Box chi-square statistic at lags 12, 24, and 36 indicating significant autocorrelation left in the residuals is a cause for concern. Figure 5.23 shows the ACF and the PACF of the residuals. From that plot, we see a significant autocorrelation at lag 11, which is likely the reason for the significant Box–Pierce/Ljung–Box chi-square statistics. Furthermore, from the summary residual checks in Figure 5.24, we can see a significant outlier corresponding to observation 74. These are all signs that perhaps the ARIMA$(1, 1, 0) \times (0, 1, 1)_{12}$ does not fit the data all that well.

5.4.4 Interlude: A Bit about the Controversy on the Sales Data

Before we proceed further, it may be helpful to briefly explain a little more about the controversy this example generated. As indicated, the ARIMA$(1, 1, 0) \times$

TABLE 5.5 **Estimation Summary for Fitting the Seasonal ARIMA$(1, 1, 0) \times (0, 1, 1)_{12}$ Model to the 77 Transformed Sales Data $z_t = y_t^{0.25}$**

Model term	Coefficient	Standard error	t	p
AR 1	-0.4870	0.1175	-4.15	0.000
SMA 12	0.7077	0.1485	4.76	0.000

Differencing: 1 regular, 1 seasonal of order 12.

Number of observations: Original series 77, after differencing 64.

Residuals: SS, 1.40755; MS, 0.02270; df, 62.

Modified Box–Pierce (Ljung–Box) chi-square statistic:

Lag	12	24	36	48
Chi square	19.7	35.9	55.1	75.2
df	10	22	34	46
p-value	0.032	0.031	0.013	0.004

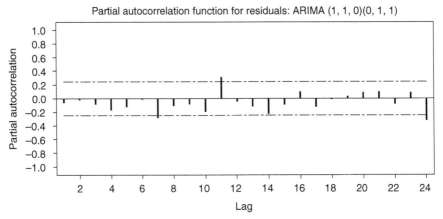

Figure 5.23 ACF and PACF of the residuals after fitting an ARIMA$(1, 1, 0) \times (0, 1, 1)_{12}$ model to $z_t = y_t^{0.25}$.

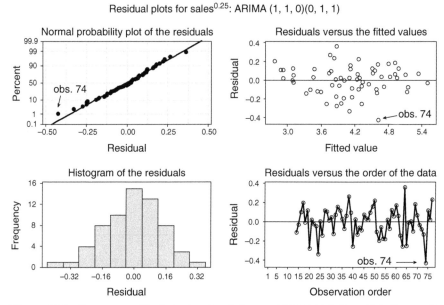

Figure 5.24   Summary residual checks after fitting an ARIMA$(1, 1, 0) \times (0, 1, 1)_{12}$ model to $z_t = y^{0.25}$.

$(0, 1, 1)_{12}$ is the model Chatfield and Prothero (1973) used, albeit fitted to $\ln(y_t)$, and not to $y_t^{0.25}$. In their response to the criticisms by Chatfield and Prothero (1973) about the poor performance of the Box–Jenkins approach, Wilson (1973), and later Box and Jenkins (1973), reanalyzed the same data and pointed out that part of the difficulties Chatfield and Prothero reported could be explained by their inappropriate "overtransformation" using the log. Instead, Wilson (1973) and Box and Jenkins (1973) suggested using the more general Box–Cox transformation that we now demonstrate. However, this approach requires that we have a tentative model in mind. Although as indicated above, the ARIMA$(1, 1, 0) \times (0, 1, 1)_{12}$ model may be problematic, this is the model used by Wilson (1973) and Box and Jenkins (1973) in their reanalysis. Therefore, we do the same here but will return to the issue of how appropriate the current model is in Section 5.4.6.

5.4.5   Box–Cox Transformation

The log transformation $\ln(y_t)$ and the power transformations $y_t^{0.25}$ and $y_t^{0.5}$ are (simplified) members of the more general class of Box–Cox transformation originally proposed by Box and Cox (1964); see Box *et al.* (2005) for a good practical discussion. We have discussed this family of transformations briefly in Chapter 3 for nonseasonal models. In this section, we provide greater detail about this very useful family of transformations. For these transformations, it is tentatively assumed that for some λ, an additive model with *normally distributed errors* and constant variance is appropriate for $y^{(\lambda)}$ defined by

$$y^{(\lambda)} = \begin{cases} \frac{(y^\lambda - 1)}{\lambda} & \text{for } \lambda \neq 0 \\ \ln(y) & \text{for } \lambda = 0 \end{cases} \qquad (5.10)$$

The Box–Cox family of transformations is continuous in λ and contains the log transformation as a special case ($\lambda = 0$). It is important to note that in the italicized statement above, it is the normality of the *errors* and not necessarily $y^{(\lambda)}$ that is referred to.

To get a feel for how the transformation works, we have, in Figure 5.25, plotted $y^{(\lambda)}$ in Equation (5.10) versus y for selected values of λ. The Box–Cox transformation affects the shape of the distribution of $y^{(\lambda)}$ as well as the relationship between the variance and the expected value of $y^{(\lambda)}$. Specifically, for $\lambda > 1$, the upper tail of the distribution of the original data, y, is stretched out, whereas for $\lambda > 1$, it is the lower tail that is drawn out. Typically, this will have a similar effect on the distribution of the residuals, but exactly *how* depends on the model.

To determine λ for a specific dataset, Box and Cox (1964) suggested using the maximum likelihood method, an approach based on a statistical argument that finds the parameter values in a model that for a given dataset is most likely. We will not go further into details, but only show how it works, which luckily is rather straightforward and easily done with standard statistical software.

To find the maximum likelihood estimate of λ and the time series parameters in an ARIMA model, we proceed in two steps. For a specific λ, we fit a given time series model and find the residual sum of squares (RSS). Note that if we use

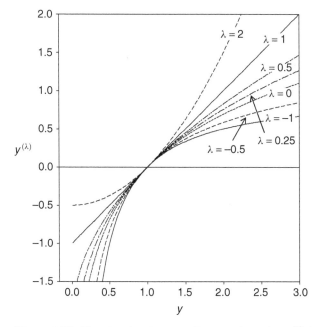

Figure 5.25 Plot showing the Box–Cox transformation $y^{(\lambda)}$ for selected values of λ.

a conditional least squares approach as MINITAB does, the results will only be approximate maximum likelihood estimates. However, those are typically close enough. Next, we repeat the fitting of the time series model and record the RSS for a range of values of λ. The (approximate) maximum likelihood value for λ is then the value for which the RSS is minimized.

The formulation of the Box–Cox transformation in Equation (5.10) is the version often used for theoretical discussions. However, the variances of the transformed data are not comparable for different values of λ and the scale depends on λ. Box and Cox (1964) therefore proposed an alternative normalized transformation that does not suffer from the scale problem given by

$$z^{(\lambda)} = \begin{cases} \frac{(y^\lambda - 1)}{\lambda \dot{y}^{\lambda-1}} & \text{for } \lambda \neq 0 \\ \dot{y} \ln(y) & \text{for } \lambda = 0 \end{cases} \tag{5.11}$$

where $\dot{y} = (y_1 y_2 \cdots y_n)^{1/n}$ is the geometric mean of the sample. Box and Cox (1964) then showed for the standardized transformation $z^{(\lambda)}$, how the appropriate λ can be estimated using maximum likelihood estimation.

The Optimal λ   In this section, we demonstrate how to find the (approximate) optimal Box–Cox transformation. Admittedly, it does involve some tedious repeated model fittings, but can be done relatively quickly with standard statistical software packages such as MINITAB and SAS JMP. It is also possible to write a simple macro to carry out the repeated computations. The steps of the computations are shown in Figure 5.26.

As indicated in Figure 5.26, we first need to compute the geometric mean, $\dot{y}$, of the data. The easiest way to compute $\dot{y}$ is by recognizing that if we take

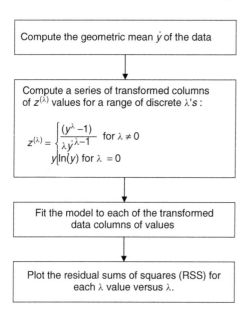

Figure 5.26   Algorithm for manually carrying out a Box–Cox transformation.

the log on both sides of $\dot{y} = (y_1 y_2 \cdots y_n)^{1/n}$, then $\ln(\dot{y}) = \frac{1}{n} \sum_{t=1}^{n} \ln(y_t)$. In other words, the log of the geometric mean is the regular arithmetic average of the log of the data. To get the geometric mean $\dot{y}$, we then take the antilog of $\ln(\dot{y})$, that is, $\dot{y} = \exp\{\ln(\dot{y})\}$. In the present case, we find that $\dot{y} = 236.567$.

Next we compute columns $\{y_t^{0.25}\}$ according to the formula in Equation (5.10) for a range of λ values. In the present case, we initially computed the columns for λ in the interval from -1 to $+1$ in steps of 0.1. However, we wanted to have better resolution close to where we suspected the optimal λ to be. Thus, we computed additional columns for $\lambda = (0.21, 0.22, \ldots, 0.29)$. Unfortunately, as we will see, that turned out to be insufficient. We therefore added additional columns for $\lambda = (0.11, 0.12, \ldots, 0.19)$. For each of these response columns, we fitted the ARIMA$(1, 1, 0) \times (0, 1, 1)_{12}$ model to all 77 observations and recorded the RSS. Finally, we plotted the RSS for the selected values of λ, which produced Figure 5.27. Now somewhat surprisingly, the minimum of the RSS is not around $\lambda = 0.25$. Instead, the optimal value is approximately $\hat{\lambda} = 0.16$.

As discussed earlier, it is always wise to provide a confidence interval for point estimates. In this case, an approximate 95% confidence interval for $\hat{\lambda}$ can be calculated graphically by finding the upper and lower confidence limits corresponding to the critical sum of squares, S, given as

$$S = S_{\hat{\lambda}} \left[1 + \frac{t_v^2(0.025)}{v} \right] \qquad (5.12)$$

where $S_{\hat{\lambda}}$ is the RSS at the minimum, $\hat{\lambda}$, and $t_v^2(0.025)$ is the lower 2.5 percentage point of the t-distribution with v degrees of freedom (i.e., the degrees of freedom of the RSS). In the present case, $v = 62$, $t_v^2(0.025) \approx 4.0$, and $S_{\hat{\lambda}} = 74,784$. Hence, we find that $S \approx 79,604$. From Figure 5.27, we see that the 95% confidence interval for $\hat{\lambda}$ stretches from about 0.0 to 0.32.

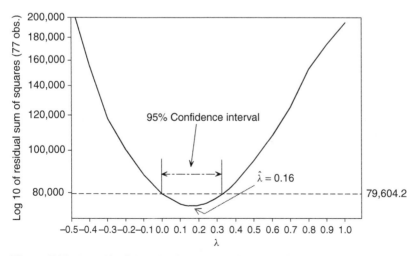

Figure 5.27   Log 10 of the RSS for a range of values of the transformation parameter λ using all 77 observations.

TABLE 5.6 Summary of Estimation Results After Fitting the ARIMA$(1, 1, 0) \times (0, 1, 1)_{12}$ Model to $\{z_t^{(0.16)}\}$ Using all 77 Observations

Model term	Coefficient	Standard error	t	p
AR 1	−0.5057	0.1124	−4.50	0.000
SMA 12	0.8005	0.1208	6.63	0.000

Differencing: 1 regular, 1 seasonal of order 12.

Number of observations: Original series 77, after differencing 64.

Residuals: SS, 74782.6; MS, 1206.2; df, 62.

Modified Box–Pierce (Ljung–Box) Chi-square statistic:

Lag	12	24	36	48
Chi square	18.0	34.3	57.2	73.6
df	10	22	34	46
p-value	0.055	0.046	0.008	0.006

Model Check for the Optimal λ With the optimal $\hat{\lambda} = 0.16$ value, we proceed to refit the ARIMA$(1, 1, 0) \times (0, 1, 1)_{12}$ model, carefully scrutinizing the fit as well as applying the usual residual checks. The estimation summary is shown in Table 5.6, the ACF and the PACF are shown in Figure 5.28, and the summary residual check is shown in Figure 5.29. We notice from Figure 5.28a that there is a relatively large residual autocorrelation left in the residuals giving a significant Box–Pierce/Ljung–Box chi-square statistic as seen in Table 5.6. We also notice from the residual plots in Figure 5.29, especially the normal plot, that we no longer see any serious outliers. Although still large and negative, the residual associated with observation 74 is no longer seriously departing from the general pattern of the residuals.

5.4.6 A Second Analysis of the Sales Data: Taking Account of Outliers

Despite the wide confidence interval for $\hat{\lambda}$ shown in Figure 5.27, this surprising discrepancy between the optimal estimate, $\hat{\lambda} = 0.16$, and the ad hoc estimate, $\tilde{\lambda} = 0.25$, based on the graphical analysis, brought back to mind that for $\tilde{\lambda} = 0.25$, observation number 74 looked like a very significant outlier. However, for the optimal transformation, it no longer appears to be one. Could this be the reason why the two λ estimates are so different? One of the unintended consequences of the Box–Cox method is that when we, by numerical optimization, find the optimal λ, we try as hard as we possibly can to find a λ value that will make even an outlier such as observation 74 fall into the normal distribution range. That means the Box–Cox approach may give undue leverage to extreme residuals and therefore is very sensitive to outliers. The outliers can also explain the relatively wider confidence intervals for the estimates of λ. However, the graphical procedures used above, although less refined, are more robust to outliers. We will now investigate this outlier issue.

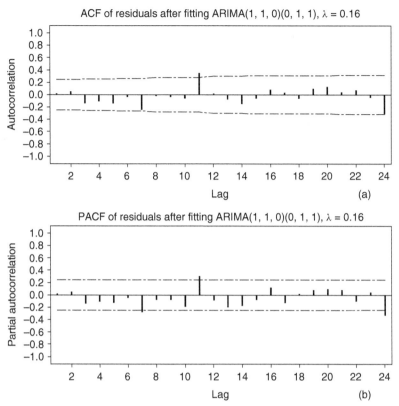

Figure 5.28 (a) ACF and (b) PACF of the residuals after fitting an ARIMA$(1, 1, 0) \times (0, 1, 1)_{12}$ model to $\{z_t^{(0.16)}\}$.

Unfortunately, outlier analysis for time series data is not as simple as for regular independent observations because the data depends critically on the particular sequence in which they appear. Therefore, we cannot simply take out a data point and reanalyze the data as we may, for example, in regression analysis. A number of approaches for outlier analysis in time series have been proposed in the literature. However, most are rather complex; for a brief overview see BJR. In our case, the outlier occurred toward the end of the sequence. Instead we simply reanalyze the first 72 observations to see what the optimal transformation based on those should be.

Following the algorithm outlined in Figure 5.26, limiting the range of λ from -0.3 to $+0.5$ in steps of 0.1, but with a finer step size between 0.1 and 0.3, we refitted the seasonal ARIMA$(1, 1, 0) \times (0, 1, 1)_{12}$ model to the 72 observations of $\{z_t^{(\lambda)}\}$. Figure 5.30 shows the plot of the RSS versus the transformation parameter λ. We now see that the optimal value is $\hat{\lambda} = 0.23$ and hence much closer to the ad hoc estimate of $\tilde{\lambda} = 0.25$. Again we also compute a confidence interval for $\hat{\lambda}$. This time we have $\nu = 57$, $t_\nu^2(0.025) \approx 4.0$, and $S_{\hat{\lambda}} = 64,261$. Hence $S \approx 68,771$. From Figure 5.30, we see that the 95% confidence interval for $\hat{\lambda}$ stretches from about 0.056 to 0.41.

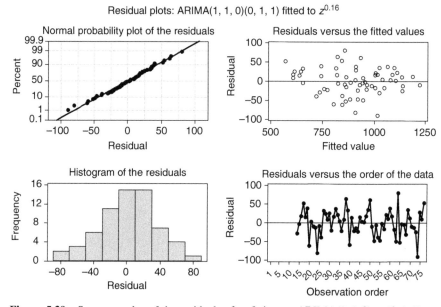

Figure 5.29 Summary plot of the residuals after fitting an ARIMA$(1, 1, 0) \times (0, 1, 1)_{12}$ model to $\{z_t^{(0.16)}\}$.

TABLE 5.7 Summary of Estimation Results After Fitting the ARIMA$(1, 1, 0) \times (0, 1, 1)_{12}$ Model to $\{z_t^{(0.25)}\}$ Using the First 72 Observations

Model term	Coefficient	Standard error	t	p
AR 1	−0.4865	0.1151	−4.23	0.000
SMA 12	0.7792	0.1396	5.58	0.000

Differencing: 1 regular, 1 seasonal of order 12.

Number of observations: Original series 72, after differencing 59.

Residuals: SS, 64310.4; MS, 1128.3; df, 57.

Modified Box–Pierce (Ljung–Box) Chi-square statistic:

Lag	12	24	36	48
Chi square	17.1	27.2	41.3	66.9
df	10	22	34	46
p-value	0.072	0.204	0.182	0.024

5.4.7 Final Model Checking

With the newly determined estimate of the transformation with $\hat{\lambda} = 0.23$, we see that when using Box–Cox transformations, we need to be observant with respect to possible outliers. Since 0.23 is an odd number and close to $0.25 = \frac{1}{4}$ and the sum of squares function in Figure 5.30 is relatively flat around the optimal value, it makes little difference if we proceed with the analysis using

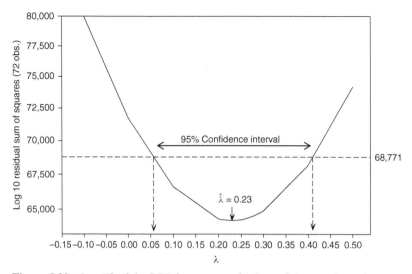

Figure 5.30 Log 10 of the RSS for a range of values of the transformation parameter λ using the first 72 observations.

$\lambda = 0.25$. Refitting the seasonal ARIMA$(1, 1, 0) \times (0, 1, 1)_{12}$ model to the first 72 observations for $\{z_t^{(0.25)}\}$, we obtain the estimation summary shown in Table 5.7. Comparing the results to those in Table 5.6, the most noteworthy difference is the change in the time series parameter estimates. However, we still see a significant Box–Pierce/Ljung–Box chi-square statistic indicating that there remains some autocorrelation in the residuals, which is a cause for some concern. This is also clear from the residual ACF and PACF as well as the summary residual check shown in Figures 5.31 and 5.32. Figure 5.33 shows the residuals plotted versus month and years. In general, the residuals look acceptable. We also notice from Figures 5.32 and 5.33 that observation number 64 now stands out as an outlier. However, it is not extreme and we will not pursue this any further except noticing that when we look at the time series plot of $z_t^{(0.25)}$ shown in Figure 5.34, it is not surprising in hindsight that observations 64 and 74 are called out as outliers; both seem to break the general pattern in the data.

5.4.8 Forecasting for the Sales Data

In the previous sections, we discussed the controversy surrounding the sales data. In this section, we look at the forecasting model proposed by Chatfield and Prothero (1973) and the alternative model proposed by Box and Jenkins (1973). We then evaluate the two models, and interpret them so we can see why Chatfield and Prothero originally felt that the ARIMA approach did not provide sensible forecasts. We also discuss in some detail an alternative seasonal time series model proposed by Wilson (1973). However, before we do that, we first digress to discuss forecasting from a very basic point of view.

With these two ARIMA$(1, 1, 0) \times (0, 1, 1)_{12}$ models, one for $\ln(y_t)$ and another for $y_t^{0.25}$ given in Table 5.8, we now produce forecasts starting at May

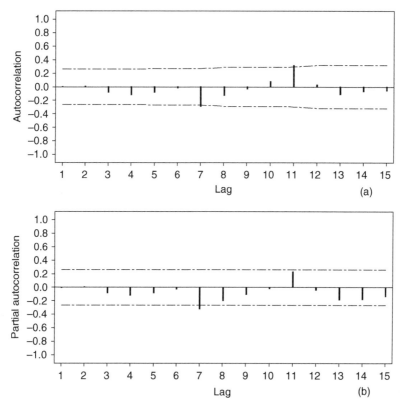

Figure 5.31 (a) ACF and (b) PACF of the residuals after fitting an ARIMA$(1, 1, 0) \times (0, 1, 1)_{12}$ model to the first 72 observations of $\{z_t^{(0.25)}\}$.

1971 or observation 77 and for the 12 months ahead. To compare the forecasts from the two models, we convert both back to the original scale. For the $\ln(y_t)$ data, we use the antilog $y_t = \exp\{\log(y_t)\}$, and for the $y_t^{0.25}$, we raise the data to the fourth power, that is, $y_t = (y_t^{0.25})^4$. The two different forecasts are shown in Figure 5.35. We see that Chatfield and Prothero were indeed correct in saying that the log-based forecast using the ARIMA$(1, 1, 0) \times (0, 1, 1)_{12}$ model provided unrealistic forecasts. In particular, we see that the forecasts for October and November of 1971 are too high. Indeed, it was those forecasts for October and November of 1971 combined with the complexity of the method that caused Chatfield and Prothero to loose faith in the Box–Jenkins approach to forecasting. Alternatively, however, we see that using the Box–Cox power transformation with $\lambda = 0.25$ as suggested by Box and Jenkins (1973) provides a much more realistic forecast more in line with the pattern seen prior to May of 1971.

5.4.8.1 An Alternative Representation of the Forecasting Model: The π Weights
We now explain how the forecasting model proposed by Box and Jenkins (1973) works and why it makes sense. The latter is an important issue

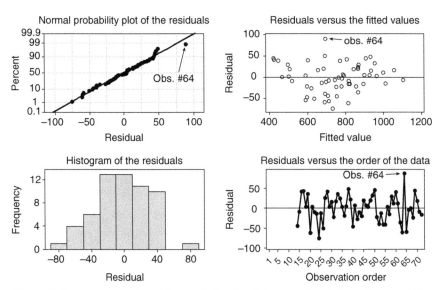

Figure 5.32 Summary plot of the residuals after fitting an $\text{ARIMA}(1,1,0) \times (0,1,1)_{12}$ model to the first 72 observations of $\{z_t^{(0.25)}\}$.

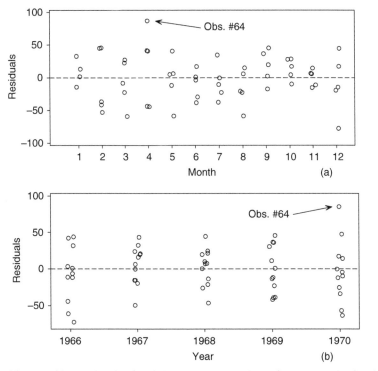

Figure 5.33 Residuals after fitting $\text{ARIMA}(1,1,0) \times (0,1,1)_{12}$ to the first 72 observations of $\{z_t^{(0.25)}\}$ plotted (a) versus month and (b) versus years. Open dots and jitter have been added to make overlapping points more visible.

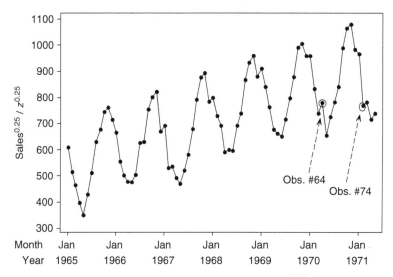

Figure 5.34 Time series plot of all 77 observations of $\{z_t^{(0.25)}\}$ with outliers highlighted.

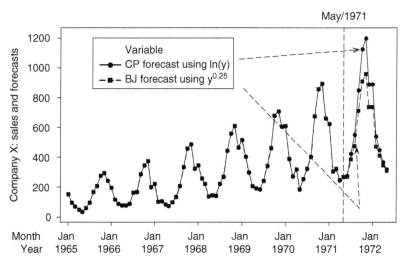

Figure 5.35 Sales data and two alternative forecasts using the transformations suggested by Chatfield and Prothero (1973) and Box and Jenkins (1973).

because it is sometimes claimed that the ARIMA-based models are difficult to interpret. There is of course some truth in that. However, as we will see, there are ways to gain insight on how they work and why they make sense. One way is to study the π weights in the EWMA forecast, which was discussed in Chapter 4. In that case, we had the representation of the forecast as the infinite weighted sum of the past observations:

$$\hat{z}_t(1) = \pi_1 z_t + \pi_2 z_{t-1} + \pi_3 z_{t-2} + \pi_4 z_{t-3} + \pi_4 z_{t-4} + \cdots$$

$$= \sum_{j=1}^{\infty} \pi_j z_{t+1-j} \tag{5.13}$$

As indicated in Chapter 4, there are several ways to express the forecasting equation for an ARIMA time series model. However, we believe that the representation given in Equation (5.13) provides insight into how the forecast works, as shown below.

5.4.8.2 *Forecasting the Seasonal ARIMA*$(1, 1, 0) \times (0, 1, 1)_{12}$ *Model*   Let us now turn to the seasonal ARIMA$(1, 1, 0) \times (0, 1, 1)_{12}$ for the sales data. This model can be written in the difference equation form as

$$z_t = (1 - \phi_1)z_{t-1} - \phi_1 z_{t-2} + z_{t-12} + (1 - \phi_1)z_{t-13}$$

$$+ \phi_1 z_{t-14} + a_t - \theta_{12} a_{t-12} \tag{5.14}$$

Admittedly, this is not an appealing equation. However, as for the IMA(1, 1) model we discussed in Chapter 4, we can, for the ARIMA$(1, 1, 0) \times (0, 1, 1)_{12}$, obtain a difference equation forecasting model similar to the EWMA, which provides the optimal forecasts for the IMA(1, 1) model. Fortunately, we do not need to write out the particular forecasting model. Most modern statistical software will automatically compute it and produce the forecasts for us as shown in Figure 5.35. However, aside from the plot, it does not facilitate much insight on how the forecast equation works. We can get this from the alternative π weight form given in Equation (5.13).

To derive the equations for computing the π weights for a given model is somewhat involved. We refer to BJR for details. Some statistical software, for example, the powerful statistical software package SCA WorkBench from Scientific Computing Associates, provides the π weights. In Figure 5.36, we simply provide a plot of the π weights for the seasonal ARIMA$(1, 1, 0) \times (0, 1, 1)_{12}$ fitted to $z_t = y_t^{0.25}$ where y_t is the sales data. We see that the forecast is a weighted average of past observations. However, because of the strong seasonality of the data, the plot of the π weights shows large positive, but exponentially declining, spikes 12 months apart. Thus, the forecast is primarily an exponentially weighted average of the data 12 months apart. For example, a forecast for the sales in December of 1971 is mostly an EWMA forecast of all the past Decembers. However, the smaller negative spikes show that the 2 months immediately before each past December are also to be taken into account. In other words, to make a forecast for December 1971, we combine an exponential moving average of all the past December sales with some additional input from the sales in the immediate past month, November 1971. We further correct it a bit by subtracting an exponential moving average of all the previous Octobers and Novembers. Since the π weights are used in a moving average manner, we can make a similar argument for the forecast of any one of the other 12 months of the year.

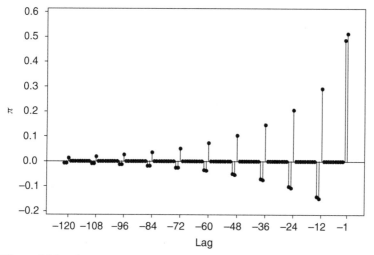

Figure 5.36 The π weights for the seasonal ARIMA $(1, 1, 0)(0, 1, 1)_{12}$ model fitted to sales$^{0.25}$.

5.4.9 An Alternative Model for the Sales Data

In the discussion of Company X's sales data, an alternative and quite different model was also discussed, in particular by Wilson (1973). He argued as follows. If we look at the original sales data (without any transformation) and examine each separate month, January, February, etc, as in Figure 5.37, we see that each of these 12 months exhibits a linear trend, each with different slope and intercept. In other words, we should be able to forecast this data by linear extrapolation for each separate month.

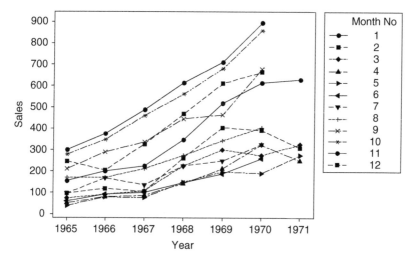

Figure 5.37 Sales for each of the 12 months of the year.

From calculus, we recall that if we differentiate a linear function $y = a + bx$ once we get $y' = b$, and if we do it twice we get $y'' = 0$. For discrete time series, the differencing operation $\nabla z_t = z_t - z_{t-1}$ plays the same role as differentiation. In this case, since we are considering a linear slope and intercept line between observations 12 months apart, we need to use a 12-month seasonal differencing $\nabla_{12} z_t = z_t - z_{t-12}$ operation, applied twice. In other words, a first order seasonal difference is $\nabla_{12} z_t = z_t - z_{t-12}$ and a second order seasonal differencing is $\nabla_{12}^2 z_t = (z_t - z_{t-12}) - (z_{t-12} - z_{t-24}) = z_t - 2z_{t-12} + z_{t-24}$. Such a second order seasonal differencing will eliminate a slope and intercept between months to make the data appear as a stationary time series fluctuating around zero. We can see how that works in Figure 5.38 showing time series plots of (a) the original (untransformed) sales data z_t, (b) the first order seasonal difference of the sales data $\nabla_{12} z_t$, and (c) the second order seasonal difference $\nabla_{12}^2 z_t$. Indeed, we see from Figure 5.37c that the second order seasonally differenced data shows the kind of stationary and variance homogenous pattern we like to see in the data before we start fitting stationary ARMA models. Of course, one unfortunate consequence of using a second order seasonal difference is that we loose the initial $2s = 2 \times 12 = 24$ observations from an already relatively short time series.

To identify a model for the second order seasonal difference $w_t = \nabla_{12}^2 z_t$ we have, in Figure 5.39a, plotted the ACF and in Figure 5.39b the PACF. From the ACF, we see a faint but nevertheless lingering pattern, suggesting an autoregressive model. The PACF shows two initial significant spikes followed by a cutoff, suggesting the need for an AR(2) model for $w_t = \nabla_{12}^2 z_t$. Without getting into the details, we simply report that the residuals from that model indicated some leftover seasonal pattern. This led us to consider an ARIMA(2, 0, 0) × (0, 0, 2)$_{12}$ model for $w_t = \nabla_{12}^2 z_t$, which is the same as an ARIMA(2, 0, 0) × (0, 2, 2)$_{12}$ for the original data z_t. This is the model Wilson (1973) suggested.

We proceeded to fit Wilson's model to the data. However, using several different software packages, we were unable to get the same estimated coefficients

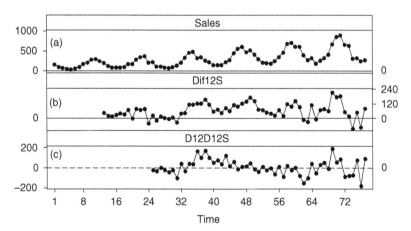

Figure 5.38   (a) Untransformed data, (b) after first order seasonal differencing, and (c) after second order seasonal differencing.

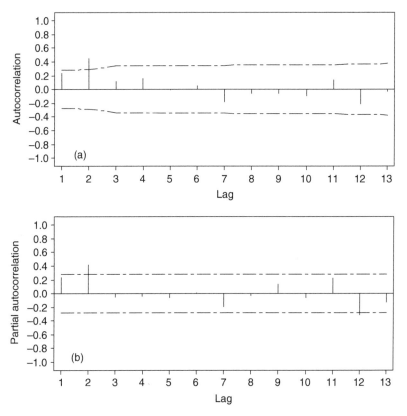

Figure 5.39 (a) ACF and (b) PACF of the second order seasonal difference, $w_t = \nabla_{12}^2 z_t$.

as he reported. We suspect that one reason is that when we start with only 77 observations and try to fit a complex 12-month seasonal double differenced model, there will be few observations left and we will therefore likely encounter unstable estimation results. The results obtained after fitting Wilson's model are shown in Table 5.9. We notice that the regular second order autoregressive term is not particularly significant. Fitting the same model with SCA WorkBench and SAS JMP, both of which use a maximum likelihood algorithm, we get the results shown in Table 5.10. It is striking (and educational) how different the estimation results are! We also notice that if we add the two SMA coefficients, 0.7726 and 0.2271, we get a number very close to one. It can be shown (BJR) that if the sum of the two SMA coefficients is one, the model is not invertible, an undesirable property as discussed in Chapter 4 because then model cannot be expressed in the form of a convergent infinite autoregressive sequence. It can further be shown that if the sum of the two moving average coefficients is close to one, it may be an indication of overdifferencing; see Chapter 7 and Abraham and Ledolter, 1983, p. 233. This may be another reason for the trouble in estimating this model. Indeed, we should always keep in mind that if ARIMA models include moving average terms, then they can be difficult to estimate. Hence, we should always

TABLE 5.8   Summary of Estimation Results After Fitting a Seasonal ARIMA$(1, 1, 0) \times (0, 1, 1)_{12}$ Model to $\{\ln(y_t)\}$ and $\{y_t^{0.25}\}$ Using all 77 Observations

Power transformation	$z = \ln(y)$	$z = y^{0.25}$
AR 1	−0.4580	−0.4870
SMA 12	0.7954	0.7077

TABLE 5.9   Estimation Summary for Fitting the Seasonal ARIMA$(2, 0, 0) \times (0, 2, 2)_{12}$ Model to the 77 Transformed Sales Data z_t Using MINITAB and Conditional Least Squares

Model term	Coefficient	Standard error	t	p
AR 1	0.4039	0.1465	2.76	0.008
AR 2	0.1961	0.1411	1.39	0.171
SMA 12	1.3271	0.2091	6.35	0.000
SMA 24	−0.5578	0.2824	−1.98	0.054

TABLE 5.10   Estimation Summary for Fitting the Seasonal ARIMA$(2, 0, 0) \times (0, 2, 2)_{12}$ Model to the 77 Transformed Sales Data z_t Using Exact Maximum Likelihood Estimation (SCA or JMP)

Model term	Coefficient	Standard error	t	p
AR 1	0.1616	0.1340	1.21	0.23
AR 2	0.4534	0.1332	3.40	0.00
SMA 12	0.7726	0.1644	4.70	0.00
SMA 24	0.2271	0.1901	1.19	0.24

be alert to convergence problems. Especially when we have few observations, maximum likelihood estimation is more reliable. An alternative interpretation of the model is the presence of a *deterministic* seasonal component, an issue we will not pursue here; see BJR for more details.

If we use the ARIMA$(2, 0, 0) \times (0, 2, 2)_{12}$ model as proposed by Wilson for forecasting using the two different sets of estimates as given in Tables 5.9 and 5.10, the forecasts turn out to be relatively close and quite reasonable. This is illustrated in Figure 5.40 where we have superimposed the two forecasts on the sales data. We notice that the model obtained by the maximum likelihood fit seems to overestimate the sales in October and November of 1971 somewhat more than the conditional least squares model, but both seem more realistic than that obtained from the seasonal ARIMA $(1, 1, 0) \times (0, 1, 1)_{12}$ model fitted to ln(sales) as proposed by Chatfield and Prothero (1973). Thus perhaps, the ARIMA$(2, 0, 0) \times (0, 2, 2)_{12}$ model merits some consideration.

Expressing the forecast equation as a generalization of an exponential moving average provides an intuitive understanding of how the forecast equation

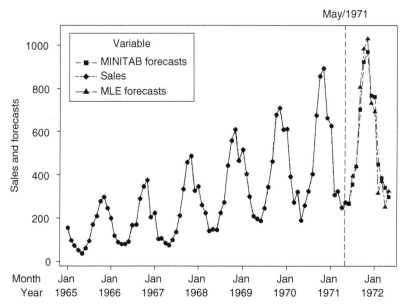

Figure 5.40 Sales and the forecasts based on model estimated by conditional least squares (MINITAB) and maximum likelihood methods.

works. There was also an alternative time series model by Wilson (1973) based on extrapolating the monthly linear trends using a somewhat unusual approach involving a second order seasonal difference. Both types of models provided what appear to be quite reasonable forecasts that were better than what CP originally reported.

The fact that two very different models can produce good forecasts is of course just another example illustrating that there is no such thing as the "correct" model. As George Box often has said "All models are wrong, but some are useful!" In this case, we found two models that appear useful for forecasting. However, regardless of the benefit of not needing to transform the data as with Wilson's model, we would like to think that the more straightforward ARIMA $(1, 1, 0) \times (0, 1, 1)_{12}$ model fitted to the double square root transformed data, $z = y_t^{0.25}$, is more preferable.

EXERCISES

5.1 Table B.8 contains the monthly residential electricity sales in the United States from January 1990 to July 2010. First consider only the data from January 1990 to December 2009. Is a transformation of the data needed as discussed in Section 5.4.1?

5.2 For the data in Exercise 5.1, fit an appropriate ARIMA model to the transformed (if needed) data.

5.3 For the data in Exercise 5.1

a. Forecast the residential electricity sales in the United States for the first 7 months of 2010 using the model you developed in Exercise 5.2.

b. The residential electricity sales from January 2010 to June 2010 are also provided in Table B.8. Using the model you developed in Exercise 5.2, make a 1-month-ahead forecast for December 2009. Then assume that the current month is January 2010. Fit a new model to original dataset extended with the observation for January. Proceed with making a 1-month-ahead forecast for February 2010. Repeat this procedure until June 2010.

c. Compare the sum of squared prediction errors $(\sum (z_t - \hat{z}_{t-1}(1))^2)$ for the forecasts in parts a and b.

5.4 Table B.8 contains the monthly average residential electricity retail price in the United States from January 1990 to July 2010. First consider only the data from January 1990 to December 2009. Is a transformation of the data needed as discussed in Section 5.4.1?

5.5 For the data in Exercise 5.4, fit an appropriate ARIMA model to the transformed (if needed) data.

5.6 For the data in Exercise 5.4

a. Forecast the residential electricity retail prices in the United States for the first 7 months of 2010 using the model you developed in Exercise 5.5.

b. The residential electricity retail prices from January 2010 to June 2010 are also provided in Table B.8. Using the model you developed in Exercise 5.5, make a 1-month-ahead forecast for December 2009. Then assume that the current month is January 2010. Fit a new model to original dataset extended with the observation for January. Proceed with making a 1-month-ahead forecast for February 2010. Repeat this procedure until June 2010.

c. Compare the sum of squared prediction errors $(\sum (z_t - \hat{z}_{t-1}(1))^2)$ for the forecasts in parts a and b.

5.7 Compare the models obtained in Exercises 5.2 and 5.5.

5.8 Table B.9 contains the monthly outstanding consumer credits provided by commercial banks in the United States from January 1980 to August 2010. First consider only the data from January 1980 to December 2009. Is a transformation of the data needed as discussed in Section 5.4.1?

5.9 For the data in Exercise 5.8, fit an appropriate ARIMA model to the transformed (if needed) data.

5.10 For the data in Exercise 5.8

a. Forecast the outstanding consumer credits in the United States for the first 8 months of 2010 using the model you developed in Exercise 5.9.

b. The outstanding consumer credits from January 2010 to August 2010 are also provided in Table B.9. Using the model you developed in Exercise 5.9, make a 1-month-ahead forecast for December 2009. Then assume that the current month is January 2010. Fit a new model to the original dataset extended with the observation for January. Proceed with making a 1-month-ahead forecast for February 2010. Repeat this procedure until August 2010.

    c. Compare the sum of squared prediction errors $(\sum (z_t - \hat{z}_{t-1}(1))^2)$ for the forecasts in parts a and b.

5.11 Consider the monthly data on sea levels in Los Angeles, CA, from 1889 to 1974 given in Table B.4. Is a transformation of the data needed as discussed in Section 5.4.1?

5.12 For the data in Exercise 5.11, fit an appropriate ARIMA model to the transformed (if needed) data. Compare your model to the model from Exercise 3.15.

CHAPTER **6**

TIME SERIES MODEL SELECTION

6.1 INTRODUCTION

One of the most difficult tasks in time series analysis is model selection—the choice of an adequate model for a given set of data. Even trained time series analysts will often find it difficult to choose a model. As we have seen in the previous chapters, in many cases, several models may seem to fit the data well. The question is then to figure out which one to choose. Some authors prefer the model that performs best according to certain numerical criteria. We prefer to exercise judgment and use such criteria as one among several considerations in choosing a model. In this chapter, we primarily use an example concerning the number of users of an internet server originally due to Makridakis *et al.* (2003) to discuss some of the issues involved in selecting adequate models for time series data.

6.2 FINDING THE "BEST" MODEL

Before we get into a detailed discussion, it is worth remembering once again that George Box famously stated that "All models are wrong but some are useful." Although this applies to all types of statistical models, it is particularly relevant for time series models. The models we fit to time series data are selected on the basis of a study of the properties of the data, in particular, sample ACF and PACF. The process of selecting an appropriate model is therefore a somewhat subjective process and, as indicated above, requires experience and judgment, something we should never be shy of exercising. Indeed, Box and Jenkins (1970) acknowledged the subjectivity when they used language such as "tentatively entertaining a model," which indicates that there is nothing definite about the models. The models may be "useful" and "adequate," but it would be inappropriate to "assume" that they are "true" or that there is anything like a "correct" model for a given time series.

One way to think of a time series model is that it is an approximation to some unknown dynamic system. Since the model is an approximation, it is possible that several models may be more or less equivalent. This raises the

Time Series Analysis and Forecasting by Example, First Edition. Søren Bisgaard and Murat Kulahci.
© 2011 John Wiley & Sons, Inc. Published 2011 by John Wiley & Sons, Inc.

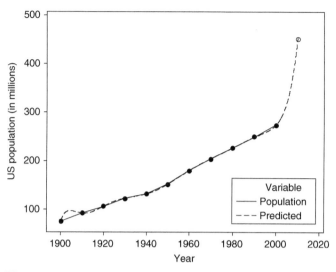

Figure 6.1 The US population according to the census versus time from 1900 to 2000. A perfect fit to the past data, but a poor and unrealistic forecast for the year 2010.

question of which one to prefer. Here, we may apply a number of model selection criteria. One would be to try to find the model that minimizes the residual sum of squares or maximizes the R^2. However, as in regression analysis, we can make the residual sum of squares as small or the R^2 as large as we like by adding more terms to the model. If a statistical model involves too many terms, the parameters will usually be poorly estimated since the information in the data will be dissipated over too many of them. Models that involve superfluous parameters will also, as a general rule, produce poor forecasts. This is illustrated in Figure 6.1 where we fitted a tenth order polynomial to the US population by the official census every 10 years from 1900 to 2000 and extrapolated to 2010. We obtained a perfect fit to the 10 observations, but a poor and unrealistic prediction for 2010. Overfitting, as this is called, often leads to models that "locally" fit nearly perfectly, but "globally" perform poorly. Therefore, we prefer *parsimonious models* that represent the data adequately while using a minimum number of parameters.

6.3 EXAMPLE: INTERNET USERS DATA

Akaike's information criterion (AIC) is often used for model selection, especially in the time series context (Akaike, 1974). This criterion includes a penalty for overparameterization similar to the adjusted R^2 in regression analysis, but is more generally applicable. As the adjusted R^2 in regression analysis, AIC should not be used mechanically as the sole criterion for model selection. Instead, we advocate the use of informed judgment where AIC or similar criteria are used when selecting a model. We will now use an example to discuss the problem of model selection and the use of model selection criteria.

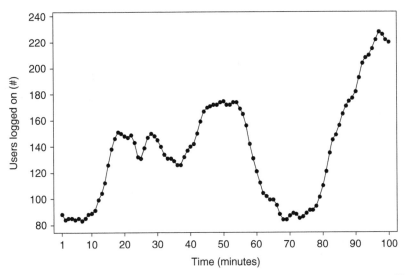

Figure 6.2   Time series plot of the number, z_t, of internet server users over a 100-minute period.

The example we will use to illustrate some of the issues associated with selecting an adequate parsimonious time series model is originally from Makridakis *et al.* (2003). The observations, z_t, are the number of users logged on to an internet server each minute for a period of 100 min. Figure 6.2 shows a plot of the original time series z_t. We immediately see that this process appears to be nonstationary. This is further corroborated by the plot of the sample autocorrelation function shown in Figure 6.3 showing persistent autocorrelations for large

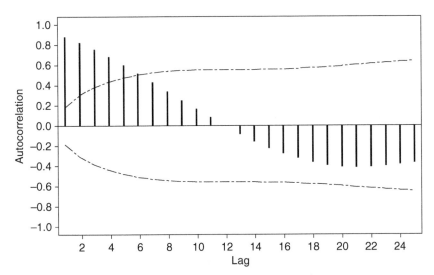

Figure 6.3   The sample autocorrelation for the number, z_t, of internet server users over a 100-minute period.

lags. Considering the context of the data—the number of users logged on to a server—it makes sense that the series is nonstationary. If a process is stationary, we expect (i.e., assume) it to have a long-term fixed mean. However, it does not seem reasonable to expect that the number of users logged on to a server would have a fixed mean at all times. More likely, the mean would gradually change over time and the number of users logged on would be a nonstationary process. Indeed, within a certain (short) period of time during regular business hours, it is more plausible that the change (i.e., the first difference) in the number of users logged on from minute to minute would be a stationary process. We therefore tentatively suggest that we should difference the data.

Figure 6.4 shows the first difference, $w_t = \nabla z_t = z_t - z_{t-1}$, the change from minute to minute of users logged on to the internet server. We see that the changes look much more stable. We also see from Figure 6.5a that ACF of the differences $w_t = \nabla z_t$ dies out relatively quickly, indicating that the differences may be a stationary process. We therefore tentatively entertain the idea that the first difference $w_t = \nabla z_t$ is a stationary process.

Next, with reference to Table 4.1, we proceed to determine a tentative model for the differences $w_t = \nabla z_t$. From Figure 6.5a and b that show the sample ACF and the PACF, respectively, we see that there are at least two different interpretations of the ACF and the PACF. Hence, at least two different models seem plausible:

1. ACF shows a damped sine wave and PACF cuts off after 3 lags. This pattern suggests that the model should be an autoregressive AR(3) model (i.e., an ARIMA(3, 0, 0) model) for the differences w_t, or an ARIMA(3, 1, 0) model for the original data z_t.

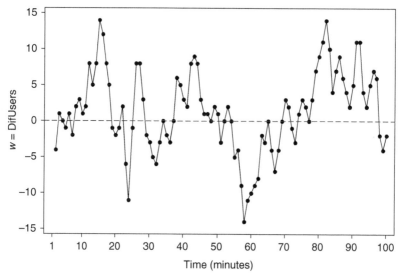

Figure 6.4 Time series plot of the difference $w_t = \nabla z_t$ (changes) of the number of internet server users over a 100-minute period.

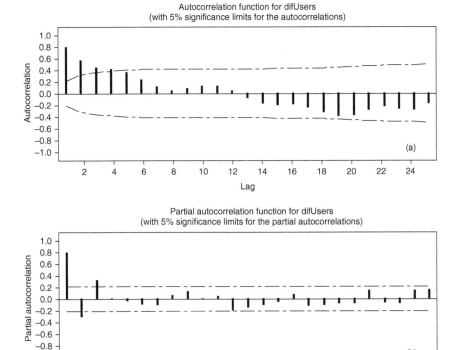

Figure 6.5 (a) The sample ACF and (b) the sample PACF of the differences (changes) of the number of internet server users over a 100-minute period.

2. Both ACF and PACF show damped sine waves. This pattern suggests a mixed ARMA model. Specifically, we decided to consider an ARIMA(1, 1, 1) for the original data z_t or an ARMA(1, 1) for the differences, w_t.

This ambiguity raises the question of which of the two models to choose. Fortunately, it is relatively simple to fit both models, see how well they fit, and compare. The estimation summary for the ARIMA(3, 1, 0) model is shown in Table 6.1 and the estimation summary for the ARIMA(1, 1, 1) model is shown in Table 6.2. The ACF and PACF of the residuals of the ARIMA(3, 1, 0) model are shown in Figure 6.6 and the corresponding residual checks in Figure 6.7. The ACF and PACF of the residuals of the ARIMA(1, 1, 1) model are shown in Figure 6.8 and the corresponding residual checks in Figure 6.9. We see from the estimation summaries that both models fit well, all the parameters are significant, and the modified Box–Pierce (Ljung–Box) chi-square statistics for both models indicate good fits. Moreover, the ACFs and PACFs in Figures 6.6 and 6.8, and the residual checks in Figures 6.7 and 6.9 for both models look good; from a diagnostic checking point of view, the two models are essentially indistinguishable from each other.

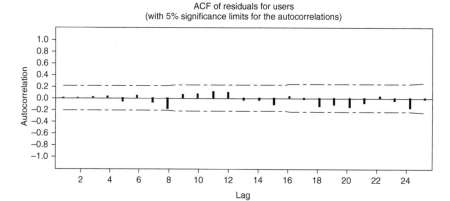

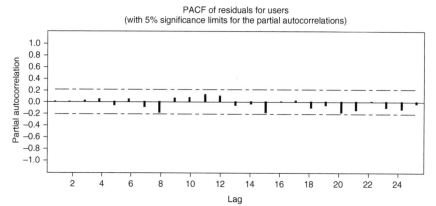

Figure 6.6 Sample ACF and sample PACF for the residuals from the ARIMA(3, 1, 0) model.

TABLE 6.1 Estimation Summary for Fitting the ARIMA(3, 1, 0) Model to the Internet Server Data z_t

Model term	Estimate	Standard error	t	p
AR 1 : ϕ_1	1.1632	0.0955	12.18	0.000
AR 2 : ϕ_2	−0.6751	0.1360	−4.96	0.000
AR 3 : ϕ_3	0.3512	0.0956	3.67	0.000

Differencing: 1 regular.

Number of observations: Original series 100; after differencing 99.

Residuals: SS $= 917.812$; MS $= 9.561$; $df = 96$.

Modified Box–Pierce (Ljung–Box) Chi–square statistic:

Lag	12	24	36	48
Chi square	7.5	20.2	31.5	46.5
df	9	21	33	45
p-value	0.587	0.509	0.539	0.411

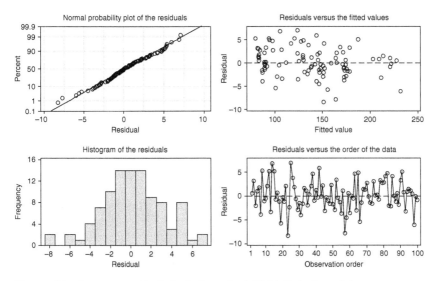

Figure 6.7 Summary residual check from the ARIMA(3, 1, 0) model.

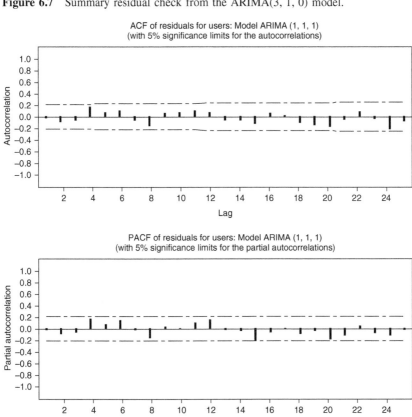

Figure 6.8 Sample ACF and sample PACF for residuals from the ARIMA(1, 1, 1) model.

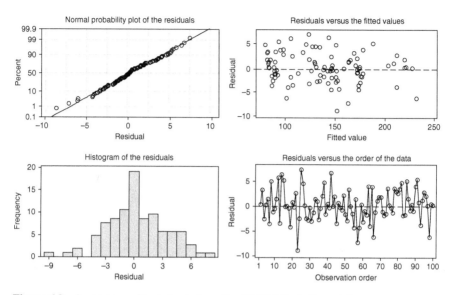

Figure 6.9   Summary residual check from the ARIMA(1, 1, 1) model.

TABLE 6.2   Estimation Summary for Fitting the ARIMA(1, 1, 1) Model to the Internet Server Data z_t

Model term	Estimate	Standard error	t	p
AR 1 : ϕ_1	0.6573	0.0868	7.57	0.000
MA 1 : θ_1	-0.5301	0.0974	-5.44	0.000

Differencing: 1 regular.

Number of observations: Original series 100; after differencing 99.

Residuals: SS = 961.617; MS = 9.914; df = 97.

Modified Box–Pierce (Ljung–Box) Chi-square statistic:

Lag	12	24	36	48
Chi square	10.3	26.3	38.1	54.2
df	10	22	34	46
p-value	0.417	0.237	0.289	0.190

The only difference between the two models is in the residual mean squares (MS), which are 9.561 for the ARIMA(3, 1, 0) model and 9.914 for the ARIMA(1, 1, 1) model. This would suggest that the ARIMA(3, 1, 0) model is slightly better than the ARIMA(1, 1, 1) model. However, the ARIMA(3, 1, 0) model involves three autoregressive parameters, whereas the ARIMA(1, 1, 1) model has two parameters: an autoregressive and a moving average parameter. Indeed, the parameters in the ARIMA(1, 1, 1) model are slightly better estimated. Yet both models seem to filter out the time series structure of the data and leave only white noise as residuals. So which of the two models is best? Which one should we choose? Moreover, what difference does it make? The following sections provide answers to these three questions.

6.4   MODEL SELECTION CRITERIA

A common approach to model selection recommended in many textbooks is to use a model fitting criterion. Some software packages even provide automatic model selection facilities using such criteria. The most popular criteria are AIC, Akaike's bias-corrected information criterion (AICC) suggested by Hurvich and Tsai (1989), and the Bayesian information criterion (BIC) introduced by Schwarz (1978). For the mathematical background for each of these criteria, see BD and BJR. Note that textbooks and software may differ in how they define and compute the different criteria and that the criteria are relative and sometimes shown with an additive constant, which is not dependent on the data, the number of observations, or the number of parameters and therefore not relevant for choosing a model for a given set of data.

For a sample size of n observations, AIC is given by

$$\text{AIC} = -2\ln(\text{maximized likelihood}) + 2r$$
$$\approx n\ln(\hat{\sigma}_a^2) + 2r \tag{6.1}$$

where $\hat{\sigma}_a^2$ is the maximum likelihood estimate of the residual variance σ_a^2; r is the number of parameters estimated in the model including a possible constant term. We prefer models with the smallest AIC. For practical purposes, if the computer software uses something other than maximum likelihood estimation, for example, conditional least squares as MINITAB does, then one can substitute in the residual MS for $\hat{\sigma}_a^2$.

To appreciate how AIC works, we can rewrite the criterion as $\text{AIC} \approx n\ln(\hat{\sigma}_a^2) + P_{\text{AIC}}$ where $P_{\text{AIC}} = 2r$ is a penalty for including superfluous parameters. Thus, as we add terms and parameters to a model, we will typically reduce the residual variance $\hat{\sigma}_a^2$ and hence $n\ln(\hat{\sigma}_a^2)$. However, we also pay a penalty, $P_{\text{AIC}} = 2r$, for the number of parameters. The optimal model is found when the additional penalty for adding a parameter exceeds the added benefit in terms of reducing the residual variance.

It turns out that for autoregressive models AR(p), Akaike's original AIC has a tendency to overestimate p. A modified criterion with a more severe penalty to avoid overfitting called the *bias-corrected Akaike's information criteria* denoted by AICC was therefore proposed by Hurvich and Tsai (1989) as

$$\text{AICC} = -2\ln(\text{maximized likelihood}) + \frac{2rn}{n-r-1}$$
$$\approx n\ln(\hat{\sigma}_a^2) + \frac{2rn}{n-r-1} \tag{6.2}$$

Likewise, the AICC can be rewritten as $\text{AICC} \approx n\ln(\hat{\sigma}_a^2) + P_{\text{AICC}}$ where $P_{\text{AICC}} = 2rn/(n-r-1)$ is the penalty function for the number of parameters.

The BIC proposed by Schwarz (1978) is yet another criterion, which attempts to correct for AIC's tendency to overfit. This criterion is given as

$$\text{BIC} = -2\ln(\text{maximized likelihood}) + r\ln(n)$$
$$\approx n\ln(\hat{\sigma}_a^2) + r\ln(n) \tag{6.3}$$

We can again rewrite this as $\text{BIC} \approx n\ln(\hat{\sigma}_a^2) + P_{\text{BIC}}$ where $P_{\text{BIC}} = r\ln(n)$ is a penalty function. As for all the criteria, the preferred model is the one with a minimum BIC.

The penalty for introducing unnecessary parameters is more severe for BIC and AICC than for AIC. Therefore, AICC and BIC tend to select simpler and more parsimonious models than AIC. Simulation studies seem to indicate that BIC performs better in large samples, whereas AICC performs better in smaller samples where the relative number of parameters is large. Thus, although AIC is the most popular criterion and available in most software packages, BIC and AICC are, by some experts, considered "better."

All three criteria are defined for stationary ARMA(p, q) models. Therefore, to use them for nonstationary and seasonal models where we may need to difference the data, we need to be very careful and define n in the above criteria as the *effective number of observations*. The effective number of observations is equivalent to the number of residuals in the series. In other words, if a time series consisting of n' original observations has been differenced d times, then the effective number of observations, n, is given by $n = n' - d$. Note also that r is the number of estimated parameters in the model (except for the residual variance) including the constant term if it is estimated. Thus, if we are modeling a nonstationary ARMA(p, d, q) time series model without a deterministic trend (i.e., without a constant term) to a series of n' observations, then the effective number of observations is $n = n' - d$ and the number of parameters is $r = p + q$. However, if the model includes a deterministic trend (i.e., a model with a constant term), then $r = p + q + 1$.

One approach to time series modeling is to fit a number of potential ARMA models to the data using the maximum likelihood estimation, choose a criterion, and select the model that has the best value according to this criterion. Although we do not advocate this approach, we will demonstrate how it is done.

Table 6.3 shows the AIC, the AICC, and the BIC values for the 36 possible models, ARIMA($p, 1, q$), with $p = 0, \ldots, 5$ and $q = 0, \ldots, 5$, fitted to the internet server data. We see that if we use AIC or AICC as the model selection criterion, then we would select the ARIMA(3, 1, 0) model. However, if we use BIC, then the ARIMA(1, 1, 1) model is the preferred model. Moreover, we see that for the AIC as well as the AICC, there is only a minor difference between the preferred ARIMA(3, 1, 0) model and the close competitor, the ARIMA(1, 1, 1) model. Using the BIC, the difference between the "winning" ARIMA(1, 1, 1) model and the "runner up" ARIMA(3, 1, 0) model seems exceedingly small.

Admittedly, this example may perhaps be a bit extreme. Often, the information criteria will provide more clear-cut answers. However, this example does indicate that relying solely on numerical criteria for model selection may not be wise.

TABLE 6.3   The AIC, AICC, and BIC Values for ARIMA($p, 1, q$), with $p = 0, \ldots, 5; q = 0, \ldots, 5$ Models Fitted to the Internet Server Data. The Numbers in Bold Face Indicate the Minimum for Each of the Information Criteria. Note That All of the Above Models are Fitted Without a Constant Term

Model	AIC	AICC	BIC
ARIMA(0, 1, 0)	628.995	628.995	628.995
ARIMA(1, 1, 0)	527.238	527.279	529.833
ARIMA(2, 1, 0)	520.178	520.303	525.368
ARIMA(3, 1, 0)	**509.994**	**510.247**	517.779
ARIMA(4, 1, 0)	511.930	512.355	522.310
ARIMA(5, 1, 0)	513.862	514.507	526.837
ARIMA(0, 1, 1)	547.805	547.847	550.401
ARIMA(1, 1, 1)	512.299	512.424	**517.490**
ARIMA(2, 1, 1)	514.291	514.544	522.077
ARIMA(3, 1, 1)	511.938	512.363	522.318
ARIMA(4, 1, 1)	510.874	511.520	523.850
ARIMA(5, 1, 1)	515.638	516.551	531.209
ARIMA(0, 1, 2)	517.875	518.000	523.065
ARIMA(1, 1, 2)	514.252	514.504	522.037
ARIMA(2, 1, 2)	515.360	515.786	525.741
ARIMA(3, 1, 2)	513.917	514.563	526.893
ARIMA(4, 1, 2)	514.179	515.092	529.750
ARIMA(5, 1, 2)	511.543	512.774	529.709
ARIMA(0, 1, 3)	518.272	518.524	526.057
ARIMA(1, 1, 3)	512.576	513.002	522.957
ARIMA(2, 1, 3)	513.773	514.418	526.749
ARIMA(3, 1, 3)	512.414	513.327	527.985
ARIMA(4, 1, 3)	517.078	518.308	535.243
ARIMA(5, 1, 3)	513.434	515.034	534.195
ARIMA(0, 1, 4)	517.380	517.805	527.760
ARIMA(1, 1, 4)	513.100	513.745	526.076
ARIMA(2, 1, 4)	511.241	512.154	526.812
ARIMA(3, 1, 4)	512.758	513.989	530.924
ARIMA(4, 1, 4)	512.808	514.408	533.569
ARIMA(5, 1, 4)	553.110	555.133	576.466
ARIMA(0, 1, 5)	516.857	517.502	529.833
ARIMA(1, 1, 5)	514.276	515.189	529.847
ARIMA(2, 1, 5)	716.050	717.281	734.216
ARIMA(3, 1, 5)	516.504	518.104	537.265
ARIMA(4, 1, 5)	515.845	517.867	539.201
ARIMA(5, 1, 5)	512.113	514.613	538.064

6.5 IMPULSE RESPONSE FUNCTION TO STUDY THE DIFFERENCES IN MODELS

Now that we have shown that several models seem to fit the data almost equally well, it would be interesting to see how different the models are. One way to investigate this is to consider the *impulse response function* for each model. We covered impulse response function and its implications in Chapter 1 where we discussed the Wold decomposition theorem in connection to the impulse response function as well.

As discussed in Chapter 1, Wold (1938) showed that a stationary time series (to be precise, a zero mean purely nondeterministic stationary stochastic process) $\{\tilde{z}_t\}$ can be represented as a linear filter with random shocks $\{a_t\}$, or white noise, as input. This concept is illustrated in Figure 6.10. Thus, the Wold representation theorem says that any stationary time series can be decomposed and represented as a weighted sum of random shocks $\{a_t\}$ where ψ_j, $j = 1, 2, \ldots$ are the weights. Mathematically, the linear representation (i.e., the linear filter) is given by

$$\tilde{z}_t = a_t + \psi_1 a_{t-1} + \psi_2 a_{t-2} + \cdots$$

$$= a_t + \sum_{j=1}^{\infty} \psi_j a_{t-j} \tag{6.4}$$

where $\sum_{j=0}^{\infty} \psi_j^2 < \infty$ and the a_t are white noise with variance σ^2. The Wold representation therefore implies that any discrete stationary time series can be represented as an infinite moving average process, MA(∞).

The weights, ψ_j, in the infinite moving average process MA(∞), are called the *impulse responses*. To appreciate the reason for this term, we go back to the pendulum example provided in Yule (1927), which was discussed in Chapter 1. Now suppose the pendulum is hit by one random shock at time zero of size $a_0 = 1$ and that the shocks at any other times are zero, $a_t = 0, t \neq 1$. If the movement $\tilde{z}_t$ around the equilibrium positions of the randomly swinging pendulum is modeled

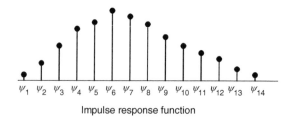

$\psi_1 \ \psi_2 \ \psi_3 \ \psi_4 \ \psi_5 \ \psi_6 \ \psi_7 \ \psi_8 \ \psi_9 \ \psi_{10} \ \psi_{11} \ \psi_{12} \ \psi_{13} \ \psi_{14}$

Impulse response function

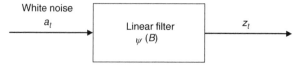

White noise a_t → Linear filter $\psi(B)$ → z_t

Figure 6.10 The representation of a stationary time series as the output from a linear filter.

using the Wold representation, then at time $t = 1$, we get

$$\tilde{z}_1 = a_1 + \psi_1 a_0 + \psi_2 a_{-1} + \cdots$$
$$= \psi_1 \tag{6.5}$$

and at time $t = 2$, we get

$$\tilde{z}_2 = a_2 + \psi_1 a_1 + \psi_2 a_0 + \psi_3 a_{-1} + \cdots$$
$$= \psi_2 \tag{6.6}$$

and so on. Thus, if the shock (i.e., impulse) is of size 1 ($a_0 = 1$) and fired at time zero, then ψ_j is the delayed response of the dynamic system to that initial impulse at time zero, j time units later. In general, the shape of the impulse response function characterizes the dynamic system and provides us with an intuitive understanding of how a dynamic system responds to a unit impulse.

To see how this works, consider the simple AR(1) process

$$\tilde{z}_t = \phi_1 \tilde{z}_{t-1} + a_t \tag{6.7}$$

Using the backshift operator, we can write the AR(1) model as $\tilde{z}_t - \phi_1 B \tilde{z}_t = a_t$ or $(1 - \phi_1 B)\tilde{z}_t = a_t$. Note that backshift operators can be manipulated like algebraic entities. Assuming $|\varphi_1| < 1$, the AR(1) process can therefore be written as

$$\tilde{z}_t = \frac{1}{(1 - \phi_1 B)} a_t \tag{6.8}$$

Using the properties of geometric series, we can write this as

$$\tilde{z}_t = \frac{1}{(1 - \phi_1 B)} a_t$$
$$= (1 + \phi_1 B + \phi_1^2 B^2 + \phi_1^3 B^3 + \cdots)a_t$$
$$= a_t + \phi_1 a_{t-1} + \phi_1^2 a_{t-2} + \phi_1^3 a_{t-3} + \cdots \tag{6.9}$$

The Wold representation of the AR(1) process is therefore given by letting the impulse responses $\psi_j = \phi_1^j$, $j = 0, 1, \ldots$. This means that the infinite sequence of ψ_j's is simply given by raising the single autoregressive parameter ϕ_1 to the jth power. In other words, with just a single parameter ϕ_1 to be estimated, we parsimoniously produce an infinite number of parameters ψ_j.

Now suppose $\phi_1 = 0.66$; then the impulse response function for the AR(1) process is shown in Figure 6.11. From that graph, we see that if a system is governed by a first order autoregressive AR(1) model with $\phi_1 = 0.66$, then a single impulse at time zero will make the system move back slowly to its equilibrium position after about 12 time units without crossing the centerline. See Appendix 6.1 for a simple way to compute the impulse response function for an ARMA(p, q) process using a spreadsheet.

Instead consider the ARMA(1, 1) process we fitted to the internet server data. This model can be written as $w_t = \phi_1 w_{t-1} + a_t - \theta_1 a_t$. Suppose further that $\phi_1 = 0.66$. In the ARMA(1, 1) process, if we let $\theta_1 = 0$, then we have the special case of an AR(1) model with the impulse response function shown in

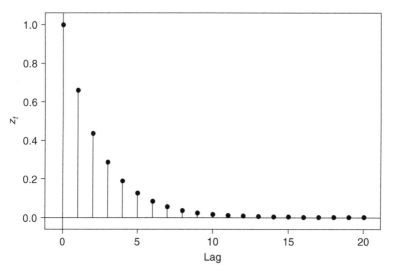

Figure 6.11 Impulse response for an AR(1) model with $\phi_1 = 0.66$.

Figure 6.11. Let us now see what happens when we change the moving average parameter θ_1 away from zero. Specifically, suppose we let θ_1 assume the values, $-0.1, -0.2, -0.3, -0.4, -0.5$; the last value is very close to that for the model we fitted for the internet server data above. The corresponding impulse response functions are shown in Figure 6.12. We see that the effect of letting θ_1 assume increasingly negative values is that a small amount is added to the basic AR(1)

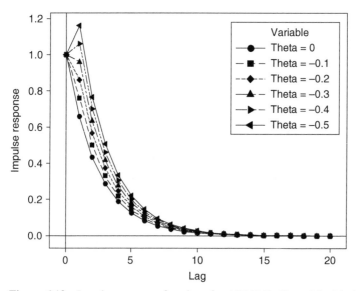

Figure 6.12 Impulse response functions for ARMA(1, 1) model with $\phi_1 = 0.66$ and $\theta_1 = 0, -0.1, -0.2, -0.3, -0.4,$ and -0.5.

impulse response function and that it fades out less rapidly. In other words, the dynamic response to an impulse for the ARMA(1, 1) system with $\phi_1 = 0.66$ and $\theta_1 = -0.5$ is stronger and longer lasting than for an AR(1) system with $\phi_1 = 0.66$.

What we have demonstrated is that by manipulating the two parameters ϕ_1 and θ_1 in the ARMA(1, 1) model, we can change the entire shape of the impulse response function so that the infinite sequence of ψ_j's can assume a range of different shapes. In general, by manipulating only $p + q$ parameters in an ARMA(p, q) process, we can generate a wide range of different impulse response functions, one of which hopefully will approximate the "true" dynamic response of the underlying system.

6.6 COMPARING IMPULSE RESPONSE FUNCTIONS FOR COMPETING MODELS

Now that we have seen that the impulse response function characterizes a time series model, let us see how different the two competing models we fitted to the internet users data are. Impulse response functions are usually computed for stationary models. We will therefore compute the impulse response functions for the corresponding stationary models, that is, for the models for $w_t = \nabla z_t$. For the coefficient estimates given in Tables 6.2 and 6.3, the impulse response functions for the ARMA(3, 0) and the ARMA(1, 1) models are shown in Figure 6.13. We see that the impulse response functions of these two seemingly different models are in fact quite similar. The main difference is the curl around lag 6 for the

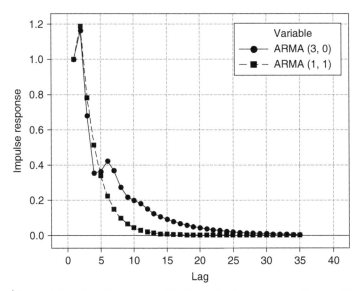

Figure 6.13 Impulse response functions for the two competing models, the ARMA(3, 0) and ARMA(1, 1).

ARMA(3, 0) model's impulse response function. We also see that the ARMA(3, 0) model's impulse response lingers around for longer time lags. However, in general, the two models look very similar. This is because these two models are essentially two different approximations of the same system dynamics. In other words, from an impulse response function perspective, the two approximations are essentially the same. However, one is more parsimonious involving only two parameters, whereas the other involves three and hence may be relatively more difficult to get good parameter estimates for.

6.7 ARIMA MODELS AS RATIONAL APPROXIMATIONS

To understand how we can have two models to be very similar approximations to the same system, we need to introduce some more time series theory. If we let $\nabla^d z_t = w_t$, then we can write the ARIMA(p, d, q) model as

$$w_t = \phi_1 w_{t-1} + \cdots + \phi_p w_{t-p} - \theta_1 a_{t-1} - \cdots - \theta_q a_{t-q} + a_t \tag{6.10}$$

Using the backshift operator introduced above, we can rewrite this as

$$w_t - \phi_1 B w_1 - \cdots - \phi_p B^p w_t = a_t - \theta_1 B a_t - \cdots - \theta_q B^q a_t \tag{6.11}$$

Treating the backshift operator B as an algebraic quantity, we can write the ARIMA(p, d, q) model as

$$(1 - \phi_1 B - \cdots - \phi_p B^p) w_t = (1 - \theta_1 B - \cdots - \theta_q B^q) a_t \tag{6.12}$$

or

$$w_t = \frac{(1 - \theta_1 B - \cdots - \theta_q B^q)}{(1 - \phi_1 B - \cdots - \phi_p B^p)} a_t \tag{6.13}$$

If we let $\phi(B) = (1 - \phi_1 B - \cdots - \phi_p B^p)$ and $\theta(B) = (1 - \theta_1 B - \cdots - \theta_q B^q)$ be two polynomials in B, then

$$w_t = \frac{\theta(B)}{\phi(B)} a_t \tag{6.14}$$

The function $\theta(B)/\phi(B)$ is a ratio of the two polynomials $\theta(B)$ and $\phi(B)$ in B and hence a rational function. Rational functions constitute a flexible class of models that, with a few adjustable parameters, $\phi_1, \ldots, \phi_p$ and $\theta_1, \ldots, \theta_q$, can approximate a wide variety of linear filter weight patterns $\{\psi_j\}$ of the Wold representation. For example, as we saw for the simple AR(1) model, the rational function with $\theta(B) = 1$ and $\phi(B) = 1 - \phi_1 B$ is

$$\begin{aligned}
\tilde{z}_t &= \frac{\theta(B)}{\phi(B)} a_t \\
&= \frac{1}{(1 - \phi_1 B)} a_t \\
&= (1 + \phi_1 B + \phi_1^2 B^2 + \phi_1^3 B^3 + \cdots) a_t \\
&= a_t + \phi_1 a_{t-1} + \phi_1^2 a_{t-2} + \phi_1^3 a_{t-3} + \cdots
\end{aligned} \tag{6.15}$$

This approximation allows us to produce an infinitive sequence of $\psi_j = \phi_1^j$ very parsimoniously with only one parameter, ϕ_1. For more general ARMA(p, q) processes, we can, likewise, with a few adjustable parameters, produce a wide variety of weight patterns $\{\psi_j\}$.

6.8   AR VERSUS ARMA CONTROVERSY

We already showed graphically that for the specific coefficient estimates, both the ARMA(3, 0) and the ARMA(1, 1) models have similar impulse response functions and hence capture the system dynamics equally well. This is not too surprising. However, it brings up an interesting issue that caused some controversy in the early days of ARIMA models. Before the seminal work on time series analysis in the late 1960s by Box and Jenkins, it was common to recommend using only AR(p) models. Depending on the data, it was then sometimes necessary to use very high order AR(p) models to get a good fit. In fact, a "heavy weight authority" in time series analysis at that time, who was a very vocal critic of the Box and Jenkins (1970, first edition of BJ) book, questioned the use of mixed ARMA models. The problem, however, is that the many parameters in higher order AR models are not always well estimated, not because of the lack of computer power, but because of the inherent difficulty in maximizing a high dimensional likelihood function for time series models. Box and Jenkins (1970) therefore promoted the use of ARMA models as parsimonious alternatives to high order AR models because the parameters in general would be better estimated. Specifically, Box and Jenkins showed that instead of a high order AR model, a mixed ARMA model with a low order AR model together with a low order MA component could capture the system dynamics equally well and in most cases would require fewer parameters. For this, Box and Jenkins at first received some harsh criticism, but today the use of mixed ARMA models is no longer controversial among statisticians. Therefore to sum up, we recommend time series analysts to use the information criteria as input to the model selection decision-making process. However, we do not advocate that such criteria be used mechanically to select the "best" model. In selecting a model, we should instead exercise judgment and in general choose one that is parsimonious. In the present case, we therefore prefer the more parsimonious and therefore easier to estimate ARIMA(1, 1, 1) model although we acknowledge that the two models are very similar in terms of the impulse response functions.

To avoid confusion, we should point out that we do not claim that the ARIMA(1, 1, 1) and the ARIMA(3, 1, 0) models in general are equivalent. We only claim that they have similar abilities in capturing the system dynamics with this particular set of coefficient estimates. In fact, the same two models with different coefficients will produce very different impulse response functions and hence model completely different dynamics. We have illustrated this in Figure 6.14. Considering only the stationary part of the model, we use the ARMA(3, 0) model with the current coefficient estimates as the base model and plotted its impulse response function as the solid line. We then plotted the impulse

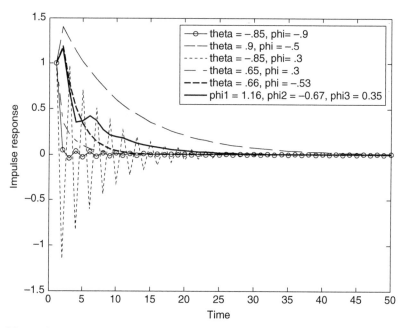

Figure 6.14 Impulse response functions of ARMA(3, 0) model together with different ARMA(1, 1) models.

response functions for a series of ARMA(1, 1) models with various values of the AR and MA coefficients. The solid dotted line with $\theta_1 = 0.66$ and $\phi_1 = -0.53$ is the best fitting ARMA(1, 1) model, which is similar to the best fitting ARMA(3, 0) model. However, from this graph, we see that for other values of θ_1 and ϕ_1, the impulse response function for the ARIMA(1, 1) model represents quite different dynamics.

6.8.1 Yet Another Model Comparison

The following is another approach to study how different the two time series models are. Let us first write the ARIMA(3, 1, 0) model as

$$(1 - 1.1632B + 0.6751B^2 - 0.3512B^3)(1 - B)y_t = a_t \qquad (6.16)$$

Similarly, we can write the ARIMA(1, 1, 1) model as

$$(1 - 0.6573B)(1 - B)y_t = (1 + 0.5301B)a_t \qquad (6.17)$$

or

$$\frac{(1 - 0.6573B)}{(1 + 0.5301B)}(1 - B)y_t = a_t \qquad (6.18)$$

If we expand the ratio $(1 - 0.6573B)/(1 + 0.5301B)$ as a Taylor polynomial in B up to order 3, we get

$$\frac{(1 - 0.6573B)}{(1 + 0.5301B)} \approx 1 - 1.1874B + 0.6279B^2 - 0.334B^3 \qquad (6.19)$$

By comparing the coefficients in the third order polynomial expansion of $(1 - 0.6573B)/(1 + 0.5301B)$ with the coefficients in the ARIMA(3, 1, 0) model, we see that the coefficients are very similar. This is yet another way to see that for the given coefficients, the ARIMA(1, 1, 1) and ARIMA(3, 1, 0) models are indeed quite similar in terms of modeling the dynamic properties of a process.

6.9 FINAL THOUGHTS ON MODEL SELECTION

One of the more difficult tasks in time series analysis is to select an adequate model for a given set of data. Often, several models may seem to fit the data equally well. In this chapter, we have discussed one popular approach using certain numerical criteria. We think such criteria are useful as inputs to the decision-making process. However, we are not in favor of using them mechanically as the sole criteria for selecting a model. Instead, we prefer to exercise judgment and use an information criterion as one among several considerations in the choice of models. First of all, we should always consider the context of the model. What is it we are trying to model, what are we going to use the model for, and so on? Besides that an important consideration in model selection should be model parsimony. In particular, we favor using parsimonious mixed ARMA models over pure higher order AR models.

Besides the estimation problems we discussed, research by Ledolter and Abraham (1981) shows that for each unnecessary parameter, the variance of the one-step-ahead forecast error increases by a factor of σ^2/n. They also showed that an approximation of a mixed ARIMA model by a nonparsimonious autoregressive model will lead to an unnecessary inflation of the forecast error. They conclude that "a model building strategy that considers only autoregressive representations will lead to nonparsimonious models and to loss in forecasting accuracy."

APPENDIX 6.1: HOW TO COMPUTE IMPULSE RESPONSE FUNCTIONS WITH A SPREADSHEET

To compute impulse response functions is simple and easy using a spreadsheet program. Suppose we need the impulse response function for an ARMA(p, q) model. We then write the model as a difference equation

$$w_t = \phi_1 w_{t-1} + \cdots + \phi_p w_{t-p} - \theta_1 a_{t-1} - \cdots - \theta_q a_{t-q} + a_t \qquad \text{(A6.1)}$$

Now suppose $p = 1$ and $q = 1$; then this general difference equation simplifies to

$$w_t = \phi_1 w_{t-1} - \theta_1 a_{t-1} + a_t \qquad \text{(A6.2)}$$

We then create columns for t, a_t, and w_t as shown in Table A6.1. The values for the column containing a_t (shown in column D in Table A6.1) should be zero for all values except for $t = 0$ where it is $a_0 = 1$. The values in the column for w_t should be zero for negative t or otherwise be computed recursively using

TABLE A6.1 Spreadsheet Computation of the Impulse Response Function

			Column	
Row	C	D	E	F
4		Model		
5		$\phi_1 =$	0.6573	
6		$\theta_1 =$	-0.5301	
7			Response	
8		Impulse		Value
9	Time	a_t	Formula	ψ_t
10	-1	0	0	0
11	0	1	=$E$5*E10+D11-$E$6*D10	1
12	1	0	=$E$5*E11+D12-$E$6*D11	1.1874
13	2	0	=$E$5*E12+D13-$E$6*D12	0.7805
14	3	0	=$E$5*E13+D14-$E$6*D13	0.513
15	4	0	=$E$5*E14+D15-$E$6*D14	0.3372
16	5	0	=$E$5*E15+D16-$E$6*D15	0.2216
17	6	0	=$E$5*E16+D17-$E$6*D16	0.1457
18	7	0	=$E$5*E17+D18-$E$6*D17	0.0958
19	8	0	=$E$5*E18+D19-$E$6*D18	0.0629
20	9	0	=$E$5*E19+D20-$E$6*D19	0.0414
21	10	0	=$E$5*E20+D21-$E$6*D20	0.0272
22	11	0	=$E$5*E21+D22-$E$6*D21	0.0179
23	12	0	=$E$5*E22+D23-$E$6*D22	0.0117
24	13	0	=$E$5*E23+D24-$E$6*D23	0.0077
25	14	0	=$E$5*E24+D25-$E$6*D24	0.0051
26	15	0	=$E$5*E25+D26-$E$6*D25	0.0033
27	16	0	=$E$5*E26+D27-$E$6*D26	0.0022
28	17	0	=$E$5*E27+D28-$E$6*D27	0.0014
29	18	0	=$E$5*E28+D29-$E$6*D28	0.0009
30	19	0	=$E$5*E29+D30-$E$6*D29	0.0006
31	20	0	=$E$5*E30+D31-$E$6*D30	0.0004

the equation $w_t = \phi_1 w_{t-1} - \theta_1 a_{t-1} + a_t$ as indicated in column E in Table A6.1. The resulting impulse responses are shown in column F in Table A6.1.

EXERCISES

6.1 In Exercise 3.16, an ARMA(1, 1) model and an appropriate AR(p) were used to fit the data given in Table B.5. Pick one of these models based on AIC and BIC.

6.2 Table B.10 contains 100 observations simulated from an ARMA(1, 1) process. Fit ARMA(p,q) models to this data where $p = 0, 1, 2$ and $q = 0, 1, 2$. Pick the best model based on AIC and BIC.

6.3 For AR(1) and MA(1) models in Exercise 6.2, consider the ACF and PACF of the residuals. Comment on whether it is possible to see from these plots how to modify the models to get a better fit—for example, for the AR(1) do the residuals indicate that there is a need for a second order AR term or an MA term in the model?

6.4 Table B.11 contains the quarterly rental vacancy rates in the United States from 1956 to the second quarter of 2010. Fit various ARIMA models using the data from 1956 to 2009 (not included) and generate a table similar to Table 6.3 for this data to find the best model based on AIC and BIC.

6.5 Using the top three best models from Exercise 6.4, make forecasts for the rental vacancy rates for the four quarters of 2009 and the first two quarters of 2010. Since you also have the true rates for those quarters, calculate the sum of squared prediction errors for the three models and comment on your results.

6.6 Table B.12 contains the famous Wölfer sunspot numbers data from 1770 to 1869 (series E of BJR). Fit various ARIMA models for this data to find the best model based on AIC and BIC.

6.7 Consider the best two models in terms of AIC and/or BIC from Exercise 6.6. Plot the impulse response functions for these models and comment on your results.

6.8 Table B.13 contains 310 viscosity readings from a chemical process collected every hour (series D of BJR). Fit various ARIMA models for this data to find the best model based on AIC and BIC.

6.9 Consider the best two models in terms of AIC and/or BIC from Exercise 6.8. Plot the impulse response functions for these models and comment on your results.

CHAPTER 7

ADDITIONAL ISSUES IN ARIMA MODELS

7.1 INTRODUCTION

ARIMA models have been used extensively in practice for the analysis of time series data and forecasting. However, it has been our experience that the analysts often have questions/concerns for which the answers are either ignored or buried in details in some textbooks. In this chapter, we aim to answer some of those questions. We do not claim the list to be comprehensive but we try to include some of the major issues. Although we attempt to provide as much intuition as possible, some theoretical results are also included. If the topic requires further theory, we as usual provide the necessary references for our advanced readers. Before discussing specific issues, we start with a section on linear difference equations, which we believe provides the theoretical foundation for the ARIMA models.

7.2 LINEAR DIFFERENCE EQUATIONS

The ARIMA(p,d,q) model represents the time series z_t in terms of the previous values $z_{t-1}, z_{t-2}, \ldots$ and current and previous disturbances, $a_t, a_{t-1}, \ldots$ The equation we used

$$(1 - \phi_1 B - \phi_2 B^2 - \cdots - \phi_p B^p)(1 - B)^d z_t$$
$$= (1 - \theta_1 B - \theta_2 B^2 - \cdots - \theta_q B^q) a_t$$

or

$$z_t - \varphi_1 z_{t-1} - \cdots - \varphi_{p+d} z_{t-p-d} = a_t - \theta_1 a_{t-1} - \cdots - \theta_q a_{t-q} \qquad (7.1)$$

is in fact a linear difference equation. In this section we discuss the solution to the linear difference equation in (7.1). A solution to a difference equation $f(t, z_t, z_{t-1}, \ldots, z_{t-n})$ is a function of t that satisfies the difference equation for all t. In other words, we are looking for a sequence $\{z_t\}_{t=-\infty}^{\infty}$ that when substituted into $f(t, z_t, z_{t-1}, \ldots, z_{t-n})$ satisfies that equation. Therefore, it is not a single number, but a discrete time curve that constitutes the solution.

Time Series Analysis and Forecasting by Example, First Edition. Søren Bisgaard and Murat Kulahci.
© 2011 John Wiley & Sons, Inc. Published 2011 by John Wiley & Sons, Inc.

By analogy with differential equations, it may be suspected that the order of a difference equation is defined by the highest order difference present in the equation. However, such a definition may lead to problems. Instead, we define the order as the difference between the largest and the smallest arguments of the function. For example, if these are z_{t+n} and z_t, then the order of the difference equation is $t + n - t = n$. However, if they are z_{t+n} and z_{t+1} for example, then the order is $t + n - (t + 1) = n - 1$. Consider, for example, the second order nonhomogenous difference equations with constant coefficients:

$$z_t + c_1 z_{t-1} + c_2 z_{t-2} = r_t \tag{7.2}$$

In physics, the function on the right-hand side is called the *forcing function*. If no external force (forcing function) is present, that is,

$$z_t + c_1 z_{t-1} + c_2 z_{t-2} = 0 \tag{7.3}$$

the equation is called homogeneous. If the right-hand side is not equal to zero for all t, then we say that it is a forced motion and that the equation is nonhomogenous. The right-hand side is sometimes called the *perturbation term*.

In the explanations below, we will closely follow BJR but skip some of the explanations provided there. Let us now go back to the ARIMA(p, d, q) model in Equation (7.1) and rewrite it as

$$\varphi(B)z_t = \theta(B)a_t \tag{7.4}$$

where $\varphi(B) = 1 - \varphi_1 B - \varphi_2 B^2 - \cdots - \varphi_{p+d} B^{p+d}$ and $\theta(B) = 1 - \theta_1 B - \theta_2 B^2 - \cdots - \theta_q B^q$.

The terms $\theta(B)a_t$ and $\varphi(B)z_t = 0$ can be considered as the (stochastic) forcing function and homogenous difference equation, respectively. To make the results more general, we derive solutions relative to a starting point $k < t$. The general solution to the linear difference equation can be written as

$$z_t = C_k(t - k) + I_k(t - k) \tag{7.5}$$

where $C_k(t - k)$, sometimes called the *complementary solution* (or *complementary function*), is the general solution to the homogenous equation and $I_k(t - k)$ is the *particular solution* (or *particular integral*). Therefore, it follows that to obtain the general solution of a nonhomogenous equation, it is sufficient to add any particular solution to the general solution of the homogeneous equation. To apply this to our ARIMA model, suppose z_t' is any particular solution to Equation (7.4) that satisfies

$$\varphi(B)z_t' = \theta(B)a_t \tag{7.6}$$

Then by subtraction, we get that

$$\varphi(B)(z_t - z_t') = 0 \tag{7.7}$$

Therefore, $z_t'' = z_t - z_t'$ satisfies the homogeneous equation $\varphi(B)z_t'' = 0$. We then have $z_t = z_t' + z_t''$ and hence we see that the general solution is the sum of the complementary function z_t'' and the particular integral z_t'. Relative to a given starting time $k < t$, we denote the complementary function z_t'' by $C_k(t - k)$ and the particular integral z_t' by $I_k(t - k)$. Now our task is to figure out what $C_k(t - k)$ and $I_k(t - k)$ are. We will first do this for a simple case and provide the general result later.

7.2.1   Linear Difference Equations and ARIMA Models

The second order homogeneous equation was given in Equation (7.3)

$$z_t + c_1 z_{t-1} + c_2 z_{t-2} = 0$$

Consider (i.e., guess) a solution of the form $z_t = m^t$ where m is constant different from zero. (Note that if we allow $m = 0$, then $z_t = 0$ for all t, which cannot be one of the solutions to constitute a fundamental set.) If we substitute this trial solution into Equation (7.3), we get $m^t + c_1 m^{t-1} + c_2 m^{t-2} = 0$. Now dividing through by the common factor m^{t-2} and assuming that the starting point $k = 1$, we get $m^2 + c_1 m^1 + c_2 = 0$. This second order polynomial is called the *characteristic equation* because it characterizes the qualitative type of solution. If m is a number satisfying the characteristic equation, that is, if it is a root of the quadratic equation, then m^t is a solution to the difference equation (7.3). The characteristic equation is a second order polynomial and has in general two nonzero solutions, say m_1 and m_2. Zero roots are excluded since we require that $c_2 \neq 0$. First we assume that these two roots are distinct and real. Then to those two roots correspond two solutions to Equation (7.3), say $z_t^{(1)}$ and $z_t^{(2)}$ given by $z_t^{(1)} = m_1^t$ and $z_t^{(2)} = m_2^t$. It can be shown that if $z^{(1)}$ and $z^{(2)}$ are any two solutions of the second order homogenous difference equation, then $C_1 z^{(1)} + C_2 z^{(2)}$ is also a solution for arbitrary constants C_1 and C_2 (see Goldberg (2010) for proof). It can be shown further that the problem of finding a general solution to this second order homogeneous difference equation amounts to finding two linearly independent solutions (Goldberg, 2010). Therefore, the general solution to Equation (7.3) can be represented as $C_1 z^{(1)} + C_2 z^{(2)}$.

Now let us go back to our ARIMA(p, d, q) model and consider the special case for which $p + d = 2$. First to find out the complementary function, $C_k(t - k)$, we consider the second order homogeneous equation given by $\varphi(B)z_t = 0$ where $(1 - \varphi_1 B - \varphi_2 B^2)z_t = 0$. Treating the backshift operator B as a variable, we can rewrite the second order homogeneous equation as

$$(1 - G_1 B)(1 - G_2 B)z_t = 0 \tag{7.8}$$

where G_1^{-1} and G_2^{-1} are the two roots of the second order equation $(1 - \varphi_1 B - \varphi_2 B^2) = 0$ in B. In other words, G_1 and G_2 are the two roots of the second order equation $(B^2 - \varphi_1 B - \varphi_2) = 0$ in B. Therefore, following the earlier arguments, the general solution for Equation (7.8) is given as

$$z_t = C_1 G_1^{t-k} + C_2 G_2^{t-k} \tag{7.9}$$

where C_1 and C_2 are constants to be determined using the first two initial values. Note that if the roots G_1 and G_2 are complex conjugates rather than real numbers, then the solution will have a sine-wave pattern.

So far, we have discussed the case for which two roots were distinct. Now let us consider when they are equal, that is

$$(1 - G_0 B)^2 z_t = 0 \qquad (7.10)$$

For that we first define $y_t = (1 - G_0 B)z_t$, then from Equation (7.10), $(1 - G_0 B)y_t = 0$, which is a first order homogeneous equation that can be solved by iteration. Suppose the process started at $t = k$ with initial value y_k. From Equation (7.10), we have $(1 - G_0 B)y_t = y_t - G_0 y_{t-1} = 0 \Rightarrow y_t = G_0 y_{t-1}$. By continuing iterations backward in time we get

$$y_t = G_0 y_{t-1} = G_0^2 y_{t-2} = G_0^3 y_{t-3} = \cdots = G_0^{t-k} y_k$$

If we set the initial value $y_k = D_1$, then we have

$$y_t = D_1 G_0^{t-k} \qquad (7.11)$$

Now substituting this solution into $y_t = (1 - G_0 B)z_t = z_t - G_0 z_{t-1}$, we get

$$
\begin{aligned}
z_t &= G_0 z_{t-1} + y_t \\
&= G_0 z_{t-1} + D_1 G_0^{t-k} \\
&= G_0 (G_0 z_{t-2} + D_1 G_0^{t-k-1}) + D_1 G_0^{t-k} \\
&\vdots \\
&= G_0^{t-k} z_k + D_1 (G_0^{t-k} + G_0 G_0^{t-k-1} + \cdots + G_0^{t-k-1} G_0) \\
&= G_0^{t-k} z_k + (t-k)D_1 G_0^{t-k} \\
&= [D_0 + (t-k)D_1] G_0^{t-k} \qquad (7.12)
\end{aligned}
$$

where $D_0 = z_k$ (the initial value).

To generalize the above results for the ARIMA(p,d,q) in Equation (7.4), all we have to do is to rewrite $\varphi(B)$ as

$$\varphi(B) = (1 - G_1 B)(1 - G_2 B) \ldots (1 - G_p B)(1 - G_0 B)^d \qquad (7.13)$$

In this notation, G_0 represents the root that is repeated d times. In our ARIMA models that are differenced d times, we have $G_0 = 1$. In Equation (7.13), we assume that G_1 through G_p are distinct. This of course may not always be the case. But the modification to Equation (7.13) is then quite straightforward. Moreover for our ARIMA models, G_1 through G_p will be less than 1 in absolute value because of the stationarity condition. Now based on Equation (7.13), the complementary function for an ARIMA(p,d,q) model is

$$C_k(t-k) = \sum_{i=1}^{p} C_i G_i^{t-k} + G_0^{t-k} \sum_{j=0}^{d-1} D_j (t-k)^j \qquad (7.14)$$

Thus the complementary function will have exponential decay pattern due to G_i's that are real, damped sine-wave pattern due to G_i's that are complex conjugates, and finally a polynomial in $(t - k)$ up to order $d - 1$ due to differencing.

We will now focus on the particular integral, $I_k(t-k)$ in Equation (7.5). It can be shown that $I_k(t-k)$ satisfying

$$\phi(B)I_k(t-k) = \theta(B)a_t, \text{ for } t-k > q \tag{7.15}$$

is given as

$$I_k(1) = a_{k+1}$$
$$I_k(2) = a_{k+2} + \psi_1 a_{k+1}$$
$$\vdots$$
$$I_k(t-k) = a_t + \psi_1 a_{t-1} + \psi_2 a_{t-2} + \cdots + \psi_{t-k-1} a_{k+1} \tag{7.16}$$

and

$$I_k(t-k) = 0 \text{ for } t \leqslant k \tag{7.17}$$

The $\{\psi_i\}$ weights in the particular integral are calculated from

$$(1 - \varphi_1 B - \cdots - \varphi_{p+d} B^{p+d})(1 + \psi_1 B + \psi_2 B^2 + \cdots)$$
$$= (1 - \theta_1 B - \cdots - \theta_q B^q) \tag{7.18}$$

as discussed in Chapter 4. We will now show how to calculate the complementary function and the particular integral for various models.

Example 1:   ARIMA(0, 1, 0) (random walk)
The ARIMA(0, 1, 0) model is given as

$$(1 - B)z_t = a_t \tag{7.19}$$

The complementary function is found from the homogeneous equation

$$(1 - B)z_t = 0 \tag{7.20}$$

From Equation (7.14), we have

$$C_k(t-k) = C_1 \tag{7.21}$$

since the characteristic equation has only one root that is equal to 1.
To calculate the particular integral, we first need to calculate the $\{\psi_i\}$. From Equation (7.17), we have

$$(1 - B)(1 + \psi_1 B + \psi_2 B^2 + \cdots) = 1 \tag{7.22}$$

We can rewrite Equation (7.22) as

$$1 + (\psi_1 - 1)B + (\psi_2 - \psi_1)B^2 + (\psi_3 - \psi_2)B^2 + \cdots = 1 \tag{7.23}$$

Since on the right-hand side of Equation (7.23), we have no terms with B, we need to have

$$(\psi_1 - 1) = 0$$
$$(\psi_2 - \psi_1) = 0$$
$$(\psi_3 - \psi_2) = 0$$
$$\vdots$$

Hence $\psi_1 = \psi_2 = \psi_3 = \cdots = 1$. Therefore, the particular integral is given as

$$I_k(t-k) = \sum_{i=0}^{t-k-1} a_{t-j} \tag{7.24}$$

This means that the solution to the random walk model at any given time is equal to a constant in addition to the sum of the current and past disturbances all the way back to the origin. We can also get to the same result from the following for $k=0$

$$z_1 = z_0 + a_1$$
$$z_2 = z_1 + a_2 = (z_0 + a_1) + a_2$$
$$z_3 = z_2 + a_3 = (z_0 + a_1 + a_2) + a_3$$
$$\vdots$$
$$z_t = z_0 + (a_1 + a_2 + a_3 + \cdots + a_t)$$
$$= z_0 + \sum_{i=0}^{t-1} a_{t-i}$$

This is, however, a relatively simple case. For more complicated models, Equations (7.14) and (7.16) provide an easy way to obtain the solution.

Example 2:   ARIMA(0, 1, 1)

The ARIMA(0, 1, 1) model is given as

$$(1-B)z_t = a_t - \theta a_{t-1} \tag{7.25}$$

The complementary function is found from the homogeneous equation

$$(1-B)z_t = 0 \tag{7.26}$$

Therefore, it is the same as in Example 1, that is,

$$C_k(t-k) = C_1 \tag{7.27}$$

For the particular integral, we calculate the $\{\psi_i\}$ from

$$(1-B)(1 + \psi_1 B + \psi_2 B^2 + \cdots) = 1 - \theta B \tag{7.28}$$

We can rewrite Equation (7.28) as

$$1 + (\psi_1 - 1)B + (\psi_2 - \psi_1)B^2 + (\psi_3 - \psi_2)B^2 + \cdots = 1 - \theta B \tag{7.29}$$

Then we have

$$(\psi_1 - 1) = -\theta$$
$$(\psi_2 - \psi_1) = 0$$
$$(\psi_3 - \psi_2) = 0$$
$$\vdots$$

Hence $\psi_1 = \psi_2 = \psi_3 = \cdots = 1 - \theta$. Therefore, the particular integral is given as

$$I_k(1) = a_{k+1}$$
$$I_k(2) = a_{k+2} + (1 - \theta)a_{k+1}$$
$$I_k(3) = a_{k+3} + (1 - \theta)a_{k+2} + (1 - \theta)a_{k+1}$$
$$\vdots$$
$$I_k(t - k) = a_t + (1 - \theta)a_{t-1} + (1 - \theta)a_{t-2} + \cdots + (1 - \theta)a_{k+1}. \quad (7.30)$$

which means $I_k(t - k) = a_t + (1 - \theta) \sum_{i=1}^{t-k-1} a_{t-i}$ for $t - k > 1$ (or $t - k > q$).

7.3   EVENTUAL FORECAST FUNCTION

In Chapter 6, we discussed in great detail the Internet users data and concluded that there were two plausible models to fit the data; ARIMA(1, 1, 1) and ARIMA(3, 1, 0). In this section, we will assume that the former model is chosen. The model with the parameter estimates is given as

$$(1 - 0.6573B)(1 - B)z_t = a_t + 0.5301a_{t-1} \quad (7.31)$$

or

$$(1 - 1.6573B + 0.6573B^2)z_t = a_t + 0.5301a_{t-1} \quad (7.32)$$

Now if we would like to make forecasts at time t for z_{t+l}, as discussed in Section 4.4, we use the following equation

$$\hat{z}_t(l) = [z_{t+l}] = \sum_{i=1}^{p+d} \varphi_i [z_{t+l-i}] + [a_{t+l}] - \sum_{i=1}^{q} \theta_i [a_{t+l-i}] \quad (7.33)$$

with

1. $[z_{t-j}] = z_{t-j}$ for $j = 0, 1, 2, \ldots$
2. $[z_{t+j}] = \hat{z}_t(j)$ for $j = 1, 2, \ldots$
3. $[a_{t-j}] = e_{t-j}(1) = z_{t-j} - \hat{z}_{t-j-1}(1)$ for $j = 0, 1, 2, \ldots$
4. $[a_{t+j}] = 0$ for $j = 1, 2, \ldots$

For that we first rearrange Equation (7.32) as

$$z_t = 1.6573z_{t-1} - 0.6573z_{t-2} + a_t + 0.5301a_{t-1}$$

Then, the l-step-ahead forecasts are given as

1-step ahead: $\hat{z}_t(1) = 1.6573z_t - 0.6573z_{t-1} + 0.5301e_t(1)$
2-step ahead: $\hat{z}_t(2) = 1.6573\hat{z}_t(1) - 0.6573z_t$
3-step ahead: $\hat{z}_t(3) = 1.6573\hat{z}_t(2) - 0.6573\hat{z}_t(1)$
4-step ahead: $\hat{z}_t(4) = 1.6573\hat{z}_t(3) - 0.6573\hat{z}_t(2)$

$\vdots$

l-step ahead:

$$\hat{z}_t(l) = 1.6573\hat{z}_t(l-1) - 0.6573\hat{z}_t(l-2) \text{ for } l = 3, 4, \ldots \quad (7.34)$$

Hence for $l = 3, 4, \ldots$, the forecast function is a homogeneous linear difference equation of the form $\hat{z}_t(l) - 1.6573\hat{z}_t(l-1) + 0.6573\hat{z}_t(l-2) = 0$ as in Equation (7.3). This suggests that eventually the autoregressive and differencing components of the ARIMA(p, d, q) model become the main driving force for forecasts with high leading times. This makes perfect sense because as the leading time increases, the error terms are replaced with our best guess for them, that is, their conditional expectation, which is equal to 0. Hence, the impact of the MA term in the forecast eventually disappears for high enough leading times. BJR aptly calls the solution to the homogeneous linear difference equation in Equation (7.34) as the *eventual forecast function*. The same results can also be extended for seasonal ARIMA models. The eventual forecast functions for various autoregressive operators and their behavior are given in BJR (pp. 382–383). Similarly in Figure 7.1, we provide the behavior of some eventual forecast functions. Note that the constants in a forecast function based on a general autoregressive operator are calculated by the number of initial values that is equal to the degree of the homogeneous difference equation. That is, the autoregressive operator $(1 - \phi B)(1 - B)(1 - B^s)$, for example, will require $s + 2$ initial values to calculate the constants in the forecast function. In Figure 7.1, those initial values are picked randomly. For each model, the initial values are indicated with a bigger square compared to the forecasted values. The seasonal periodicity s is set to be equal to 4 as, for example, in a quarterly financial data. Also, the forecasts are stopped at the end of the fourth period. Without loss of generality, both seasonal and nonseasonal autoregressive parameters are set to 0.5 as in Figure 9.8 of BJR.

Models (1) through (4) are all nonseasonal models. Only model (1) is stationary. This is also apparent in the behavior of the eventual forecast function, which exhibits an exponential decay pattern and reaches 0. Note that for the stationary series (nonseasonal or seasonal), we assume that the data is demeaned (or mean-corrected). That is, we take the constant mean, μ, from the data and use $\tilde{z}_t = z_t - \mu$, of which the mean is 0. It should of course be expected that the eventual forecast function would asymptotically reach μ instead of 0 if z_t were used instead.

The forecast for model (2) is simply the initial value for all lead times. In (3), we see that the forecasts reach an asymptote different from the mean as the impact of the stationary component wears off. The model in (4) corresponds to data that needed to be differenced twice. The eventual forecast function therefore has a linear trend behavior in time.

Models (5) through (12) have seasonal components. Model (5) is a stationary seasonal ARIMA model. Therefore, the eventual forecast converges asymptotically to the mean. In model (6), however, we have a nonstationary purely seasonal model. As in (2), we can see that the forecasts remain constant and equal to the initial values used for each period. That is, the forecast for the first period, for example, remains the same and is equal to the first of the four initial values. The model in (7) is a stationary model with both seasonal and nonseasonal components and as expected, the forecasts do eventually reach the mean value, which once again is 0 in this case. In model (8), we have a stationary nonseasonal component and nonstationary seasonal component. If we refer back to model (6),

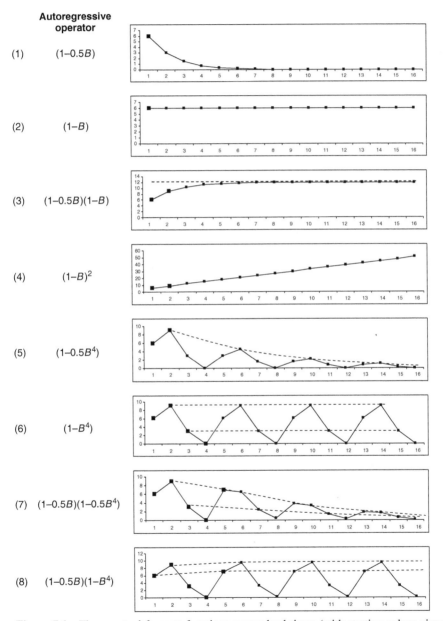

Figure 7.1 The *eventual* forecast functions versus lead times (with starting values given in bigger square boxes) for various ARIMA models.

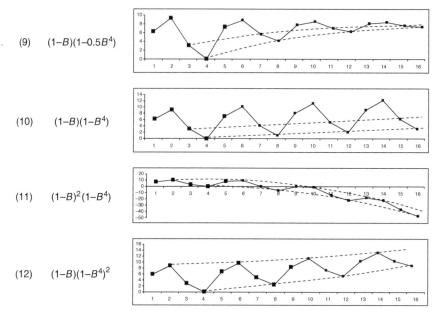

(9) $(1-B)(1-0.5B^4)$

(10) $(1-B)(1-B^4)$

(11) $(1-B)^2(1-B^4)$

(12) $(1-B)(1-B^4)^2$

Figure 7.1 (*continued*)

the nonstationary component is expected to imply that the forecasts remain the same. However, as we can see from the figure, the forecasts do go up a little bit with the effect of the stationary nonseasonal component but eventually reach an asymptote for each of the four periods. As opposed to model (8), model (9) has a nonstationary nonseasonal component and a stationary seasonal component. Because of the latter, we expect the forecasts to eventually converge toward the mean but because of the former, they reach a different asymptote than the mean. Model (10) implies a nonstationary series both at the seasonal and nonseasonal components. What we see in the forecasts in fact is that the seasonal forecast pattern is repeated with an yearly increase due to the nonseasonal nonstationarity.

Models (11) and (12) correspond to double differencing at nonseasonal and seasonal components, respectively. Since both involve the double differencing, we expect to see a trend upward or downward in time. The difference between the models is that in model (11), the rate of decrease (which is simply a result of the initial values and with a different set of initial values, we could have an increase rather than a decrease) in the quarters is constant because of the fact that the trend is due to the nonseasonal component. However, the trend in model (12) is due to the seasonal component, which makes the rate of increase to be different from one quarter to the next.

Figure 7.1 is by no means comprehensive in covering all possible ARIMA models but it does nonetheless provide a general overview on what to expect as the behavior of the eventual forecast functions for some of the most common autoregressive operators. It should be noted that the forecasts in Figure 7.1 are obtained for randomly chosen initial values. With different initial values, different behaviors such as downward trends rather than upward trends could very well be obtained while the general conclusions drawn above would remain the same.

7.4   DETERMINISTIC TREND MODELS

An alternative way of modeling time series data is based on the decomposition of the model into trend, seasonal, and noise components

$$z_t = T_t + S_t + N_t \tag{7.35}$$

where T_t is the trend component often expressed as a polynomial in time, t, S_t is the seasonal component often expressed as trigonometric functions in time, t, and N_t represents the usual uncorrelated, constant variance, zero mean noise. Forecasting based on this decomposition can be performed using the methodologies developed by Holt (1957) and Winters (1960). For further details on this methodology, see MJK Chapter 4. Admittedly, the representation in Equation (7.35) is being extensively used in practice and can in certain cases provide same or even better results than the models we discussed up to this point (see BD pp. 328–330). We nonetheless have some reservations about this approach. Consider, for example, the modification of Equation (7.35) with only a linear trend model

$$z_t = b_0 + b_1 t + N_t \tag{7.36}$$

What is the implication of such a model? If we intend to use Equation (7.36), we expect our system to produce outcomes that follow a deterministic line with constant intercept b_0 and constant slope b_1 with some noise sprinkled on top. Now, for some physical systems, this may very well be appropriate; however, it is nonetheless quite restrictive. First of all, how can we be sure that, based on the period for which we observe the system, it follows a fixed, deterministic trend line? Second, can we believe that in that model noise is only added to the trend line and does not necessarily behave like a true disturbance that may have a direct impact on the system? Recording errors due to imperfections in measurement devices can be considered as an example of the type of noise assumed in Equation (7.36). However, as Yule (1927) pointed out, would we not in general expect the external disturbances to affect the system directly? In fact, it has been the main premise of this book so far that each system is constantly bombarded with various disturbances that are absorbed and "eliminated" by the system on the basis of its dynamics. Depending on the system dynamics, the impact of the individual disturbances may be short lived or linger for a while. Therefore, it is more realistic to think of disturbances as having a direct impact on the system and not merely sprinkled on top of the deterministic trend as in Equation (7.36).

As a second counterargument, consider once again the linear trend model in Equation (7.36). The data produced by such a model will undoubtedly show an upward or downward linear trend depending on the slope b_1. If we were to model this data with an ARIMA model, we would obviously consider differencing to get rid off the trend in the data. If we apply the first difference to the linear trend component, we obtain

$$(b_0 + b_1 t) - (b_0 + b_1(t - 1)) = b_1 \tag{7.37}$$

Hence, the linear trend can be eliminated by taking the first difference of the data which then becomes stationary. Note that by taking the first difference, we also reduce the number of parameters to be estimated from $2(b_0, b_1)$ to $1(b_1)$. This is in fact true in general, providing yet another reason for us to favor ARIMA models over models given in Equation (7.35) referring back to our discussion in Section 1.7.

We can also show that there is a very close relationship between the deterministic trend models and the ARIMA models when it comes to the forecasts made on the basis of these models. For an extensive and great overview of this subject, we refer to Abraham and Ledolter (1983), Chapter 7. We illustrate it using a very simple example. Consider the linear trend model in (7.36) and the ARIMA(0, 2, 2) model given as

$$(1 - B)^2 z_t = (1 - \theta_1 B - \theta_2 B^2) a_t \tag{7.38}$$

It can be easily shown that the forecast function for both models is the same and is equal to

$$\hat{z}_t(l) = b_0^{(t)} + b_1^{(t)} l \tag{7.39}$$

where $b_0^{(t)}$ and $b_1^{(t)}$ are the estimates of the intercept and slope at time t. Since both models give exactly the same forecasts, how similar are they? We will be particularly concerned with two issues: (1) prediction variance and (2) updating of the parameter estimates.

1. It can be shown that the variance of the l-step forecast error for the deterministic linear trend model in Equation (7.37) is constant for all l's and is equal to

$$\text{Var}(e_t(l)) = \sigma_a^2 \tag{7.40}$$

We have already shown in Chapter 4 that for the ARIMA(0, 2, 2) model in Equation (7.38), we have

$$\text{Var}(e_t(l)) = \sigma_e^2(l) = (1 + \psi_1^2 + \psi_2^2 + \cdots + \psi_{l-1}^2)\sigma_a^2 \tag{7.41}$$

where the weights $\{\psi_t\}$ can be calculated from the following relationship

$$(1 - B)^2 (1 + \psi_1 B + \psi_2 B^2 + \cdots) = (1 - \theta_1 B - \theta_2 B^2) \tag{7.42}$$

There is a striking difference between the variances in Equations (7.40) and (7.41). As mentioned earlier, the variance of the forecast errors for the deterministic trend model remains constant for all lead times, whereas the variance for the ARIMA model in Equation (7.41) obviously gets larger as l gets bigger, implying that when we attempt to make forecasts further in the future, the uncertainty about our forecasts becomes larger. Which case is more realistic? We should intuitively expect the variation in the error that we make in our forecast to get larger and larger with increasing lead times, which is the case for the forecasts made using the ARIMA model.

2. The forecast function in Equation (7.39) implies that the origin of forecast (or current time) is at t. Let us consider that we are now at $t + 1$ and that the new data is available. We are then interested in updating the parameter estimates b_0 and b_1. For the deterministic trend model, the updating equations are given as

$$b_0^{(t+1)} = b_0^{(t)} + b_1^{(t)}, \quad b_1^{(t+1)} = b_1^{(t)} \tag{7.43}$$

As can be seen, the slope has not changed, whereas the intercept is updated to the new origin by simply adding the slope to the earlier intercept.

The updating equations for the ARIMA model are given as

$$b_0^{(t+1)} = b_0^{(t)} + b_1^{(t)} + (1 + \theta_2)e_t(1), \quad b_1^{(t+1)} = b_1^{(t)} + (1 - \theta_1 - \theta_2)e_t(1) \tag{7.44}$$

(For the derivations of these equations, see BJR.)

There is once again a striking difference between these updating equations. The updating equations for the ARIMA model use the one-step forecast error to update the intercept and the slope enabling them to evolve in time. Needless to say, we certainly favor the updating equations that will take into account the forecast errors that we have committed along the way.

It can be argued that there are ways to work around the rigidity of the deterministic trends, for example, by simply allowing the trend and seasonal components to follow stochastic models. A detailed discussion on these models together with their relationship to ARIMA models can be found in BJR who call these *structural component models*. But it is beyond the scope of this book to go further into this topic. We therefore conclude this section by reiterating that ARIMA models offer flexible, reasonable, sound, and easy-to-understand alternatives for many situations encountered in practice.

7.5 YET ANOTHER ARGUMENT FOR DIFFERENCING

Let us assume that the stochastic process $\{z_t\}$ has a level μ_t that is nonstationary. However, we want to consider an ARMA model for *a function of* z_t such that after this transformation, the process is mean stationary or the mean is independent of time. Now consider the ARMA $(p + 1, q)$ process written in terms of z_t (the non-mean-adjusted process). The model is then

$$\varphi(B)z_t = \theta_0 + \theta(B)a_t \tag{7.45}$$

where the constant term θ_0 is given by

$$\theta_0 = \varphi(B)\mu_t$$
$$= (1 - \phi_1 B - \phi_2 B^2 - \cdots - \phi_{p+1}B^{p+1})\mu_t \tag{7.46}$$

We seek a transformation of z_t such that the θ_0 is independent of time. Specifically, suppose the mean μ_t is a polynomial function $g(t)$ of time. For example, $\mu_t = g(t) = \beta_0 + \beta_1 t$. Then

$$\theta_0 = (1 - \phi_1 B - \phi_2 B^2 - \cdots - \phi_{p+1} B^{p+1}) g(t)$$
$$= (1 - \phi_1 B - \phi_2 B^2 - \cdots - \phi_{p+1} B^{p+1})(\beta_0 + \beta_1 t) \qquad (7.47)$$

Now, suppose we factor the autoregressive polynomial $\phi(B)$ such that

$$\varphi(B) = (1 - \varphi B)(1 - \phi_1 B - \phi_2 B^2 - \cdots - \phi_p B^p)$$
$$= (1 - \varphi B)\phi(B) \qquad (7.48)$$

where $\phi(B) = (1 - \phi_1 B - \phi_2 B^2 - \cdots - \phi_p B^p)$. We can then write

$$\theta_0 = (1 - \varphi B)(1 - \phi_1 B - \phi_2 B^2 - \cdots - \phi_p B^p)(\beta_0 + \beta_1 t) \qquad (7.49)$$

Multiplying the right-hand side of Equation (7.49) out we get

$$\theta_0 = \phi(B)(1 - \varphi B)(\beta_0 + \beta_1 t)$$
$$= \phi(B)[(1 - \varphi B)\beta_0 + (t - \varphi B t)\beta_1] \qquad (7.50)$$

Now if we let $\varphi \to 1$ from below, we get

$$\theta_0 = \phi(B)[(1 - B)\beta_0 + (t - Bt)\beta_1]$$
$$= \phi(B)\beta_1 \qquad (7.51)$$

We can see that this latter expression is independent of time. Hence, we conclude that a first order difference function $\nabla = (1 - B)$ of z_t obtained by letting $\varphi \to 1$ in $(1 - \varphi B)$ provides the function we were seeking.

For higher order polynomials, i.e., $g(t) = \beta_0 + \beta_1 t + \cdots + \beta_d t^d$, we would still seek a function of z_t for which the level is independent of time. To do so, we choose an autoregressive operator that includes $(1 - B)^d$. Therefore, if

$$\varphi(B) = (1 - B)^d (1 - \phi_1 B - \phi_2 B^2 - \cdots - \phi_p B^p) \qquad (7.52)$$

then

$$\theta_0 = (1 - \varphi B)^d (1 - \phi_1 B - \phi_2 B^2 - \cdots - \phi_p B^p)(\beta_0 + \beta_1 t + \cdots + \beta_d t^d)$$
$$= \phi(B)(1 - B)^d (\beta_0 + \beta_1 t + \cdots + \beta_d t^d)$$
$$= \phi(B)\beta_d \, d! \qquad (7.53)$$

which is a constant independent of t. Note that the above argument is made for $\mu_t = g(t)$. However results will be similar for a randomly changing μ_t.

7.6   CONSTANT TERM IN ARIMA MODELS

In Section 7.3, we showed that for the linear trend model $z_t = b_0 + b_1 t + N_t$, differencing reduced the linear trend to a constant as in Equation (7.37). This is directly related to the issue we briefly discussed in Chapter 4 about adding a constant term in an ARIMA model. We indicated specifically that the need for a constant term in an ARIMA(p, d, q) model with $d > 0$ indicates a deterministic trend. For the general ARIMA model

$$\varphi(B)z_t = \theta_0 + \theta(B)a_t \tag{7.54}$$

if the constant term θ_0 is assumed to be zero and $d > 0$, then the general ARIMA model is capable of representing a time series with stochastic trend, where there are random changes in the level or slope. For example, we see that if $d = 1$, then for an arbitrary constant added to the process, we have $\varphi(B)(z_t + c) = \phi(B)(1 - B)(z_t + c) = \phi(B)\nabla z_t = \varphi(B)z_t$. Therefore, adding a constant to the process does not alter the behavior of the process, but just shifts its level. This property is referred to as *homogenous nonstationary*.

In general, we may, for some applications, want to allow for a deterministic trend $g(t)$ in the model. In particular, if $g(t)$ is a polynomial of degree d, we may use a model with $\theta_0 \neq 0$. For example, if $d = 1$, then we allow for a linear trend in the presence of nonstationary noise in the model. Since we allow $\theta_0 \neq 0$, it follows that

$$E\{w_t\} = E\{\nabla^d z_t\} = \mu_w = \theta_0/(1 - \phi_1 - \cdots - \phi_p) \neq 0 \tag{7.55}$$

In other words, $\tilde{w}_t = w_t - \mu_w$ follows a stationary invertible ARMA process.

In many applications, there would be no obvious physical reason to suggest a deterministic component in the time series. We will therefore often assume that θ_0 or μ_w is zero unless the data suggests otherwise. Indeed, in many applications, it is more realistic to assume a stochastic trend. Specifically, a stochastic trend does not necessitate that the entire series follows the same pattern. Thus, if $d > 0$, we will most often assume that θ_0 or μ_w is zero unless the physical circumstances of the problem or the data indicates that this is an unrealistic assumption.

7.7   CANCELLATION OF TERMS IN ARIMA MODELS

In Section 7.4, we discussed the similarities (or more appropriately the differences) between the linear trend model

$$z_t = b_0 + b_1 t + N_t \tag{7.56}$$

and the ARIMA(0, 2, 2) model

$$(1 - B)^2 z_t = (1 - \theta_1 B - \theta_2 B^2)a_t \tag{7.57}$$

We saw that while the l-step-ahead forecast for both models could be obtained using

$$\hat{z}_t(l) = b_0^{(t)} + b_1^{(t)}l \tag{7.58}$$

the updating equations for linear trend model were

$$b_0^{(t+1)} = b_0^{(t)} + b_1^{(t)}, \quad b_1^{(t+1)} = b_1^{(t)} \tag{7.59}$$

whereas for the ARIMA model, they were

$$b_0^{(t+1)} = b_0^{(t)} + b_1^{(t)} + (1+\theta_2)e_t(1), \quad b_1^{(t+1)} = b_1^{(t)} + (1-\theta_1-\theta_2)e_t(1) \tag{7.60}$$

Now suppose $\theta_2 \to -1$ and $\theta_1 \to 2$ simultaneously. Then we can see from Equation (7.60) that the factors in front of the 1-step-ahead forecast error, $e_t(1)$ approach 0. This means that the updating Equations (7.59) and (7.60) will be exactly the same. Let us consider the MA term in Equation (7.57) at the limiting case of $\theta_2 = -1$ and $\theta_1 = 2$,

$$(1 - 2B + B^2) = (1 - B)^2$$

Hence for $\theta_2 \to -1$ and $\theta_1 \to 2$, $(1 - \theta_1 B - \theta_2 B^2) \to (1 - B)^2$. This will lead to cancellation of the terms on both sides of Equation (7.57). In general, when the parameters in $\theta(B)$ approach values that will result in such a cancellation in $\varphi(B)$, we will then need a deterministic trend component in the model as in Equation (7.56). To show this more theoretically, let $\varphi(B) = 1 - \varphi_1 B - \varphi_2 B^2 - \cdots - \varphi_{p+1} B^{p+1}$ and $\theta(B) = 1 - \theta_1 B - \theta_2 B^2 - \cdots - \theta_p B^q$. Then, the ARIMA$(p,d,q)$ model is

$$\varphi(B)z_t = \theta_0 + \theta(B)a_t \tag{7.61}$$

From Section 7.1, the general solution $\{z_t\}$ of the linear difference equation (7.61) relative to the time origin $k < t$ is

$$z_t = C_k(t-k) + I_k(t-k) \tag{7.62}$$

where $C_k(t-k)$ is the complementary function and $I_k(t-k)$ is a particular function. The complementary function is the solution to the homogenous difference equation $\varphi(B)z_t = 0$.

Note that the constant θ_0 can be considered as a constant value forcing function. Now suppose $G(B)$ is the common factor such that $\varphi(B) = G(B)\varphi_1(B)$ and $\theta(B) = G(B)\theta_1(B)$ where $G(B) = 1 - g_1 B - \cdots - g_r B^r$. Hence, the model is

$$G(B)\varphi_1(B)z_t = \theta_0 + G(B)\theta_1(B)a_t \tag{7.63}$$

Furthermore, suppose the polynomial $G(B) = 1 - g_1 B - \cdots - g_r B^r$ has roots $G_1^{-1}, \ldots, G_r^{-1}$. We will assume that the roots are distinct and greater than one in absolute value. It now follows that we can write Equation (7.63) as

$$\varphi_1(B)z_t = \theta_0/G(B) + \theta_1(B)a_t \tag{7.64}$$

Hence, $\theta_0/G(B)$ is a deterministic forcing function, and $\theta_1(B)a_t$ is a stochastic forcing function. Now let $\{x_t\}$ be a deterministic function that satisfies $G(B)x_t = \theta_0$. Then we may write the reduced time series model as $\varphi_1(B)z_t = x_t + \theta_1(B)a_t$. As can be seen, $\{z_t\}$ is a combined solution of a linear difference equation $\varphi_1(B)z_t$ with a deterministic driving function $\{x_t\}$ and a stochastic driving function $\{u_t\}$ where $u_t = \theta_1(B)a_t$.

The absolute values of the roots of $G(B)$ are important. If $|G_i| < 1$, the contribution of the deterministic trend will quickly die out. We can therefore often ignore such terms in practice. However, if roots $|G_i| = 1$, then these roots correspond to nonstationary differencing or a simplifying operator such as $(1 - B)^d$ or $(1 - B^s)$ and they should not be ignored but need to be included in the model as deterministic functions, either as polynomials or as sine and cosine functions depending on the roots of $G(B)$.

Let k be the starting point of the time series $k < t$. If the polynomial $G(B)$ has d equal roots G_0^{-1}, then it contains a factor $(1 - G_0 B)^d$. The deterministic trend is then given by

$$x_t = [b_0 + b_1(t - k) + b_2(t - k)^2 + \cdots + b_{d-1}(t - k)^{d-1}]G_0^{t-k} \qquad (7.65)$$

In the special case when $G_0 = 1$, we have

$$x_t = b_0 + b_1(t - k) + b_2(t - k)^2 + \cdots + b_{d-1}(t - k)^{d-1} \qquad (7.66)$$

which is a $(d - 1)$th order polynomial.

As in dealing with ordinary linear differential equations with constant coefficients, we need to consider three cases. If the polynomial $G(B)$ has distinct roots G_i^{-1} then

$$x_t = \alpha_1 G_1^{t-k} + \alpha_2 G_2^{t-k} + \cdots + \alpha_p G_p^{t-k} \qquad (7.67)$$

If the roots are real, then they contribute exponential terms G_j^{t-k}. However, if the roots are complex, then they contribute pairs of sine functions $D^{t-k}\sin(2\pi ft + F)$. In the special case where the roots are complex but $|G_j| = 1$, that is, the roots are on the unit circle in the complex plane, we get from De Moivre's theorem that the roots of unity $x^n = 1$ are given by

$$\varepsilon_1 = \varepsilon = \cos(2\pi/n) + i\sin(2\pi/n),$$
$$\varepsilon_2 = \varepsilon^2 = \cos(2 \times 2\pi/n) + i\sin(2 \times 2\pi/n), \ldots$$
$$\varepsilon_{n-1} = \varepsilon^{n-1} = \cos((n-1) \times 2\pi/n) + i\sin((n-1) \times 2\pi/n),$$
$$\varepsilon_n = \varepsilon^n = 1$$

We will now consider two simple examples to show how the deterministic component is calculated.

Example 3: Suppose the model is

$$(1 - B)z_t = \theta_0 + (1 - B)\theta_1(B)a_t \qquad (7.68)$$

Then

$$z_t = x_t + \theta_1(B)a_t \qquad (7.69)$$

where $(1 - B)x_t = \theta_0$. The general solution of this deterministic first order difference equation is $x_t = b_0 + b_1 t$. Hence, the solution to the time series model is

$$z_t = x_t + \theta_1(B)a_t = b_0 + b_1 t + \theta_1(B)a_t \qquad (7.70)$$

Example 4: For the model

$$(1 - \sqrt{3}B + B^2)z_t = \theta_0 + (1 - \sqrt{3}B + B^2)\theta_1(B)a_t \qquad (7.71)$$

we have $G(B) = (1 - \sqrt{3}B + B^2)$. The roots are

$$G_j = \frac{\sqrt{3} \pm \sqrt{3-4}}{2} = \frac{\sqrt{3} \pm \sqrt{-1}}{2} = \sqrt{3}/2 \pm i/2$$

Now $r = \sqrt{(\sqrt{3}/2)^2 + (1/2)^2} = 1$. Thus $|G_j| = 1$ and hence the solution to the time series model is

$$z_t = b_0 + b_1 \cos(2\pi t/12) + b_2 \sin(2\pi t/12) + \theta_1(B)a_t \qquad (7.72)$$

It should be noted that so far in this section, we assumed that the true values of the parameter estimates are known. However, in practice, we will have estimates of those parameters. These estimates of course are not fixed and have a variance due to sampling errors. This makes the determination of the common factor $G(B)$ in Equation (7.62) quite complicated. In those circumstances, we may have terms that are close in value but not exactly the same, resulting in near-cancellation. We therefore strongly recommend, as usual, using judgment and in near-cancellation situations seriously considering the alternative model with a deterministic trend.

7.8 STOCHASTIC TREND: UNIT ROOT NONSTATIONARY PROCESSES

First order nonstationary processes are sometimes called *unit root nonstationary processes*. The random walk $\nabla z_t = a_t$ is a simple example of a nonstationary unit root process. Given a starting value z_0 and the sequence of noise $\{a_t\}$, the current value can be found iteratively from $z_j = z_{j-1} + a_j$. The forecast is $\hat{z}_h(\ell) = z_p$. Therefore, for all forecast horizons ℓ, the forecast is simply the value at the forecast origin. In other words, there is no mean reverting as there would be the case for a stationary process.

The MA representation of the random walk is $z_t = a_t + a_{t-1} + a_{t-2} + \cdots$. Thus, the forecast error is

$$e_h(\ell) = a_{h+\ell} + a_{h+\ell-1} + \cdots + a_{h+1} \qquad (7.73)$$

From this we see that $\text{Var}\{e_h(\ell)\} = \ell\sigma^2$, which diverges to infinity as $\ell \to \infty$. In other words, the forecast interval will continue to increase as the forecast horizon increases.

Another property of the unit root process is that it has a strong memory. In other words, the shocks all have a permanent impact.

Let us now consider a slightly different process, the random walk with drift, written as $\nabla z_t = \theta_0 + a_t$ where $\theta_0 \neq 0$. This model has a deterministic linear trend and can also be written as

$$z_t = \theta_0 + z_{t-1} + a_t \qquad (7.74)$$

Now, assuming that the process started at time $t = 0$ with initial value z_0, then

$$z_1 = \theta_0 + z_0 + a_1$$
$$z_2 = \theta_0 + z_1 + a_2 = 2\theta_0 + z_0 + a_1 + a_2$$
$$\vdots$$
$$z_t = t\theta_0 + z_0 + a_t + a_{t-1} + \cdots + a_1 + a_2 \qquad (7.75)$$

Thus we see that $E\{z_t\} = t\theta_0$ (for $z_0 = 0$) and this process exhibits a linear deterministic trend $\theta_0 t$ and a pure random walk process $\sum_{j=1}^{t} a_j$. Moreover, we see that $\text{Var}\left\{\sum_{j=1}^{t} a_j\right\} = t\sigma^2$ and hence $\text{Var}\{z_t\} = t\sigma^2$.

A closely related model to the random walk with drift is given by

$$z_t = \beta_0 + \beta_1 t + \varepsilon_t \qquad (7.76)$$

where ε_t follows a stationary time series, for example, AR(p) such that $\phi(B)\varepsilon_t = a_t$. This model is called *trend stationary* in the economics literature; in the statistics literature, we say that it has a deterministic trend. Like the random walk with drift, this model increases linearly in time due to $\beta_1 t$. However, the two models are fundamentally different. To see that, consider first the random walk with drift. For this process, the expected value is $E\{z_t\} = t\theta_0$ and the variance is $\text{Var}\{z_t\} = t\sigma^2$. They are both time dependent. However, the trend stationary (deterministic trend) model has a time-dependent mean $E\{z_t\} = \beta_0 + \beta_1 t$ and variance $\text{Var}\{\varepsilon_t\} = \sigma_\varepsilon^2$, which is constant.

If we know that the data follows a trend stationary model, then we can remove the trend with linear regression. As discussed earlier, we can also difference the data and get rid of the trend. Which method is better? For the stochastic trend model, we know that removing the trend by fitting a polynomial will obscure the model and create false residuals. However, that need not be the case for the trend stationary model (deterministic model). A nice discussion on trend stationary and random walk models for macroeconomic data can be found in Nelson and Plosser (1982).

7.9   OVERDIFFERENCING AND UNDERDIFFERENCING

In Chapter 4, we applied differencing to a nonstationary time series to transform it into a stationary series. We also suggested using differencing several times until the stationarity is achieved. It should be noted that we also added that differencing more than twice is very rarely needed. The question is of course when and how many times to use differencing. First of all, we strongly recommend considering the characteristics of the process from which the data is collected. For example, do we expect the data to have a fixed level? The answer to this question may easily give a clue for the need for differencing. It may, however, be the case that we have limited information about the process. We then suggest considering two very basic and simple tools, plotting the data and the sample ACF. The former will often reveal whether the data has time-independent, constant level and spread. Therefore, it is often possible to check for stationarity by simply looking at the plot of the data. We should, however, keep in mind that for certain (often short) periods of time, perfectly nonstationary data may exhibit stationary behavior. If there are doubts about the stationarity of the data, we also recommend considering differencing the data and observing the behavior of the differenced series. With today's powerful software packages, looking into alternative models is very easily done with the least hassle to the analysts.

As the second tool, we suggest examining the sample ACF. A slow, almost linear, decaying sample ACF value is a good indication of nonstationarity. The sample ACF of a properly differenced time series should exhibit a rapidly decreasing behavior. A common misconception about the sample ACF of a nonstationary time series is that the first lag sample ACF, $r(1)$, is expected to be close to 1. However, as shown in Wichern (1973), the starting value of the sample ACF can be as low as 0.5 or less for a nonstationary series because of sampling variability. Therefore, it is the behavior of the decay (slow and almost linear) in sample ACF values that is the indicator for nonstationarity and not the value of $r(1)$ to be close to 1.

These plots and the ability of trying various alternative models with ease using today's user-friendly software would in most cases be more than enough to make the right decision of differencing a time series to achieve stationarity. However, a more formal test to check for unit roots in autoregression, which is the indication for nonstationarity and hence for the need for differencing by Dickey and Fuller (1979), is also available. For further details of this test, we refer the readers to BD. But as mentioned earlier, we truly believe that in most cases, the visual inspection of the time series together with the sample ACF is a very simple yet effective tool to avoid underdifferencing.

On the flip side of the coin lies overdifferencing. Remember that the main goal of differencing is to achieve stationarity. If we overdifference a time series, for example, take the first difference of an already stationary time series, the new differenced series will remain stationary. Therefore, seemingly we would do no harm in overdifferencing. However, as we can see below, overdifferencing will often complicate the estimation of the model parameters and inflate the variance of the overdifferenced series.

Consider the following MA(1) model:

$$z_t = (1 - \theta B)a_t \qquad (7.77)$$

Now suppose we take the difference of both sides. We then get

$$(1 - B)z_t = (1 - B)(1 - \theta B)a_t \qquad (7.78)$$

Multiplying out the right-hand side we get

$$(1 - B)z_t = (1 - B)(1 - \theta B)a_t$$
$$= (1 - (1 + \theta)B + \theta B^2)a_t \qquad (7.79)$$

If we were to fit this model to $\nabla z_t = (1 - B)z_t$ but not knowing that it was an overdifferenced MA(1) process, we would likely fit an MA(2) model to ∇z_t and get

$$\nabla z_t = (1 - \theta_1 B - \theta_2 B^2)a_t \qquad (7.80)$$

where to a close approximation $\hat{\theta}_1 \approx 1 + \theta$ and $\hat{\theta}_2 \approx -\theta$. Hence, we would expect to get $\hat{\theta}_1 + \hat{\theta}_2 \approx 1$. We can see that there is a close relationship between the two MA parameters, which will certainly affect their estimation. Also from the 1-parameter MA(1) model, we now move to a less parsimonious 2-parameter MA(2) model. Moreover, since one of the roots in $(1 - B)(1 - \theta B)$ is equal to 1, we now have a noninvertible MA model. Finally, Abraham and Ledolter (1983) showed that the variance of the overdifferenced series is larger than the variance of the original series.

Since overdifferencing can have such serious consequences, how can we detect it? As in the unit root tests in autoregression we discussed above, we can also test for unit roots in moving averages. For a detailed discussion, see BD. As for simple tools, however, one can consider the sample variances as suggested in Tintner (1940). Since the variance of the overdifferenced series gets larger (Abraham and Ledolter, 1983), we can use sample variances of successively differenced series to decide on when to stop differencing. At each differencing of the nonstationary series, the sample variance will get smaller until the stationary series is reached. Further differencing will then cause the sample variance to increase. As another simple tool, once a model is obtained for differenced data, we can check whether the model is invertible. If not, we would then suspect overdifferencing.

7.10   MISSING VALUES IN TIME SERIES DATA

Figure 7.2 shows the annual sea levels in millimeters for Copenhagen, Denmark, from 1889 to 2006. We can clearly see that toward the end, there are a few data points missing. The data for years 1983, 1991, 1999, and 2001 were not available. The data is available in Appendix A. We obtain this data from www.psmsl.org, a great website to obtain global sea level data. A number of datasets, however, suffer from the same problem of having some missing observations. Unfortunately, this is not at all uncommon in time series data. There are some advanced methods based on state space models in dealing with missing data as given in BJR Section 13.3. However, a simple way to deal with these missing values is to replace the missing values with some sort of an average value based on the observations before and after the missing value and then proceed with the analysis as usual. But before we even attempt to do so, we first and foremost have to ask why and how often we have missing values. The answers to these questions may affect the choice of methods used to deal with the missing data.

In our discussions so far, we assumed that the time series data we dealt with were collected at equal time intervals. However if, for example, the data is missing with a systematic pattern, for example, every third observation is

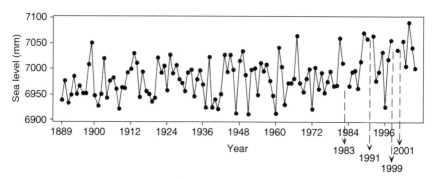

Figure 7.2   The annual sea levels in millimeters for Copenhagen, Denmark, with missing values on 1983, 1991, 1999, and 2001. (*Source:* www.psmsl.org.)

missing or the data is collected whenever it becomes available and not at fixed time intervals, then we can no longer make the assumption that we have equally spaced data. The methods to analyze this type of data are beyond the scope of this book and instead we refer the readers to Jones (1985). We would rather prefer to focus on the issue of a few observations missing seemingly at random as in Figure 7.2.

As mentioned earlier, the common practice in dealing with missing data is to insert an estimated value. We can, for example, simply use a smoothing method such as EWMA and insert the EWMA prediction for the missing value. BJR discuss the use of a weighted average of p observations before and after the missing value for an AR(p) model. For that, the best estimate of the missing value is

$$\hat{z}_t = -d_0^{-1} \sum_{j=1}^{p} d_j \left(z_{T-j} + z_{T+j} \right) \tag{7.81}$$

where $d_j = \sum_{i=j}^{p} \phi_i \phi_{i-j}, \phi_0 = -1$, and $d_0 = 1 + \sum_{i=1}^{p} \phi_i^2$

We also recommend interpolation techniques such as spline regression and Kriging to estimate the missing value. The former is based on continuous piece-wise linear polynomials to make the interpolations. For further details see, for example, Montgomery *et al.* (2006). The latter is a method originated from geo-statistics developed by Krige and improved by Matheron (1963). Kriging is commonly used to model the data obtained from deterministic computer experiments for which the deterministic output $y(x)$ is assumed to be described by the random function

$$Y(x) = f(x)'\beta + Z(x) \tag{7.82}$$

where $f(x)'\beta$ is a parametric trend component, which often consists of a constant and $Z(x)$ is a zero mean Gaussian random field assumed to be second order stationary with a covariance function $\sigma^2 R(x_i, x_j)$. Kriging, similar to any other interpolation technique, uses a weighted average of the surrounding data to predict the response at a new location. The weights depend on the predetermined correlation structure, the fitted model, and also the location for which the prediction is to be made. For further details on Kriging, see Santner *et al.* (2003).

For the data in Figure 7.2, we used DACE, a Matlab Kriging toolbox available at http://www2.imm.dtu.dk/~hbn/dace/ to fit an ordinary Kriging model where the parametric trend component is simply a constant. The predictions from the Kriging model for years 1983, 1991, 1999, and 2001 are 6996, 7012, 7030, and 7034, respectively. ACF and PACF plots of the "complete" dataset where the missing observations are replaced with these predictions are given in Figures 7.3 and 7.4.

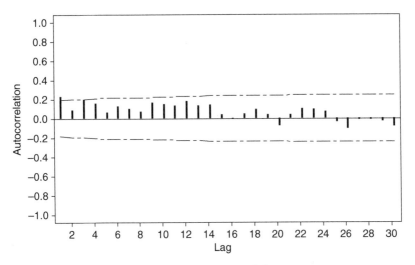

Figure 7.3   ACF of the complete annual sea level dataset.

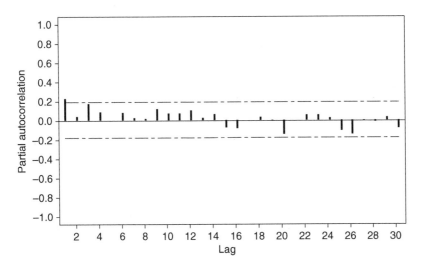

Figure 7.4   PACF of the complete annual sea level dataset.

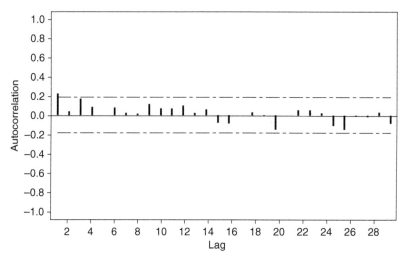

Figure 7.5 ACF of the residuals of the AR(1) model for the annual sea level data.

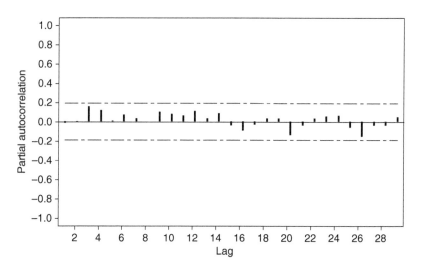

Figure 7.6 PACF of the residuals of the AR(1) model for the annual sea level data.

Both ACF and PACF show only one significant correlation at the first lag. However, we observe a slight decaying pattern in ACF and therefore we decide to fit an AR(1) model. The model estimates are given in Table 7.1, which suggests that both terms in the model (AR(1) parameter and the constant term) are significant. ACF and PACF of the residuals given in Figures 7.5 and 7.6 show no sign of remaining autocorrelation. Therefore, we deem the model adequate.

TABLE 7.1   AR(1) Model Estimates for the Annual Sea Level Data

Type	Coefficient	Standard error	t	p
AR	0.2204	0.0906	2.43	0.017
Constant	5445.82	3.57	1524.21	0
Mean	6985.5	4.58	—	

EXERCISES

7.1 Show how to calculate the complementary function and the particular integral for ARIMA(1, 0, 0), ARIMA(1, 1, 0), ARIMA(0, 1, 1), and ARIMA(1, 1, 1) processes as in the examples given in Section 7.2.

7.2 Reproduce Figure 7.1 and use arbitrary initial values when needed. Compare your plots of the eventual forecast functions to the ones given in Figure 7.1.

7.3 Table B.14 shows the mid-year population data in the United Kingdom from 1950 to 2010. Using the data from 1950 to 2005, fit appropriate ARIMA models with and without the intercept and make forecasts for 2006 to 2010. Compare your results based on mean-squared prediction errors.

7.4 Repeat Exercise 7.3 using the linear trend model given in Equation (7.36). Compare your results with the ones you obtained in Exercise 7.3.

7.5 In Chapter 3, we used an AR(2) model for temperature readings from a ceramic furnace. The data is given in Table A.1. Fit an ARMA(3, 1) model to this data. Comment on the estimates of the parameters of this new model.

7.6 Consider again the data given in Table A.1. Assume that the observations 70, 73, and 78 are missing. Estimate the missing values using the following techniques and compare your results.

a. Estimate the missing value by taking the average of the observations before and after.

b. Estimate the missing value using Equation (7.81).

c. Estimate the missing value using EWMA with an appropriate θ value.

d. Estimate the missing value using an interpolation technique such as Kriging.

7.7 Consider sea level data shown in Figure 7.2 and given in Table A.7. Estimate the missing values using the following techniques and compare your results with the ones obtained using Kriging provided in Section 7.10.

a. Estimate the missing value by taking the average of the observations before and after.

b. Estimate the missing value using Equation (7.81).

c. Estimate the missing value using EWMA with an appropriate θ value.

7.8 Consider temperature readings data given in Table A.2. Take the successive differences of the data and for each differencing calculate the variance. Compare the variances to decide the right degree of differencing as discussed in Section 7.9. We used this data in Chapter 4. Compare your resulting degree of differencing with the one used in that chapter.

7.9 Show that the forecast function for the linear trend model and an ARIMA(0, 2, 2) is
$$\hat{z}_t(l) = b_0^{(t)} + b_1^{(t)} l.$$

7.10 Show that the variance of the l-step forecast error for the deterministic linear trend model is $\text{Var}(e_t(l)) = \sigma_a^2$.

7.11 Show that the variance of the l-step forecast error for the ARIMA(0, 2, 2) model is
$$\text{Var}(e_t(l)) = \sigma_e^2(l) = (1 + \psi_1^2 + \psi_2^2 + \cdots + \psi_{l-1}^2)\sigma_a^2.$$

CHAPTER *8*

TRANSFER FUNCTION MODELS

8.1 INTRODUCTION

We have so far discussed the analysis of univariate time series. In this chapter and the next, we build on these concepts and apply and extend them to facilitate the analysis of two or more time series. In this chapter, we will primarily focus on the unidirectional relationship between an input and an output, both of which are time series. For the purpose of forecasting, it is often important to be able to take into account the information from certain inputs also called *leading indicators*. For example, early information about housing starts can provide early indications about how many economic variables will develop such as the use of materials, the labor market and employment, and so on. Therefore, the simultaneous analysis of two or more time series is important in many areas of applications.

8.2 STUDYING INPUT–OUTPUT RELATIONSHIPS

In this section, we discuss how to analyze the relationship between an input and an output of a process. Understanding such relationships is important not only for engineering but also for a wide range of problems including many economic, financial, and managerial applications of time series analysis such as forecasting with leading indicators. The input–output relationships in the engineering context will be studied first because it provides a simple understanding that can be generalized to other processes and applications that are typically much more complex.

We denote the input by x and the output by y, and represent the process by a box as in Figure 8.1. The functional relationship between the input and the output is expressed as $y = f(x)$. In the simplest case, the relationship is a linear regression equation $y = a + bx + \varepsilon$. If so, the implication is that the effect is immediate; a change in x is immediately followed by a proportional change in y. In the regression literature, there is typically no reference to time—it is simply understood to be a contemporaneous relationship. However, it would be

Time Series Analysis and Forecasting by Example, First Edition. Søren Bisgaard and Murat Kulahci.
© 2011 John Wiley & Sons, Inc. Published 2011 by John Wiley & Sons, Inc.

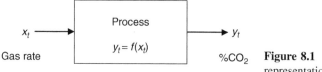

Figure 8.1 A diagrammatic representation of a process.

more correct to express the relationship as $y_t = f(x_t)$ and in the specific case of a linear relationship as $y_t = a + bx_t + \varepsilon_t$.

Input–output relationships sometimes get more complicated. For example, the response may be delayed and distributed over a period of time. Such relationships are called *dynamic transfer functions*. A familiar example is the shower. If we manipulate the shower control knob, it may take a few seconds before the temperature of the water coming out of the nozzle changes. Many economic and manufacturing processes exhibit similar dynamic relationships with a delayed and distributed response to changes. Especially, delays make it difficult to control a process. For example, imagine how troublesome it would be to drive a car if there was even a few seconds delay in the steering mechanism.

In this chapter, we demonstrate how to assess delayed dynamic relationships using the theory originally developed by Box and Jenkins (1969). To provide an impression of the scope of this method, we use an example to demonstrate step-by-step how to assess the form of the input–output relationship for a relatively complex process.

8.3   EXAMPLE 1: THE BOX–JENKINS' GAS FURNACE

To develop methods for controlling processes, Box and Jenkins had a laboratory scale gas furnace built to demonstrate how to assess the nature of a dynamic relationship between an input and an output. The furnace combined air and methane gas to form a mixture containing carbon dioxide, CO_2. In one of their experiments, they kept the air feed constant. However, the input methane gas rate, x_t, was varied and the output CO_2 concentration, y_t, was measured. The CO_2 concentration was sampled every 9 s. It is important to note that there was no feedback loop in the process. A set of 296 pairs of simultaneous observations of x_t and y_t provided as series J in BJR are shown in Figure 8.2.

If we make a scatter plot of contemporaneous values of the input and the output as in Figure 8.3, we see that there is a negative relationship between the gas feed rate, x_t, and the CO_2 concentration, y_t. The estimated correlation coefficient is $r_{xy} = -0.484$. It makes physical sense that the correlation is negative; an increase in gas feed rate causes a reduction in the CO_2 concentration. If a linear regression line is naively fitted to the data, the relationship would be $y_t = 53.43 - 1.446x_t$. This regression line is superimposed on the scatter plot in Figure 8.3. However, we see below that this is not a good idea.

In Figure 8.3, we plotted simultaneous values of x_t and y_t. However, if we suspect a dynamic behavior in the process, we could also compute the correlation

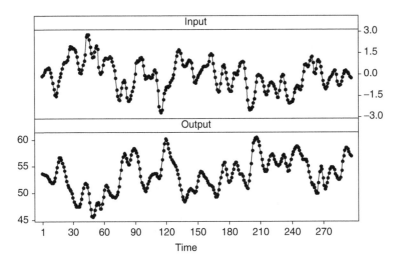

Figure 8.2   Time series plots of Box and Jenkins' gas furnace data.

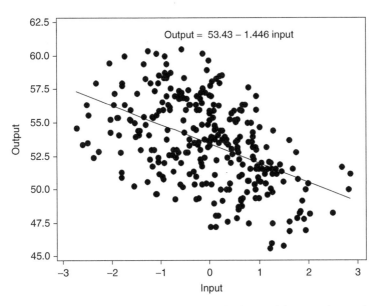

Figure 8.3   Scatter plot of CO_2 versus gas feed rate with a superimposed regression line.

coefficient between x_t and the response y_{t+1} 1 time unit later. This is called the *lag 1 cross correlation*, which in this case is $r_{xy}(1) = -0.598$. It shows an even stronger correlation than the contemporaneous cross correlation. We can, of course, also calculate the correlation between the input at time t and the output k time units later. This is called the *cross correlation for lag k* and is given by

$$r_{xy}(k) = \frac{c_{xy}}{s_x s_y} \tag{8.1}$$

where

$$
c_{xy} = \begin{cases}
\dfrac{1}{n} \displaystyle\sum_{t=1}^{n-k} (x_t - \bar{x})(y_{t+k} - \bar{y}), & k = 0, 1, 2, \ldots \\[4mm]
\dfrac{1}{n} \displaystyle\sum_{t=1}^{n+k} (y_t - \bar{y})(x_{t-k} - \bar{x}), & k = 0, -1, -2, \ldots
\end{cases}
\tag{8.2}
$$

$\bar{x}$ and $\bar{y}$ are the sample means of x_t and y_t, respectively, and s_x and s_y are the sample standard deviations of x_t and y_t, respectively. The cross correlation between x_t and y_t from lags $k = -30$ to $k = 30$ is shown in Figure 8.4.

Notice that the cross correlation function is not symmetric for positive and negative time lags. The strongest cross correlation occurs for lag $k = 5$. However, we also note that the cross correlation function is "smeared out" over a wide range of lags, both positive and negative.

The most surprising observation from Figure 8.4 is that there seems to be large cross correlations at negative lags. But how can that be? If there is a true cause-and-effect relationship, a change in x_t should precede a change in y_t, not the other way around. We should expect to see significant cross correlations for positive, but not for negative, lags. This is indeed one of the more puzzling aspects of time series relationships. As is shown below, the reason is the autocorrelation of the input x_t.

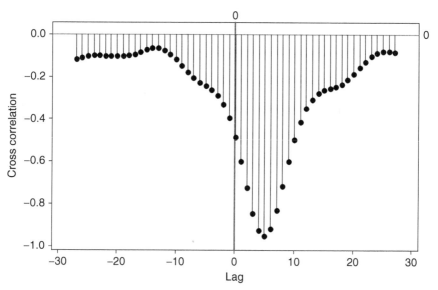

Figure 8.4 The estimated cross correlation function between the input gas feed rate and the output CO_2.

8.4   SPURIOUS CROSS CORRELATIONS

If the input is autocorrelated, the effect of any change in the input itself will take some time to play out. Therefore, the subsequent effects of the input on the output tend to linger around. When a new change in x_t occurs, the current effect gets mixed up with the lingering after-effects of previous changes in x_t. We may therefore observe spurious effects of changes in y_t, which appear to have occurred before the change in x_t.

To see more precisely how this can happen, consider an extremely simple transfer function $y_t = x_t$ where the output is exactly the same as the input with no time delay. We may intuitively expect a strong cross correlation at lag zero and no other cross correlations at any other lags. However, that is not so. If we substitute $y_t = x_t$ in the cross correlation formula above, we see that it becomes identical to the ACF.

Now remember we always plot the autocorrelation for positive lags. This is because the autocorrelation is symmetric in time. It looks exactly the same for negative lags. Suppose the input to our simple process, $y_t = x_t$, is the gas feed rate data. Figure 8.6 shows the autocorrelation for x_t. We see that it is positively auto-correlated for a considerable number of lags. Therefore, since $y_t = x_t$, the cross correlation function between x_t and y_t will look exactly like Figure 8.6 for posi-tive lags. But because of the symmetry of the ACF, the cross correlation function for negative lags will look like the mirror image of Figure 8.6. There will be sig-nificant cross correlations for both positive and negative lags. Therefore, because of the autocorrelation in the input, even if the input/output relationship is instan-taneous, the cross correlation will be smeared out over a considerable number of lags. Indeed, it may give the spurious appearance of effects before the cause.

8.5   PREWHITENING

What can we do to alleviate this spurious effect in the estimated cross correlation function caused by the autocorrelation in the input? The answer is to *prewhiten* the input. The idea is to make the input look like white noise. In other words, we remove the autocorrelation in the input series that caused the spurious cross correlation effect. The steps involved in prewhitening are outlined in Figure 8.5. We now demonstrate each step for the gas furnace data. Each of these steps can be performed easily with standard statistical software.

8.5.1   Time Series Modeling of the Input

The first step in prewhitening involves identifying and fitting a time series model to the input data x_t.

The autocorrelation for the input data x_t is shown in Figure 8.6. We see a damped sine wave pattern typical for an autoregressive model. Further, the

Prewhitening

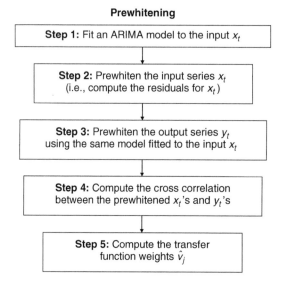

Step 1: Fit an ARIMA model to the input x_t

Step 2: Prewhiten the input series x_t (i.e., compute the residuals for x_t)

Step 3: Prewhiten the output series y_t using the same model fitted to the input x_t

Step 4: Compute the cross correlation between the prewhitened x_t's and y_t's

Step 5: Compute the transfer function weights $\hat{v}_j$

Figure 8.5 Prewhitening step-by-step.

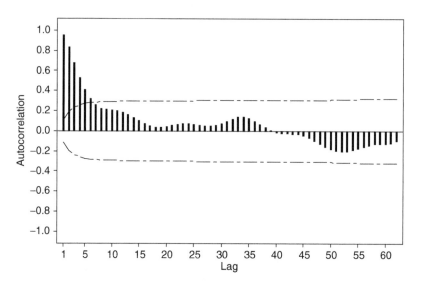

Figure 8.6 Autocorrelation of the input, gas feed rate.

partial autocorrelation shown in Figure 8.7 cuts off after three lags. These two plots, when combined, indicate that an appropriate model might be a third order autoregressive AR(3) model

$$\tilde{x}_t = \phi_1 \tilde{x}_{t-1} + \phi_2 \tilde{x}_{t-2} + \phi_3 \tilde{x}_{t-3} + \alpha_t \tag{8.3}$$

where $\tilde{x}_t = x_t - \mu$ and α_t is an error term usually assumed to be white noise. The input data x_t has already been scaled so that the mean is zero. Thus, we will

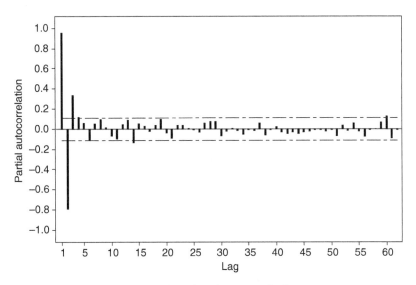

Figure 8.7 Partial autocorrelation of the input, gas feed rate.

assume that $\mu_x = 0$. Fitting Equation (8.3) to the data, we obtain the estimated coefficients summarized in Table 8.1. Rewriting Equation (8.3), we see that the residuals $\hat{\alpha}_t$ are given by

$$\hat{\alpha}_t = \tilde{x}_t - (\hat{\phi}_1 \tilde{x}_{t-1} + \hat{\phi}_2 \tilde{x}_{t-2} + \hat{\phi}_3 \tilde{x}_{t-3}) \qquad (8.4)$$

The residuals after fitting an appropriate model should look like "white noise." For the gas furnace input data, the residuals, or the prewhitened input as we will call them, are shown in Figure 8.8. Even though ACF and PACF plots of the residuals in Figure 8.9 show two borderly significant autocorrelations at higher lags, we deem the model to be appropriate.

8.5.2 Prewhitening the Output

After prewhitening the input, the next step (step 3) is to prewhiten the output, y_t. This involves filtering the output data through the same model with the same coefficient estimates that we fitted to the input data. In other words, we first subtract the estimated mean $\hat{\mu}_y = 53.509$ from the y_t data to obtain $\tilde{y}_t$. Next we substitute $\tilde{y}_t$ for $\tilde{x}_t$ in Equation (8.4) to get the filtering equation for the

TABLE 8.1 Summary Statistics from Fitting an AR(3) Model to the Gas Rate Data

Term	Estimated coefficient	Standard error	t-Ratio	p-Value
$\hat{\phi}_1$	1.9762	0.0549	36.01	0.000
$\hat{\phi}_2$	−1.3752	0.0995	−13.82	0.000
$\hat{\phi}_3$	0.3434	0.0549	6.25	0.000

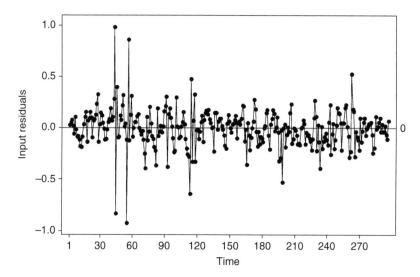

Figure 8.8   Residual plot of the input also called the prewhitened input.

prewhitened output

$$\hat{\beta}_t = \tilde{y}_t - (\hat{\phi}_1 \tilde{y}_{t-1} + \hat{\phi}_2 \tilde{y}_{t-2} + \hat{\phi}_3 \tilde{y}_{t-3})$$
$$= \tilde{y}_t - (1.9762 \tilde{y}_{t-1} - 1.3752 \tilde{y}_{t-2} + 0.3434 \tilde{y}_{t-3}) \qquad (8.5)$$

The $\hat{\beta}_t$'s are the prewhitened output. This recursive computation can be accomplished very easily with a spreadsheet. To get the recursion started, we simply set the first three terms $\tilde{y}_t, t = 1, 2, 3$ equal to zero. In the rest of the discussion below, unless otherwise specified, we simply use y_t to represent the demeaned value of the output, $\tilde{y}_t$.

8.5.3   Determining the Input–Output Relationship

From the flow chart in Figure 8.5, we see that the next step (step 4) in the prewhitening process is to compute the cross correlation function between the prewhitened input $\hat{\alpha}_t$ and the prewhitened output $\hat{\beta}_t$. This cross correlation function is shown in Figure 8.10. A rough estimate of the standard error of the cross correlations is given by $1/\sqrt{n}$ where n is the number of observations. We have superimposed two dotted horizontal lines at $\pm 2\hat{\sigma}_r \approx \pm 2/\sqrt{n} \approx \pm 0.1168$ to help see which of the cross correlations then seem to be significantly different from zero.

Figure 8.10 provides a much clearer picture of the input–output relationship than Figure 8.4. Now, all the cross correlations for negative lags appear to be insignificant. This fits better with our physical understanding of the process; a change in the input causes a subsequent change in the output, but not the other way around. Further, we see that the cross correlations for lags 0, 1, and 2 seem insignificant. This indicates that there is a time delay of about 3 time units in

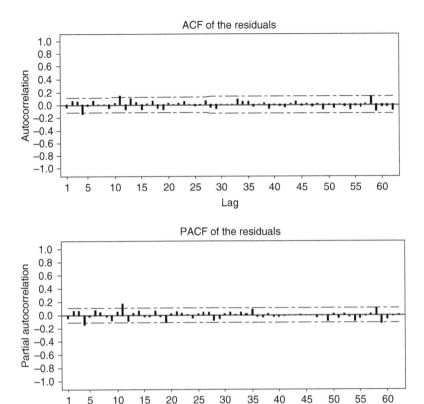

Figure 8.9 ACF and PACF of the residuals (prewhitened input).

the system. The cross correlations are significant for lags 3 to 7 after which they again appear to be insignificant. This indicates a dynamic relationship stretching over five time periods that initially builds up and then fades out.

A dynamic relationship between input and output can often be approximated by a linear transfer function

$$y_t = v_0 x_t + v_1 x_{t-1} + \cdots + v_k x_{t-k} + n_t \tag{8.6}$$

where the v_j's are transfer function weights and n_t is a noise term. When that is the case, BJR showed that rough estimates of the transfer function weights can be computed from

$$\hat{v}_j = \frac{r_{\alpha\beta}(j)s_\beta}{s_\alpha}, \quad j = 0, \ldots, k \tag{8.7}$$

where s_α is the estimated standard deviation of the prewhitened input, s_β is the estimated standard deviation of the prewhitened output, and $r_{\alpha\beta}(j)$ is the estimated cross correlation between the prewhitened input and the prewhitened output. Insignificant weights have been rounded off to zero. The weights are summarized in Table 8.2 and shown graphically in Figure 8.11.

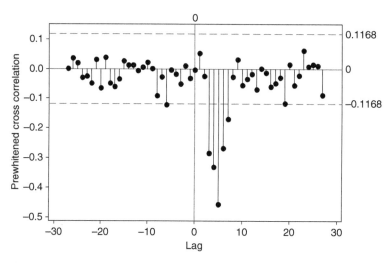

Figure 8.10   The estimated cross correlation between prewhitened input and output.

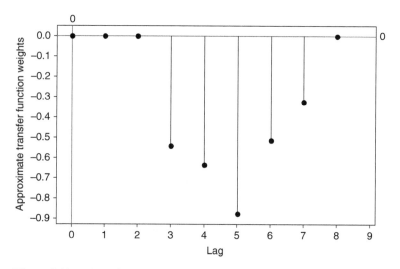

Figure 8.11   Approximate transfer function weights.

From the discussion above, we now see that an approximate transfer function for the gas furnace is of the form

$$y_t = v_3 x_{t-3} + v_4 x_{t-4} + v_5 x_{t-5} + v_6 x_{t-6} + v_7 x_{t-7} + n_t \qquad (8.8)$$

where the error term n_t is not necessarily white noise.

What we have done so far is to identify the general shape of the transfer function. It may be tempting to use standard regression methods to fit the linear model in Equation (8.8) to the data. However, that is fraught with other problems. The errors n_t will likely be autocorrelated. This autocorrelation violates the

TABLE 8.2 Approximate Transfer Function Weights

Lag, j	Cross correlation	Weight, $\hat{v}_j$
0	0	0
1	0	0
2	0	0
3	−0.282	−0.542
4	−0.330	−0.634
5	−0.455	−0.875
6	−0.267	−0.513
7	−0.168	−0.323
8	0	0

standard assumption for the error in regression analysis. Furthermore, a model such as the one in Equation (8.8) with several parameters fitted to lagged relationships is very likely ill-conditioned or collinear. This can cause serious bias in the estimated coefficients. BJR therefore explained that it is better to fit a dynamic transfer function model that requires a small number of parameters. They further recommended that the autocorrelated errors be modeled with an appropriate time series model. We show how to do this in the next section.

8.6   IDENTIFICATION OF THE TRANSFER FUNCTION

We define the dynamic relationship between the output Y_t and the input X_t using

$$Y_t = v(B)X_t + N_t \qquad (8.9)$$

where N_t is assumed to be independent of X_t and generated by an ARIMA process. However, if the series exhibits nonstationarity, we would take an appropriate degree of differencing to obtain stationary x_t, y_t and n_t corresponding to nonstationary X_t, Y_t and N_t respectively. Hence, the transfer function–noise model becomes

$$y_t = v(B)x_t + n_t \qquad (8.10)$$

$$v(B) \approx \frac{\omega(B)}{\delta(B)}B^b \qquad (8.11)$$

This suggests that Equation (8.10) can be written as

$$y_t = v(B)x_t + n_t$$

$$\approx \frac{\omega(B)B^b}{\delta(B)}x_t + n_t$$

$$\approx \frac{\omega(B)}{\delta(B)}x_{t-b} + n_t \qquad (8.12)$$

where $\omega(B) = \omega_0 - \omega_1 B + \cdots - \omega_s B^s$ and $\delta(B) = 1 - \delta_1 B - \cdots - \delta_r B^r$. It should be noted that with even a fairly low order of $\delta(B)$ ($r = 1$ or 2), it is possible to capture dynamics that would require a $v(B)$ of very high order. It should also be noted that the stability of the transfer function in Equation (8.12) depends on $\delta(B)$ as the stationarity of an ARMA process depends on the AR component. That is, $v(B) = \frac{\omega(B)}{\delta(B)} B^b$ is said to be stable if all the roots of $m^r - \delta_1 m^{r-1} - \cdots - \delta_r$ are less than 1 in absolute value.

To show that infinite number of parameters in $v(B)$ can be computed from a finite number of parameters in $w(B)$ and $\delta(B)$, consider

$$\delta(B)v(B) = \omega(B)$$

or

$$\left(1 - \delta_1 B - \cdots - \delta_r B^r\right)\left(v_0 + v_1 B + v_2 B^2 + \cdots\right) \qquad (8.13)$$

$$= \left(\omega_0 - \omega_1 B - \cdots - \omega_s B^s\right) B^b \qquad (8.14)$$

This means that

$$v_j - \delta_1 v_{j-1} - \delta_2 v_{j-2} - \cdots - \delta_r v_{j-r} = \begin{cases} -\omega_{j-b} & j = b+1, \ldots, b+s \\ 0 & j > b+s \end{cases}$$

with $v_b = \omega_0$ and $v_j = 0$ for $j < b$.

Here are a few observations about the equations in Equation (8.14)

1. We expect the weights in $v(B)$ to be 0 up to the lag that determines the delay b between the input and the output.

2. At lag b, the value of the corresponding weight v_b is simply equal to ω_0.

3. If the degree of $w(B)$, $s = 0$, then Equation (8.14) simply becomes a homogeneous difference equation. For a stable transfer function, this would mean that the weights in $v(B)$ follow a mixture of exponential decay and damped sine wave pattern.

4. For $s > 0$, the weights in $v(B)$ after lags $b + s$ will once again follow a mixture of exponential decay and damped sine wave pattern, but for lags between $b + 1$ and $b + s$, the weights will be adjusted by weights in $w(B)$ and do not necessarily follow any obvious pattern.

Consider, for example, the approximate transfer function weights for the gas furnace data in Figure 8.11. Clearly, the first nonzero weight starts at lag 3, indicating that the delay is 3 time units, that is, $b = 3$. For the nonzero weights, one plausible interpretation is that the weights exhibit exponential decay after lag 5 which suggests that $s = 5 - 3 = 2$. The exponential decay after lag 5 could be due to a single exponential decay term or sum of more than one exponential decay terms. The more plausible and simpler choice would be a single or sum of two exponential decay terms. This suggests that $r = 1$ or 2. At this point, it should be noted that we are not searching for the "right" r value (whatever that may be). We are at an early stage of model identification and whichever model we pick will be a tentative one. Moreover, the standard diagnostics checks will usually reveal if a low value is used for r. Nonetheless, one can certainly try both models. A sensible choice, however, would be to start with $r = 2$ and check whether the second δ weight (δ_2) is necessary once the overall model is fit. Therefore, our tentative transfer function is

$$v(B) = \frac{\omega_0 - \omega_1 B - \omega_2 B^2}{1 - \delta_1 B - \delta_2 B^2} B^3 \tag{8.15}$$

We estimate the parameters in the transfer function using Equation (8.14) and the estimates for the v weights from Table 8.2

$$\hat{v}_3 = \hat{\omega}_0 = -0.542$$
$$\hat{v}_4 = \hat{\delta}_1 \hat{v}_3 + \hat{\delta}_2 \hat{v}_2 - \hat{\omega}_1 = -0.542\hat{\delta}_1 - \hat{\omega}_1 = -0.634$$
$$\hat{v}_5 = \hat{\delta}_1 \hat{v}_4 + \hat{\delta}_2 \hat{v}_3 - \hat{\omega}_2 = -0.634\hat{\delta}_1 - 0.542\hat{\delta}_2 - \hat{\omega}_2 = -0.875$$
$$\hat{v}_6 = \hat{\delta}_1 \hat{v}_5 + \hat{\delta}_2 \hat{v}_4 = -0.875\hat{\delta}_1 - 0.634\hat{\delta}_2 = -0.513$$
$$\hat{v}_7 = \hat{\delta}_1 \hat{v}_6 + \hat{\delta}_2 \hat{v}_5 = -0.513\hat{\delta}_1 - 0.875\hat{\delta}_2 = -0.323 \tag{8.16}$$

Using the last two equations in Equation (8.16), we have $\hat{\delta}_1 = 0.554$ and $\hat{\delta}_2 = 0.044$. We then have

$$\hat{\omega}_1 = 0.634 - 0.542\hat{\delta}_1 = 0.334$$
$$\hat{\omega}_2 = 0.875 - 0.634\hat{\delta}_1 - 0.542\hat{\delta}_2 = 0.5$$

Hence, the transfer function in Equation (8.15) is estimated as

$$\hat{v}(B) = \frac{-0.542 - 0.334B - 0.5B^2}{1 - 0.554B - 0.044B^2} B^3 \tag{8.17}$$

Note that the estimate for $\hat{\delta}_2$ is quite small somehow, confirming our initial guess about the possibility of r being equal to 1. But at this stage, we can simply ignore this fact and proceed with the modeling of the noise. We will come back to this issue once we fit the overall model.

8.7 MODELING THE NOISE

In the previous section, we used the transfer function–noise model in Equation (8.10) to describe the dynamic relationship between the output and the input with added noise. In that model, we assumed that the noise term was most likely autocorrelated. In this section, we discuss how to fit a model just for the noise and then incorporate it into the overall model in Equation (8.10).

So far, we have developed two versions of the transfer function–noise model

$$y_t = v(B)x_t + n_t$$

and

$$y_t = \frac{\omega(B)B^b}{\delta(B)} x_t + n_t$$

Consequently, we have two ways of estimating the noise

$$\hat{n}_t = y_t - \hat{v}(B)x_t$$

$$= y_t - \hat{v}_0 x_t - \hat{v}_1 x_{t-1} - \hat{v}_2 x_{t-2} - \cdots \tag{8.18}$$

or

$$\hat{n}_t = y_t - \frac{\hat{\omega}(B)}{\hat{\delta}(B)} x_{t-b} \tag{8.19}$$

Even though the number of significant terms in $\hat{v}(B)$ sometimes can be small as in the gas furnace example where we only had four significant terms, as discussed earlier, a more parsimonious model can usually be obtained using the rational polynomial representation of $v(B)$ as in Equation (8.11). Therefore, we prefer the model in Equation (8.19) to obtain the estimate of the noise series. We define

$$\hat{\delta}(B)\hat{y}_t = \hat{\omega}(B)x_{t-b}$$

or

$$\hat{y}_t = \hat{\delta}_1 \hat{y}_{t-1} + \hat{\delta}_2 \hat{y}_{t-2} \ldots + \hat{\delta}_r \hat{y}_{t-r} + \hat{\omega}_0 x_{t-b} - \hat{\omega}_1 x_{t-b-1} - \cdots - \hat{\omega}_s x_{t-b-s} \tag{8.20}$$

Then the estimate of the noise in Equation (8.19) is given by

$$\hat{n}_t = y_t - \hat{y}_t$$

$$= y_t - (\hat{\delta}_1 \hat{y}_{t-1} + \hat{\delta}_2 \hat{y}_{t-2} \cdots + \hat{\delta}_r \hat{y}_{t-r} + \hat{\omega}_0 x_{t-b} - \hat{\omega}_1 x_{t-b-1} - \cdots - \hat{\omega}_s x_{t-b-s}) \tag{8.21}$$

Once the parameter estimates in $w(B)$ and $\delta(B)$ are calculated, it is fairly easy to calculate the estimate of the noise series in Equation (8.21).

For the gas furnace data, the estimate for the noise series is calculated from

$$\hat{n}_t = y_t - (0.554\hat{y}_{t-1} + 0.044\hat{y}_{t-2} - 0.542x_{t-3} - 0.334x_{t-4} - 0.5x_{t-5}) \tag{8.22}$$

Note that in Equation (8.22), the calculation of $\hat{y}_t$ requires previous values of $\hat{y}_t$ and x_t. Therefore, we start t from 6 in our calculations for $\hat{n}_t$ and set $\hat{y}_5$ and $\hat{y}_4$ to 0, which is the average value of the demeaned y_t since as we mentioned at the end of Section 8.4.2, we demeaned the output by subtracting $\hat{\mu}_y = 53.509$ from each observation. Shifting of the starting time from 1 to 6 causes loss of information, which can be important for small datasets. But the gas furnace data is large enough for us to overlook this issue.

Figure 8.12 shows the estimate of the noise. Figures 8.13 and 8.14 are the ACF and PACF plots of the estimate of the noise, respectively. From all these plots, it is safe to assume that the series is indeed stationary. The PACF plot cuts off after lag 2. Together with the observed behavior of ACF in Figure 8.12, it seems that an AR(2) model would be appropriate for modeling the noise. It could also be argued that lag 3 of the PACF seems significant. Therefore, we can

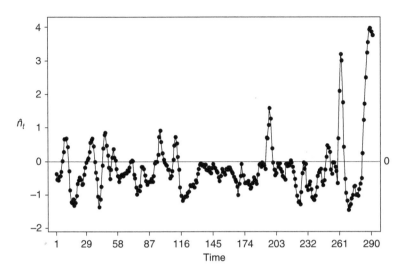

Figure 8.12 Time series plot of $\hat{n}_t$.

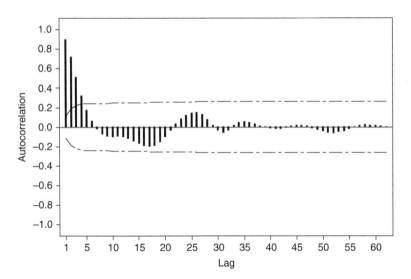

Figure 8.13 ACF of $\hat{n}_t$.

also consider an AR(3) model and let the model diagnostics determine whether the third order term is needed.

The parameter estimates for the AR(3) model are given in Table 8.3. Clearly, the third order autoregressive term is insignificant. We therefore fit the AR(2) model and summarize the model in Table 8.4. As a final check, we consider the residuals from the AR(2) model. ACF and PACF plots of the residuals in Figures 8.15 and 8.16, respectively, reveal no sign of autocorrelation. Hence,

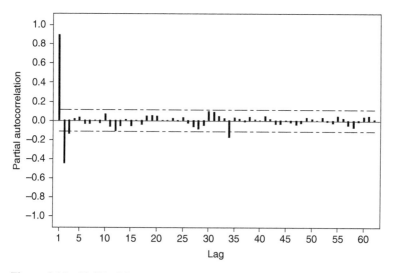

Figure 8.14 PACF of $\hat{n}_t$.

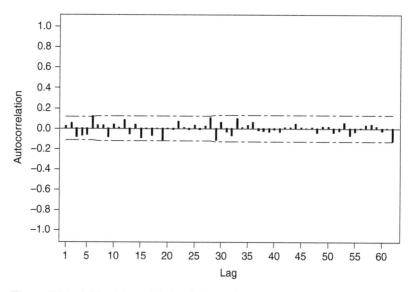

Figure 8.15 ACF of the residuals of the AR(2) model for $\hat{n}_t$.

we conclude that the noise can be modeled using an AR(2) model

$$(1 - \phi_1 B - \phi_2 B^2) n_t = a_t \tag{8.23}$$

or

$$n_t = \frac{1}{(1 - \phi_1 B - \phi_2 B^2)} a_t \tag{8.24}$$

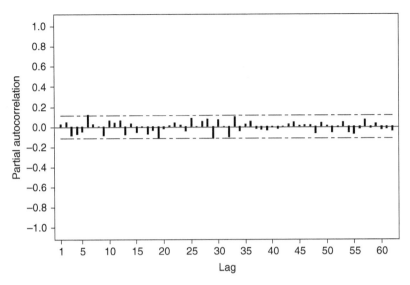

Figure 8.16   PACF of the residuals of the AR(2) model for $\hat{n}_t$.

TABLE 8.3   The Parameter Estimates of the AR(3) Model for $\hat{n}_t$

Type	Estimate	Standard error	t-Ratio	p-Value
AR(1)	1.5754	0.0591	26.66	0
AR(2)	−0.7139	0.1024	−6.97	0
AR(3)	0.0473	0.0616	0.77	0.443

TABLE 8.4   The Parameter Estimates of the AR(2) Model for $\hat{n}_t$

Type	Estimate	Standard error	t-Ratio	p-Value
AR(1)	1.5471	0.0461	33.56	0
AR(2)	−0.6445	0.0478	−13.47	0

Therefore, the transfer function−noise model for the gas furnace data is

$$
\begin{aligned}
y_t &= \frac{\omega(B)B^b}{\delta(B)} x_t + n_t \\
&= \frac{\omega(B)B^b}{\delta(B)} x_t + \frac{\theta(B)}{\phi(B)} a_t \\
&= \frac{\omega_0 - \omega_1 B - \omega_2 B^2}{1 - \delta_1 B - \delta_2 B^2} x_{t-3} + \frac{1}{(1 - \phi_1 B - \phi_2 B^2)} a_t
\end{aligned}
\tag{8.25}
$$

TABLE 8.5   The Parameter Estimates for the Model in Equation (8.25)

Parameters	Estimate	Standard error	t-Ratio	p-Value
ω_0	−0.53	0.07	−7.11	0.000
ω_1	0.37	0.14	2.63	0.009
ω_2	0.51	0.15	3.52	0.000
δ_1	0.57	0.19	3.04	0.003
δ_2	−0.01	0.13	−0.11	0.912
ϕ_1	1.53	0.05	32.59	0.000
ϕ_2	−0.62	0.05	−12.64	0.000

Unfortunately, not all standard statistical software packages have the capability of fitting the model in Equation (8.25). In this case, we used SAS JMP. The parameter estimates are given in Table 8.5.

The parameter estimate of δ_2 and its standard error suggest that we can omit that parameter from the model as we suspected earlier. Thus, we fit the following model

$$y_t = \frac{\omega_0 - \omega_1 B - \omega_2 B^2}{1 - \delta_1 B} x_{t-3} + \frac{1}{(1 - \phi_1 B - \phi_2 B^2)} a_t \qquad (8.26)$$

The new parameter estimates are given in Table 8.6. For diagnostics checks, we first consider the cross correlation between the residuals from the model in Equation (8.26) and the residuals of the model we had for the input, that is, the prewhitened input. If the transfer model is correctly identified, then there should be no significant cross correlation left between the prewhitened input series and the residuals from the final model. If that is not the case, that would mean that there is still some information left in the residuals that can be explained by the input series. Figure 8.17, generally speaking, shows no significant cross correlation. Hence, we move forward with the second step in diagnostics checks—the ACF and PACF plots of the residuals. Assuming that the transfer function and the noise model are correctly identified, the ACF and PACF of the residuals should show no significant auotocorrelation, which is the case in Figures 8.18 and 8.19. Hence we conclude that the model in Equation (8.26) with the parameter estimates in Table 8.6 is capable of explaining the dynamic relationship between the output and the input.

With this example, we have demonstrated how to assess the nature of a dynamic relationship between two time series when both the input and the output are autocorrelated. This is important in many applications including economics, engineering, and management where such relationships frequently occur. Our primary objective was to show that simply computing a regression equation or cross correlations may yield a misleading picture of the real functional relationship. Further, the cross correlations between the original input and the original output showed a spurious relationship where the effect appeared to precede the cause. The reason for these spurious relationships was the autocorrelation in the

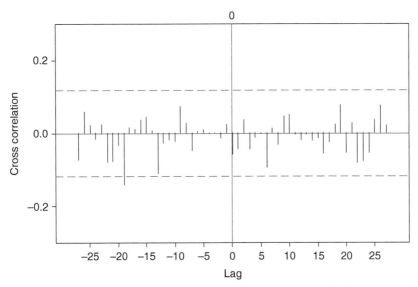

Figure 8.17 Cross correlation between the residuals of the model in Equation (8.26) and the prewhitened input series.

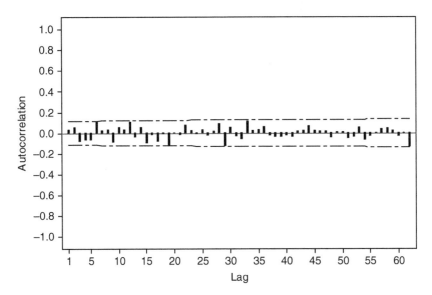

Figure 8.18 ACF of the residuals of the model in Equation (8.26).

input. Therefore, we used prewhitening. This procedure clarified the true nature of the cause-and-effect relationship. The analysis based on prewhitening showed that the process had a time-delayed dynamic response spread over several lags. Once the delay and the general structure of the transfer function are identified, the final model is shown to provide a good fit. In the next section, we summarize

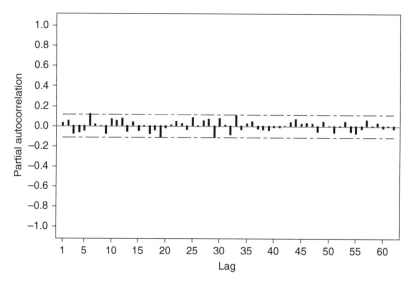

Figure 8.19 PACF of the residuals of the model in Equation (8.26).

TABLE 8.6 The Parameter Estimates for the Reduced Model in Equation (8.26)

Parameters	Estimate	Standard error	t-Ratio	p-Value
ω_0	−0.53	0.07	−7.14	0.000
ω_1	0.38	0.10	3.74	0.000
ω_2	0.52	0.11	4.80	0.000
δ_1	0.55	0.04	13.78	0.000
ϕ_1	1.53	0.05	32.59	0.000
ϕ_2	−0.62	0.05	−12.65	0.000

the steps we followed in the analysis of the gas furnace data in order to provide a general overview of the methodology proposed by BJR.

8.8 THE GENERAL METHODOLOGY FOR TRANSFER FUNCTION MODELS

The steps we covered in the previous section in identifying, fitting, and making the diagnostic checks for the transfer function models are as follows:

1. *Plot the data.* As in any statistical analysis, the first step should always start with plotting the data. In this case, time series plots of the input and the output will give us a preliminary overview of the general behavior of these variables. The plot of the cross correlation function should be studied with caution because of possible spurious correlations.

2. *Prewhitening.* To get a better picture of the cross correlation between the input and the output, we first fit a model to the input and obtain the residuals. We then apply exactly the same model with the same parameter estimates to the output to obtain "pseudo-residuals." The cross correlation function between these two sets of residuals allows for the tentative determination of the delay between the input and the output as well as the estimates of the terms in the transfer function $v(B)$. Note that suitable differencing is applied to the series to achieve stationarity. If the series is stationary, it is recommended that the series is also demeaned by subtracting a fixed value equal to the average of the series from each observation.

3. *Rational polynomial representation of the transfer function.* The transfer function in step 2 will often have a large number of terms. Instead, a rational polynomial representation of the transfer function, that is, $v(B) = \omega(B)/\delta(B)$, can greatly reduce the number of parameters to be estimated.

4. *Fitting a model for the noise.* The noise in the general transfer function model is expected to be autocorrelated. The estimate of the noise is obtained by subtracting the tentative model in step 3 from the observations, which is then modeled using an ARIMA model. A tentative model for the estimate of the noise provides the missing component to be used to fit an overall model.

5. *Fitting the transfer function–noise model.* Once the degrees of the rational polynomial of the transfer function and a model for the estimate of the noise are identified, the combined model is fit. Unfortunately, this option is not available in all statistical software packages. SAS, SAS JMP, and SCA are today among the few that have the capability of fitting such models.

6. *Diagnostics checks.* If the transfer function is not correctly identified, there will still be information left in the residuals that can be explained by the input series. Therefore, we should first check the cross correlation between the residuals and the prewhitened input series. Note that we use the prewhitened input series rather than the original input series for reasons we already discussed. If no significant cross correlation is present, we then proceed with checking whether there is any autocorrelation left in the residuals, that is, to see whether the noise is modeled properly. We accomplish this with the usual ACF and PACF plots of the residuals as well as tests such as Ljung and Box mentioned in Chapter 4. Any violations observed in the diagnostic checks will suggest the reevaluation of the noise model and/or transfer function model.

8.9 FORECASTING USING TRANSFER FUNCTION–NOISE MODELS

In the previous chapters, we have developed forecasts for the univariate time series. This type of univariate forecasting is often powerful because the past history of a time series includes information not just coming from the time series itself, but indirectly contains information from other sources that may influence the present and future values of the time series we are interested in forecasting. To see that, suppose we are interested in forecasting y_t one or several steps ahead. Now further suppose x_t is a leading indicator that provides early information about the changes in y_{t+1}. More specifically, suppose changes in x_t tell us something about what will happen to y_{t+1}. That is, x_t is a leading indicator for y_t one time unit later. Therefore, a simple relationship will be $y_{t+1} = v_0 + v_1 x_t + a_t$. That means, going one step back, we have $y_t = v_0 + v_1 x_{t-1} + a_{t-1}$ and going one step further back, $y_{t-1} = v_0 + v_1 x_{t-2} + a_{t-2}$. Thus we see that y_t contains information input coming from x_{t-1}. Similarly, y_{t-1} contains information input from x_{t-2}. Therefore, if we use y_t, y_{t-1}, y_{t-2}, and so on to forecast y_{t+1}, then those quantities already incorporate information coming from x_{t-1}, x_{t-2}, and so on. However, if we can identify directly those leading indicators, then that will allow us to build better forecasting models, for example, by using the delay between the leading indicator and the output. For that, we consider in this section another standard example from the time series literature.

8.9.1   Example 2: Sales with Leading Indicators

The data shown in Figure 8.20 is typical for many business applications of time series analysis and originally given as Series M in BJR. We are interested in making forecasts for future sales and after careful thinking about the business and analysis of date, we have been able to identify a certain leading indicator for our sales. For example, building permit applications is a leading indicator for many sectors of the economy that is influenced by construction activities. In the present case, we are not informed about the specifics of the sales Y_t and the leading indicator X_t. However, we are provided with the two sets of time series data shown in Figure 8.20. We see that both Y_t and X_t are clearly nonstationary series trending up over the given time period. We also see that the two time series show very similar trends. Indeed, it seems plausible that one may be the leading indicator for the other.

In the following, we will go through the steps we presented in Section 8.8 to come up with a transfer function–noise model. Our goal is to use this model to make forecasts. We will use the first 125 observations out of 150 to build the model and the last 25 observations as a validation test to see how well our forecasts will be. Therefore, all the model building efforts in the following sections are based on the first 125 observations. Note that the entire dataset is given in Appendix A.

Now, as we already discussed in the previous section, we need to be careful when trying to assess the cross correlation between two nonstationary time series and not let ourselves be confused by spurious correlation. Thus, since both time series are nonstationary, we apply the first order difference to obtain $x_t = \nabla X_t = X_t - X_{t-1}$ and $y_t = \nabla Y_t = Y_t - Y_{t-1}$. Time series plots of the differenced data are shown in Figure 8.21. We see that both look much more stationary. This impression is further confirmed by looking at the ACFs of $x_t = \nabla X_t$ and $y_t = \nabla Y_t$

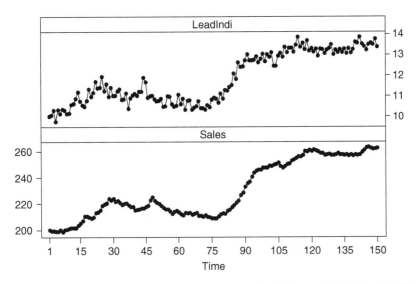

Figure 8.20   Time series plots of sales and a leading indicator (Series M of BJR).

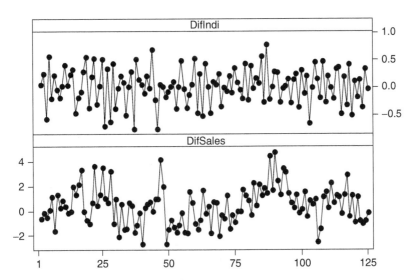

Figure 8.21   First difference of sales $y_t = \nabla Y_t$ and the leading indicator $x_t = \nabla X_t$.

in Figures 8.22 and 8.23, respectively. In both cases, the ACFs die out quickly, indicating that taking the first difference is sufficient to achieve stationarity.

   To get a clear view of the cross correlation between the two time series, we need to prewhiten the series. However, even taking the first difference goes a long way in clarifying the cross correlation between the two time series and can provide a preliminary indication if it is worth the trouble of going through the prewhitening of the series. Figure 8.24 shows the cross correlation between

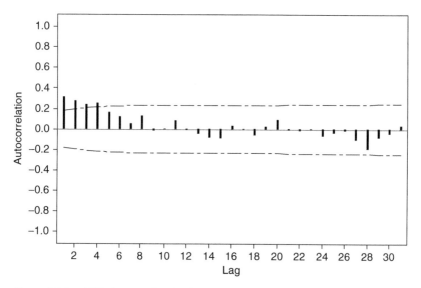

Figure 8.22 ACF of $y_t = \nabla Y_t$, the first difference of sales.

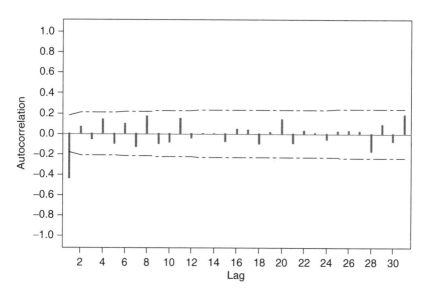

Figure 8.23 ACF of $x_t = \nabla X_t$, the first difference of the leading indicator.

$y_t = \nabla Y_t$ and $x_t = \nabla X_t$. We see that it seems plausible that there is significant cross correlations at lag 2 and 3, lending credence to the conjecture that X_t is a leading indicator for Y_t. With this preliminary assessment of the relationship between sales and the leading indicator, we now proceed with a full prewhitening of the two series.

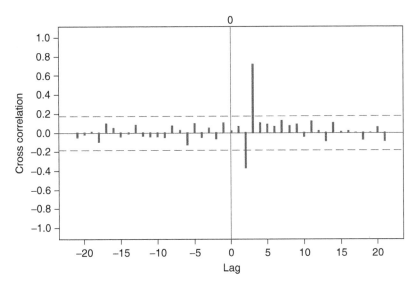

Figure 8.24   Cross correlation between $y_t = \nabla Y_t$ and $x_t = \nabla X_t$.

8.9.2   Modeling the Leading Indicator Series

To prewhiten the two time series, we first need to develop a time series model for the input series, in this case, the leading indicator series. From Figure 8.23 showing the ACF of $x_t = \nabla X_t$, we see that the autocorrelation cuts off after lag 1, indicating a first order moving average process, MA(1). The PACF of $x_t = \nabla X_t$ is shown in Figure 8.25. In that plot, we see a decay providing further indication that $x_t = \nabla X_t$ may tentatively be modeled as a first order moving average MA(1). In other words, we may tentatively try to model the original time series X_t as an IMA(1, 1) process.

Table 8.7 provides a summary of the estimation output after fitting an IMA(1, 1) model to the leading indicator data. We see that the moving average coefficient is quite significant with a t-value of 5.45. Further, the Ljung–Box chi-square statistics looks good for all lags with no apparent significant autocorrelations. The ACF of the residuals in Figure 8.26 and the PACF in Figure 8.27 also show no particular worrisome patterns. Our tentative model for the leading indicator is $X_t - X_{t-1} = \alpha_t - \theta\alpha_{t-1}$ where the estimate of the moving average parameter is $\hat{\theta} = 0.4408$.

8.9.3   Prewhitening the Sales Series

As discussed earlier, without taking into account the autocorrelation in both series, it is not possible to assess the dependency between the two time series from the sample cross correlation function. However, if we prewhiten both series, we get a better picture.

Although we already explained how to prewhiten a pair of time series in Section 8.5, the basic equations are reproduced here. Let the time series model for

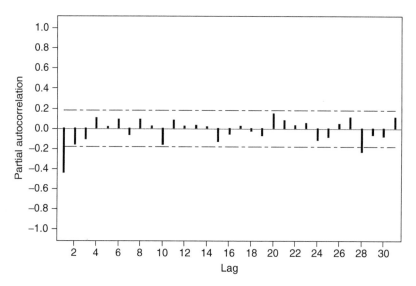

Figure 8.25   PACF of $x_t = \nabla X_t$, the first difference of the leading indicator.

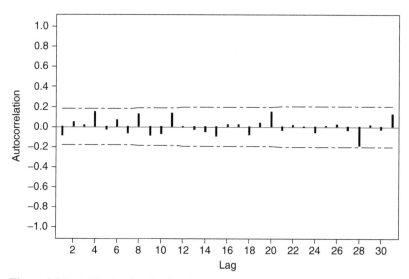

Figure 8.26   ACF of residuals after fitting an IMA(1, 1) model to the leading indicator X_t data.

the input time series x_t be given by $\phi_x(B)x_t = \theta_x(B)\alpha_t$ where x_t is the suitably differenced series. Then, we may write the prewhitening filter for the input series $\{x_t\}$ as

$$\hat{\alpha}_t = [\hat{\theta}_x(B)]^{-1}\hat{\phi}_x(B)x_t \tag{8.27}$$

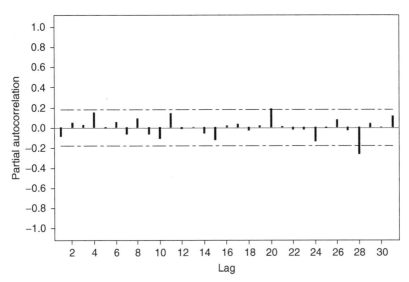

Figure 8.27   PACF of residuals after fitting an IMA(1, 1) model to the leading indicator X_t data.

TABLE 8.7   Estimates of an IMA(1, 1) Model for the Leading Indicator, X_t

Type		Estimate	Standard error	t-Ratio	p-Value
MA	1	0.4408	0.0809	5.45	0
Modified Box–Pierce (Ljung–Box) chi-square statistic					
Lag	12	24	36	48	—
Chi square	11.8	18.9	46.6	61.6	—
df	11	23	35	47	—
p-value	0.381	0.709	0.092	0.075	—

where $\{\hat{\alpha}_t\}$ is the residual or whitened input series. The sample variance of the whitened input series is given by $\hat{\sigma}_{\hat{\alpha}}^2$. Now we apply the same filter $[\hat{\theta}_x(B)]^{-1}\hat{\phi}_x(B)_t$ with the same coefficients to the suitably differenced output series $\{y_t\}$. That yields the prewhitened output series $\{\hat{\beta}_t\}$ given by

$$\hat{\beta}_t = [\hat{\theta}_x(B)]^{-1}\hat{\phi}_x(B)y_t \qquad (8.28)$$

In our case, the model for the differenced input is $x_t = (1 - \hat{\theta}_x B)\hat{\alpha}_t$ where $x_t = \nabla X_t = X_t - X_{t-1}$. Therefore, the prewhitening filter is $[\hat{\theta}_x(B)]^{-1}\hat{\phi}_x(B)_t = [(1 - \hat{\theta}_x B)]^{-1}$. For computations, it is, however, easier if we write the model as $X_t - X_{t-1} = \hat{\alpha}_t - \hat{\theta}_x\hat{\alpha}_{t-1}$ or in the recursive form as

$$\hat{\alpha}_t = X_t - X_{t-1} + \hat{\theta}_x\hat{\alpha}_{t-1} \qquad (8.29)$$

Equation (8.29) can directly be used to compute the residuals. Now with the filter written in this form for the input, we then simply substitute Y_t for X_t and β_t for α_t to get the prewhitening filter for the output series

$$\hat{\beta}_t = Y_t - Y_{t-1} + \hat{\theta}_x \hat{\beta}_{t-1} \tag{8.30}$$

where $\hat{\theta}_x = 0.4408$.

8.9.4 Identification of the Transfer Function

After we have prewhitened the output series, we then compute the cross correlation between the prewhitened input and the output as shown in Figure 8.28 where we can see that the significant correlation starts at lag 3. This indicates that the delay is equal to 3. Furthermore, we see at lag 3 a strong positive cross correlation immediately followed by an exponentially decaying set of cross correlations for larger lags. This suggests a first order transfer function of the form

$$v(B) = \frac{\omega_0 B^3}{(1 - \delta_1 B)} \tag{8.31}$$

The estimates of terms in $v(B)$ given in Table 8.8 are calculated using Equation (8.7) with $s_\alpha = 0.29$ and $s_\beta = 2.05$ as the estimated standard deviations of the prewhitened input and the prewhitened output, respectively. We then estimate the parameters in the rational polynomial in Equation (8.31) using Equation (8.14)

$$\hat{v}_3 = \hat{\omega}_0 = 4.8$$
$$\hat{v}_4 = \hat{\delta}_1 \hat{v}_3 = 4.8 \hat{\delta}_1 = 3.16 \Rightarrow \hat{\delta}_1 = 0.66 \tag{8.32}$$

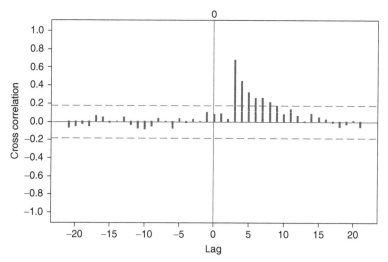

Figure 8.28 Cross correlation of the prewhitened sales $\{\hat{\beta}_t\}$ and leading indicator series $\{\hat{\alpha}_t\}$.

TABLE 8.8   Approximate Transfer Function Weights

Lag j	Cross correlation	Weight $\hat{v}_j$
0	0	0
1	0	0
2	0	0
3	0.679	4.800
4	0.447	3.160
5	0.325	2.297
6	0.261	1.845
7	0.263	1.859
8	0.213	1.506
9	0.173	1.223
10	0.086	0.608
11	0.135	0.954

8.9.5   Transfer Function–Noise Model

The next step is to find a tentative model for the noise series. Previously, we showed how to do this using the recursive equation in Equation (8.21). In this section, we instead use a slight trick and simply fit the transfer function–noise model without specifying any model for the noise. Therefore, we use the following model

$$(1 - B)Y_t = \delta_0 + \frac{\omega_0 B^3}{1 - \delta_1 B}(1 - B)X_t + n_t \tag{8.33}$$

Note that we added a constant term, δ_0, to the transfer function–noise model, which gives a more general representation of the transfer function–noise models.

The ACF and PACF plots of the residuals from the model in Equation (8.33) are given in Figures 8.29 and 8.30. ACF of the residuals cuts off after the first lag indicating a possible MA(1) model, that is, $n_t = (1 - \theta B)a_t$, and the (approximately) exponential decay behavior of the PACF confirms that. Therefore, we decide to skip the step of modeling the residuals and directly fit the transfer function–noise model of the form

$$(1 - B)Y_t = \delta_0 + \frac{\omega_0 B^3}{1 - \delta_1 B}(1 - B)X_t + (1 - \theta B)a_t \tag{8.34}$$

The estimates of the parameter of the model in Equation (8.34) are given in Table 8.9. All parameters seem to be significant.

For diagnostics checks, we start with the cross correlation between the prewhitened input and the residuals of the model in Equation (8.34). Figure 8.31 shows no sign of correlation between these two series. Furthermore, from the ACF and PACF plots of the residuals in Figures 8.32 and 8.33, we can confirm

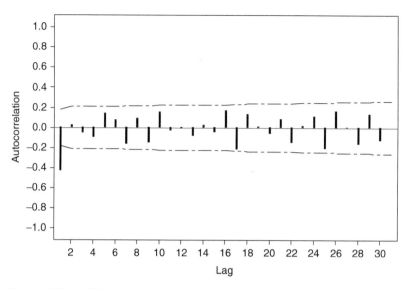

Figure 8.29   ACF of the residuals ($\hat{n}_t$) of the model in Equation (8.33).

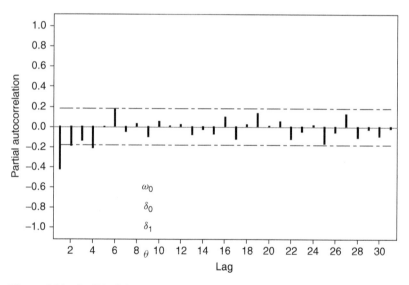

Figure 8.30   PACF of the residuals ($\hat{n}_t$) of the model in Equation (8.33).

that there is no autocorrelation left in the residuals and therefore we declare the model in Equation (8.34) as our final model.

8.9.6   Forecasting Sales

Forecasting using transfer function–noise models is quite similar to forecasting using ARIMA models, with the exception of the additional input variable in the model. The general equation for forecasting can get complicated because of the

TABLE 8.9   The Parameter Estimates of the Model in Equation (8.34)

Parameters	Estimate	Standard error	t-Ratio	p-Value
ω_0	4.73	0.057	82.56	0.000
δ_0	0.03	0.010	2.87	0.005
δ_1	0.72	0.004	174.39	0.000
θ	0.58	0.079	7.42	0.000

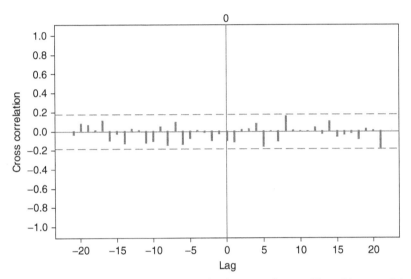

Figure 8.31   The cross correlation function between the prewhitened input and the residuals of the model in Equation (8.34).

terms in the transfer function as well as the model for the noise. For that, we refer the reader to BJR Chapter 12 or MJK Chapter 6. We will use the following model to show how to make forecasts

$$(1 - B)Y_t = \frac{\omega_0 - \omega_1 B}{1 - \delta_1 B}B^3(1 - B)X_t + \frac{1 - \theta B}{1 - \phi B}a_t \qquad (8.35)$$

Now let us rewrite this model as

$$(1 - \phi B)(1 - \delta_1 B)(1 - B)Y_t = (1 - \phi B)(\omega_0 - \omega_1 B)B^3(1 - B)X_t$$
$$+ (1 - \delta_1 B)(1 - \theta B)a_t \qquad (8.36)$$

In other words

$$(1 - (1 + \phi + \delta_1)B + (\phi + \delta_1 + \phi\delta_1)B^2 - \phi\delta_1 B^3)Y_t$$
$$= (\omega_0 B^3 - (\omega_0 + \omega_1 + \phi\omega_0)$$
$$B^4 + (\omega_1 + \phi\omega_0 + \phi\omega_1)B^5 - \phi\omega_1 B^6)X_t$$
$$+ (1 - (\theta + \delta_1)B + \theta\delta_1 B^2)a_t \qquad (8.37)$$

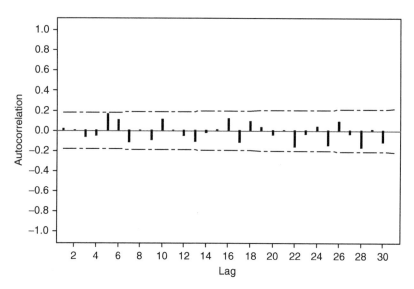

Figure 8.32 ACF of the residuals of the model in Equation (8.34).

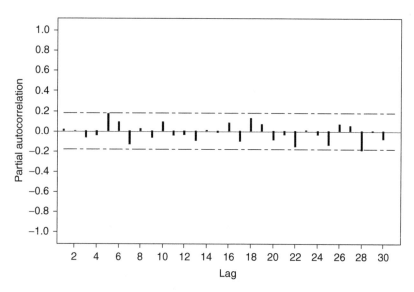

Figure 8.33 PACF of the residuals of the model in Equation (8.34).

or

$$(1 - \phi_1^* B - \phi_2^* B^2 - \phi_3^* B^3) Y_t = (v_0^* B^3 + v_1^* B^4 + v_2^* B^5 + v_3^* B^6) X_t$$
$$+ (1 - \theta_1^* B - \theta_2^* B^2) a_t \qquad (8.38)$$

where

$$\phi_1^* = 1 + \phi + \delta_1$$
$$\phi_2^* = -(\phi + \delta_1 + \phi\delta_1)$$
$$\phi_3^* = \phi\delta_1$$
$$v_0^* = \omega_0$$
$$v_1^* = -(\omega_0 + \omega_1 + \phi\omega_0) \qquad (8.39)$$
$$v_2^* = \omega_1 + \phi\omega_0 + \phi\omega_1$$
$$v_3^* = -\phi\omega_1$$
$$\theta_1^* = \theta + \delta_1$$
$$\theta_2^* = -\theta\delta_1$$

Thus, the output at $t + 1$ is given as

$$Y_{t+1} = \phi_1^* Y_t + \phi_2^* Y_{t-1} + \phi_3^* Y_{t-2} + v_0^* X_{t-2} + v_1^* X_{t-3} + v_2^* X_{t-4}$$
$$+ v_3^* X_{t-5} + a_{t+1} - \theta_1^* a_t - \theta_2^* a_{t-1} \qquad (8.40)$$

Note that if the constant term δ_0 is added to Equation (8.35), then on the right-hand side of Equation (8.40), we would add $(1 - \delta_1 B)(1 - \phi B)\delta_0 = \delta_0(1 - \delta_1 - \phi + \delta_1\phi)$. The minimum mean square error forecast of Y_{t+1} at t is obtained by taking the conditional expectations in Equation (8.40)

$$[Y_{t+1}] = \hat{Y}_t(1) = \phi_1^*[Y_t] + \phi_2^*[Y_{t-1}] + \phi_3^*[Y_{t-2}]$$
$$+ v_0^*[X_{t-2}] + v_1^*[X_{t-3}] + v_2^*[X_{t-4}] + v_3^*[X_{t-5}]$$
$$+ [a_{t+1}] - \theta_1^*[a_t] - \theta_2^*[a_{t-1}] \qquad (8.41)$$

Note that we once again used square brackets to denote the conditional expectations at time t. The right-hand side of Equation (8.41) is calculated based on the rules we had in Chapter 5.

1. For the output, for present and past values we use the observations. Although we do not need it in Equation (8.41), since it is for lead time of 1 only, for lead times bigger than 1, we will also need to have future values of the output. For those, we use their forecasts at time t. In other words,

$$[Y_{t+j}] = \begin{cases} Y_{t+j} & j \le 0 \\ \hat{Y}_t(j) & j > 0 \end{cases} \qquad (8.42)$$

2. For the input, for present and past values, we use the observations. If future values are needed, we use the univariate ARIMA model we obtained in the prewhitening step and the forecasts from that model. This is possible primarily because of the independence assumption between the input and noise series. Hence, we have

$$[X_{t+j}] = \begin{cases} X_{t+j} & j \le 0 \\ \hat{X}_t(j) & j > 0 \end{cases} \qquad (8.43)$$

3. For the noise, for present and past values, we can use the residuals of the transfer function–noise model or for $b \geq 1$, the one-step-ahead forecast error, for example, $\hat{a}_t = Y_t - \hat{Y}_t(1)$. The future values are set to 0. That is,

$$\left[a_{t+j}\right] = \begin{cases} \hat{a}_{t+j} & j \leq 0 \\ 0 & j > 0 \end{cases} \tag{8.44}$$

It is quite straightforward to generalize Equation (8.41) to l-step-ahead forecasts, which we leave to the reader.

For the calculation of the variance of the forecast error, we write the general transfer function–noise model as

$$\begin{aligned} Y_t &= \frac{\omega(B)}{\delta(B)} X_{t-b} + \frac{\theta(B)}{\phi(B)(1-B)^d} a_t \\ &= \frac{\omega(B)}{\delta(B)} X_{t-b} + \frac{\theta(B)}{\varphi(B)} a_t \\ &= v(B)\alpha_t + \psi(B)a_t \end{aligned} \tag{8.45}$$

where the prewhitened input $\alpha_t = \varphi_X(B)\theta_X(B)^{-1}X_t$, $v(B) = \delta(B)^{-1}\varphi_X(B)^{-1}$ $\omega(B)\theta_X(B)B^b$, and $\psi(B) = \varphi(B)^{-1}\theta(B)$. Then the l-step-ahead forecast error variance is

$$\text{Var}(Y_{t+l} - \hat{Y}_t(l)) = \sigma_\alpha^2 \sum_{i=b}^{l-1} v_i^2 + \sigma_a^2 \sum_{i=0}^{l-1} \psi_i^2 \tag{8.46}$$

One striking feature of Equation (8.46) is that the forecast error variance from a transfer function–noise model is a function of not only the variance of the errors of the transfer function–noise model but also the variance of the errors from the univariate model of the input.

For the sales data, we use SAS JMP to provide the 1- to 25-step-ahead forecasts. For comparison purposes, we also fit a univariate ARIMA model to the sales data to better understand the impact of adding the exogenous variable, X_t on the forecasts. We follow the usual model building approach given in Chapter 6 and decide on an ARIMA(2, 1, 0) model for the sales data. At first, it is not clear whether a constant term in the univariate model is needed, so we try both. The constant term turns out to be borderline significant. However, the implication of having that constant term is revealed in the forecasts in Figure 8.34. The forecasts from the univariate model with the constant term included labeled as "Univariate" in Figure 8.34 gives forecasts following a linear trend as expected. We discussed this issue in great detail in Chapter 7. Obviously, adding the constant term in the univariate model yields progressively bad forecasts as lead time gets larger. We then fit an ARIMA(2, 1, 0) without the constant term, which is labeled as "Univariate_NoCst" in Figure 8.34. Clearly, these forecasts are better. However, the best forecasts are obtained from the transfer function–noise model as can be seen in Figure 8.34. This is particularly true for short lead times. As the lead time gets big, the forecasts from the transfer function–noise model get closer to

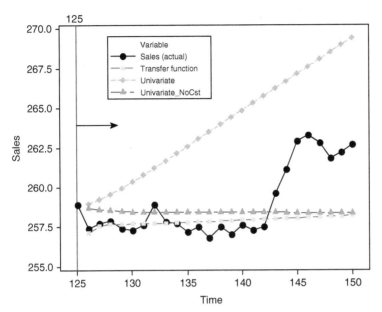

Figure 8.34 1- to 25-step-ahead forecasts from the transfer function model (Transfer function), univariate model for the output including a constant term (Univariate), and univariate model for the output including a constant term (Univariate_NoCst).

the forecasts made by the univariate model and at those big leads, we no longer have sensible forecasts. However, particularly for nonseasonal models, we do not necessarily expect the forecasts made for big lead times to be as accurate as the ones made for small lead times and this is generally true irrespective of the model we use. Another possible reason for the discrepancy between the actual values and their forecasts for large lead times is that the forecasts made using transfer function–noise model require forecasted values of the input as well. In this example, we have an ARIMA(0, 1, 1) model for the input, which resulted in a constant forecast of 13.234 for all lead times. Hence, inaccuracies in these forecasts will be somehow projected in the forecasts of the transfer function–noise model.

We also notice that the behavior of the output changes right around observation 143. After that point, both models fail to keep up and start producing large forecast errors. Of course, it is conjectured that if we make 1-step-ahead casts and update our forecast, each time new data is available, we will with both models have better forecasts for even observations 143 and on. In Figure 8.35, we clearly see the change in behavior in the output series at around the 143rd observation. In the same figure, we also show the input series for the same time period. We can also clearly see that there is a change in the behavior of the input series and it starts at around the 140th observation. The difference corresponds exactly to the delay of 3 time units we observed earlier. Hence, we expect that the transfer function–noise model with this appropriate delay between the input and

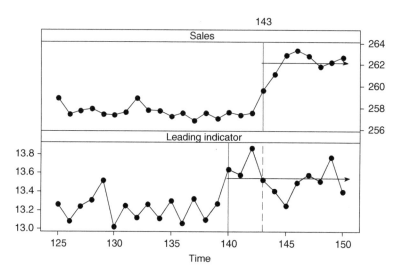

Figure 8.35 Sales and leading indicator data from 125 to 150.

the output would be able to accurately predict the change in the output behavior and produce good 1-step-ahead forecasts even after the 143rd observation.

8.10 INTERVENTION ANALYSIS

The natural course of time series is often interrupted by events or changes such as process adjustments or changes in raw materials for manufacturing processes or policy changes, strikes, natural disasters in economic time series, the effect of advertisement promotions in business time series, or changes in regulations for environmental time series. Typically, these events after some transition period will cause a level change in the time series. Often, the timing of the event is known. If that is the case, we can use intervention analysis to analyze and model the change. It is often of primary interest how much of a change in the level of the process did the intervention lead to. If the process is a manufacturing process and the intervention is a process adjustment, the engineers would like to know how much the action changed the output. As another example, legislators may want to know the effect of seatbelt legislation on the number of death and injuries on the nation's highways. In this section, we will show two examples of intervention analysis. The first example is a simulated manufacturing process where a step change was applied midway in the process. The point of this simulation where we know the exact change is to show how poor and biased a t-test can be if the data is a nonstationary time series. For the second example, we use the well-known dataset about the Colgate and Crest fight for market share of the toothpaste market in light of the endorsement of the American Dental Association (ADA) of Crest. Here, we want to assess the effect of the ADA's endorsement on the sales.

TABLE 8.10   Simulated Process Data with $\delta = 1.5$ and $\theta = 0.7$ (Box and Tiao (1965))

4.393	5.296	6.546	6.655	5.561	5.853	6.977	7.300	7.139	8.582
7.231	7.046	5.176	6.148	8.000	7.273	6.072	7.026	8.254	7.585
5.817	7.880	7.021	7.657	7.615	9.000	9.624	8.972	8.169	8.620
9.073	8.312	6.879	9.192	9.016	7.789	8.070	9.252	9.400	9.449
9.238	10.842	10.897	8.842	10.618	10.401	10.679	10.462	10.348	9.531

8.10.1   Example 3: Box and Tiao Level Change

In this section, we provide an example of how to estimate a simple shift in level of a nonstationary process. Suppose $z_t, t = 1, 2, \ldots, n$ is the daily output of a chemical process and that an action was taken at time $T, 1 < T < n$ to change the process. If the observations z_t could be assumed to be independently, normally distributed with variance σ^2 with a fixed mean μ before the change to the process and with a fixed mean $\mu + \delta$ after the change, then we could simply use a two-sample t-test to test and estimate the change δ. However, these are not realistic assumptions. Daily outputs of a chemical process are much more likely to be dependent and nonstationary. As we will illustrate, even moderate dependence between the observations will make the t-test an inappropriate and misleading procedure for assessing the level change. A more realistic model for the observations is the IMA(1, 1) process given by $z_t = z_{t-1} + a_t - \theta a_{t-1}$, which has often been found to be useful in representing a variety of industrial processes.

Now to demonstrate the approach to estimating a level change when the underlying process is nonstationary, we use the simulated data provided by Box and Tiao (1965) so that we know the true level change and can compare the estimate we get from using the t-test inappropriately as compared to the intervention model approach developed by Box and Tiao (1965). The simulated level change was $\delta = 1.50$ and the moving average parameter was $\theta = 0.7$. There are 50 observations and the level change happened between observation 25 and 26. The data is shown in Table 8.10.

Figure 8.36 shows a time series plot of the process data. It is possible with the naked eye to see that the process jumped up a small amount after observation 25. Another obvious pattern is that the process looks nonstationary both before and after observation 25.

An Inappropriate t-Test   We now show what not to do. That is, if we suspect the data is dependent and nonstationary, we should not use a two-sample t-test to compare before and after a level change. However, it is of interest for later purposes to see how misleading the approach can be. Table 8.11 provides the standard output from a computer-generated two-sample t-test. We see that the estimated difference is $\hat{\delta} \approx 2.50$ with a 95% confidence interval of $[1.91, 3.09]$. That is about one unit larger than the true value. Now it is intuitively easy to see why the t-test will be misleading. Since the process is nonstationary and appears to be drifting upward during the initial phase before the change, the

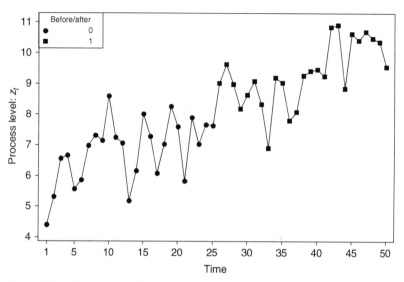

Figure 8.36 Time series plot of the simulated process data with a step change intervention after observation 25.

t-test is based on the assumption that the process is stable and hence provides an estimate of the mean that is based on taking a regular average of the first 25 observations, which is 6.80. The estimate of the level after the change is computed as a regular average of the second set of 25 observations after the change. The average of those numbers is 9.31. These averages are superimposed on the time series plot in Figure 8.37. However, if the process is nonstationary, then it does not have a constant mean. Rather, the mean keeps changing. Therefore, since the process before the change was drifting upward, computing a straight average of the process given equal weights to all 25 observations before the change is not appropriate. The current mean around the time observation 25 was generated would have drifted somewhat higher than 6.80. In other words, the straight average will in this case underestimate the current mean. Now reasoning heuristically, a more realistic estimate of the current mean would have been generated from an exponentially weighted moving average (EWMA) with $\theta = 0.8$ that would put more weight on the most recent observations in computing an estimate of the current mean around observation 25.

Similarly, since the process after the change continued to drift upward, the average of the remaining 25 observations after the change will tend to be an overestimate in this case of the current mean around observation 26. A more realistic estimate of the current mean around observation 26 could be produced if an EWMA is calculated by going backward in time from observation 50 back toward observation 26. The EWMA computed for time $t = 26$ in the backward manner would then be an estimate of the current level right after the event. Therefore, a more realistic estimate of the change δ would be produced by subtracting these two EWMA estimates of the means before and after the change. This is

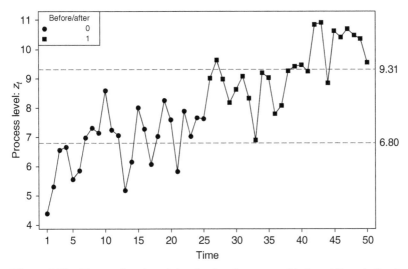

Figure 8.37 Time series plot of the simulated process with dotted lines indicating the arithmetic averages computed before and after the intervention.

TABLE 8.11 Summary of a Two-Sample t-Test for the Simulated Process Data

```
Two-sample t-test for z_t

Indi     N     Mean    StDev    SE Mean

0        25    6.80    1.04     0.21
1        25    9.31    1.03     0.21

Difference = mu (0)- mu (1)
Estimate for difference:-2.50286
95% CI for difference: (-3.09110,-1.91462)
T-Test of difference = 0 (vs not =): T-Value = -8.55 P-Value =
0.000 DF = 48
Both use Pooled StDev = 1.0344
```

illustrated in Figures 8.38 and 8.39. From Figure 8.39, we see that a rough estimate of the change based on subtracting the forward iterating EWMA at time $t = 25$ from the backward moving EWMA going backward from observation 50 to observation 26 would be $\hat{\delta} \approx 8.84 - 7.20 = 1.64$, which is much closer to the true value of $\delta = 1.5$.

This, of course, does bring up the philosophical issue of what a level change means in the context of a nonstationary process. However, it seems reasonable to say that it should be an estimate of the level change computed by subtracting an estimate of the current mean before the change from an estimate of the current mean right after the change. In the next section, we show how intervention

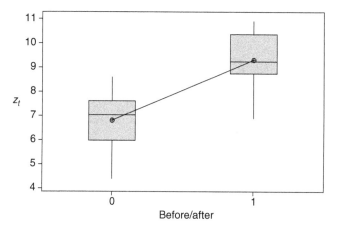

Figure 8.38   Box plot of the data before and after the simulated step change intervention.

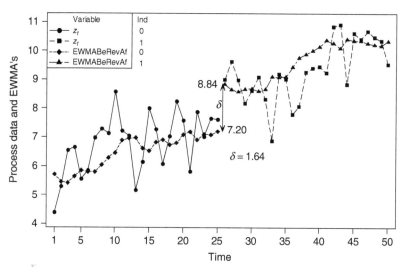

Figure 8.39   Time series plot of the simulated process with a forward iterating EWMA superimposed for the first 25 observations and a similar backward iterating EWMA superimposed on the second set of 25 observations after the intervention.

analysis can more formally provide a procedure for the unbiased estimation of a level change for nonstationary time series.

8.10.2   Intervention Analysis of a Level Change

We now consider the intervention modeling approach that allows for a wide variety of process changes as the result of an intervention. In the case of the chemical process, we will consider the simplest possible change, namely, a simple step change that takes effect immediately.

The general intervention model is given by

$$z_t = v(B)x_t + n_t \tag{8.47}$$

here $v(B) = v_0 + v_1 B + v_2 B^2 + \cdots$ is a linear transfer function, x_t is the intervention driving the change, and n_t is the noise process, usually modeled as an ARIMA (p, d, q) process $\phi(B)(1 - B)^d n_t = \theta(B)a_t$. Typically, the intervention x_t is a step change at time T. We therefore often denote this by an indicator variable $I_t(T)$ that takes the values

$$I_t(T) = \begin{cases} 0 & t < T \\ 1 & t \geq T \end{cases} \tag{8.48}$$

Thus, we can model our simple step change as

$$\begin{aligned} z_t &= v_0 I_t(T) + n_t \\ &= \delta I_t(T) + n_t \end{aligned} \tag{8.49}$$

where we have used the more intuitive symbol δ rather than the generic symbol v_0. Now in this case, we need not identify the noise model since we simulated the process as an IMA(1, 1) process. That is, $(1 - B)n_t = (1 - \theta B)a_t$, which can be rewritten as $n_t = \frac{(1 - \theta B)}{(1 - B)}a_t$. Therefore, the intervention model is

$$z_t = \delta I_t(T) + \frac{(1 - \theta B)}{(1 - B)}a_t \tag{8.50}$$

This model formulation can be surprising as one may expect that the model should be written as $(1 - B)z_t = \delta I_t(T) + (1 - \theta B)a_t$. However, that would not be correct. To see that, consider the situation where $t < T$. In that case, $I_t(T) = 0$ and hence the model in Equation (8.50) is given by $z_t = \frac{(1 - \theta B)}{(1 - B)}a_t$ or $(1 - B)z_t = (1 - \theta B)a_t$. That is the correct model and would be the same for both formulations. However, after the intervention, that is, $t \geq T$ and $I_t(T) = 1$, the model in Equation (8.50) gives $z_t = \delta + \frac{(1 - \theta B)}{(1 - B)}a_t$ or $(1 - B)(z_t - \delta) = (1 - \theta B)a_t$. However, if we used the other model formulation, we would get $(1 - B)z_t = \delta + (1 - \theta B)a_t$, which is a different model. Indeed this is a model with a deterministic trend, which we do not want. Thus, the model formulation in Equation (8.50) is the correct one for a step change to a nonstationary process; this subtlety should always be kept in mind.

We define the pulse function as

$$P_t(T) = \begin{cases} 0 & t \neq T \\ 1 & t = T \end{cases} \tag{8.51}$$

It is then readily seen that if we difference the indicator function $I_t(T)$, then we get

$$\nabla I_t(T) = P_t(T) \tag{8.52}$$

TABLE 8.12 The Parameter Estimates of the Intervention Model

Parameters	Estimate	Standard error	t-Ratio	p-Value
δ	1.56	0.074	2.10	0.041
θ	0.61	0.120	5.06	0.000

Thus, the model we need in this case for the noise is given by $(1 - B)n_t = (1 - \theta B)a_t$ or $n_t = \frac{(1-\theta B)}{(1-B)} a_t$ and the transfer function is given by

$$z_t = v(B)x_t + n_t$$

$$= \delta I_t(T) + \frac{(1 - \theta B)}{(1 - B)}a_t \qquad (8.53)$$

Now rewriting this by multiplying throughout by $(1 - B)$ we get

$$(1 - B)z_t = (1 - B)\delta I_t(T) + (1 - \theta B)a_t \qquad (8.54)$$

or

$$\nabla z_t = \delta P_t(T) + (1 - \theta B)a_t \qquad (8.55)$$

In other words, we need to difference both z_t and $I_t(T)$ before we engage in the estimation of the model. We have elaborated this in great detail because this is one of the subtleties that in practice often lead to mistakes.

To fit the model in Equation (8.55) in SAS JMP, we use the transfer function option for which the rational polynomial simply consists of a single term in the numerator and the input series consists of 0's except at the 26th observation, which is 1. The parameter estimates are given in Table 8.12. The estimate of δ, $\hat{\delta} = 1.56$ is much closer to the true value of 1.5 compared to the estimate we obtained from the t-test. In Figure 8.40, we can see that the fitted values clearly track the change in the level of the process.

As for diagnostic checks, we consider the ACF and PACF of the residuals from model (Equation 8.55) given in Figures 8.41 and 8.42, respectively. Note that at lag 2, both ACF and PACF show a slightly significant value. However, this could be overlooked since both plots were in general consistent with the assumption of uncorrelated errors and therefore we declare the model to be adequate.

8.10.3 Example 4: Crest Versus Colgate Market Share Fight

Marketing managers would in general like to know the effectiveness of their marketing expenditures. Intervention analysis provides a tool that can be helpful in assessing the impact of initiatives that the firm, the competitors, or third parties may make on the market mix. In this section, we consider the fight for market share between Colgate–Palmolive's Colgate Dental Cream and Proctor and Gamble's Crest toothpaste. Proctor and Gamble introduced Crest into the

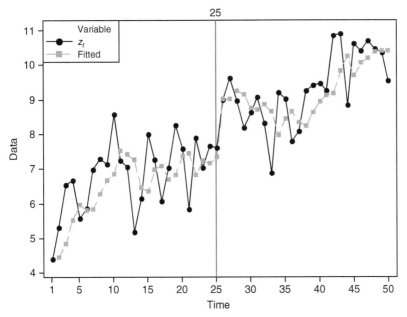

Figure 8.40 Time series plot of simulated process data and the fitted values with the simple step function intervention model.

US market in 1956. Prior to that time, Colgate enjoyed a market leadership with a close to 50% market share. For the following 4 years, Colgate remained a dominant competitor and Crest only captured a relatively modest but stable 15% market share. However, on August 1, 1960, the Council on Dental Therapeutics of the ADA endorsed Crest as an "important aid in any program of dental hygiene." Meanwhile, Proctor and Gamble reinvigorated its marketing campaign to take advantage of the ADA endorsement resulting in an almost immediate jump in Crest's market share to the detriment of Colgate's. We now proceed to analyze the effect of the ADA intervention and in particular how large the Crest's increase in market share and the decrease in the Colgate's market share were. The following analysis is based on the article by Wichern and Jones (1977) who originally presented and analyzed this data.

8.10.3.1 Preliminary Graphical Analysis Figure 8.43 shows 250 weekly observations of the Colgate z_{1t} and Crest z_{2t} market shares spanning the period from 1958 to 1963. The ADA endorsement occurred on August 1, 1960, which corresponds to week 135. It is quite visible from the two graphs in Figure 8.43 that the market shares changed dramatically in the weeks that followed the ADA endorsement.

8.10.3.2 Time Series Modeling of the Noise Process: Colgate Since the ADA intervention happened at observation 135, we use the first 134 observations

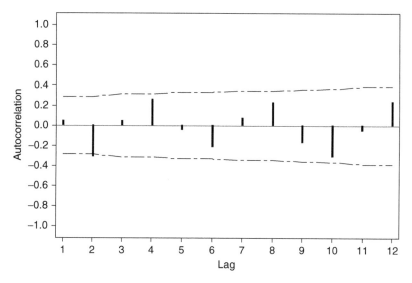

Figure 8.41 ACF of the residuals after fitting the intervention model in Equation (8.55).

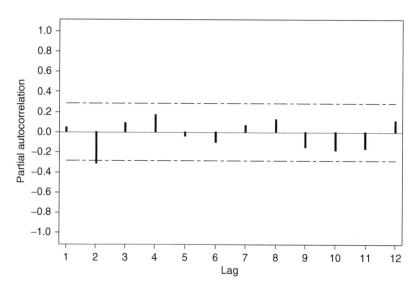

Figure 8.42 PACF of the residuals after fitting the intervention model in Equation (8.55).

to identify the models for the time series. We first analyze the Colgate market share data.

The time series plot in Figure 8.44 suggests that the Colgate market share z_{1t} may be nonstationary. This impression is further indicated by the lingering, nonfading ACF in Figure 8.45. Therefore, we consider a first difference of the data, ∇z_{1t}. Figure 8.46 shows the time series plot of the first differences. The

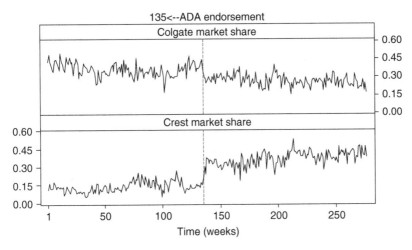

Figure 8.43   Time series plot of the weekly Colgate and Crest market shares.

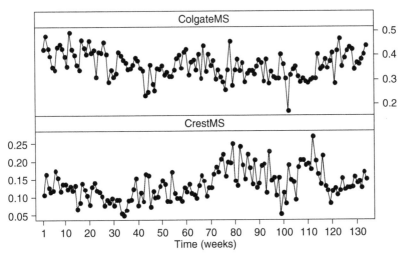

Figure 8.44   Time series plot of the first 134 weekly market share observations of Colgate and Crest before the ADA endorsement. ColgateMS, Colgate market share; CrestMS, Crest market share.

ACF shown in Figure 8.47 dies out relatively quickly. The time series plot of ∇z_{1t} provides an impression of stationary variation. Further, the ACF in Figure 8.47 shows a significant autocorrelation at lag 1 and then a cut off. This suggests that the first difference can be modeled as first order moving average model. The PACF of ∇z_{1t} shown in Figure 8.48 shows a cyclical decaying pattern also consistent with a first order moving average model. Thus, the original time series z_{1t} can be modeled as a first order integrated moving average process ARIMA(0, 1, 1) or IMA(1, 1), $z_{1t} - z_{1t-1} = a_t - \theta a_t$.

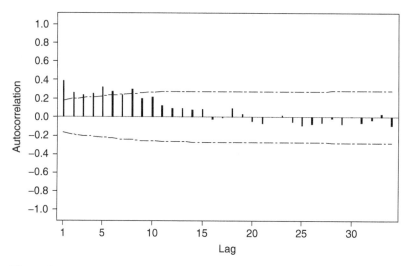

Figure 8.45 ACF of the first 134 weekly observations of Colgate's market share.

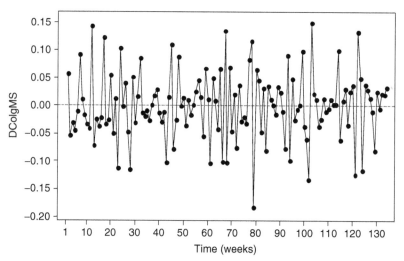

Figure 8.46 Time series plot of the first difference of the first 134 weekly observations of Colgate's market share. DColgMS, Difference of Colgate market share.

From Table 8.13, the estimate of the moving average parameter is $\hat{\theta} = 0.8095$. The ACF and PACF plots of the residuals in Figures 8.49 and 8.50 suggest that the model is indeed adequate.

8.10.3.3 Time Series Modeling of the Noise Process: Crest

We now turn to the Crest data, z_{2t}. Since the two time series exhibit very similar dynamics, we will not provide as much details as we did for the Colgate modeling. Figure 8.51 shows the ACF of the Crest market share data for the first 134 weeks. We

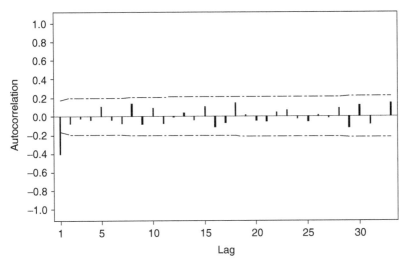

Figure 8.47   ACF of the first difference of the first 134 weekly observations of Colgate's market share.

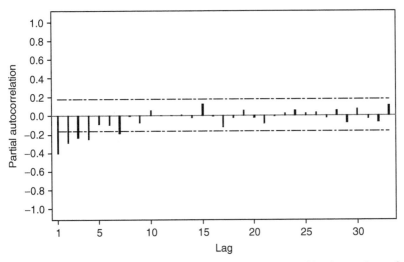

Figure 8.48   PACF of the first difference of the first 134 weekly observations of Colgate's market share.

see a very persistent autocorrelation indicating the need for a first difference. Figure 8.52 shows the ACF of the first difference ∇z_{2t}. The ACF seems to die out relatively quickly, indicating that the first difference is stationary. This impression is further reinforced by looking at the time series plot of ∇z_{2t} in Figure 8.53. Indeed, it looks like the ∇z_{2t} has a zero mean after differencing, which implies that there does not seem to be a deterministic trend. Furthermore, we see from the ACF of ∇z_{2t} that the lag one autocorrelation is significant but

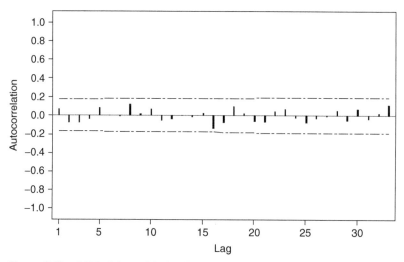

Figure 8.49 ACF of the residuals after fitting an IMA(1, 1) model of the first 134 weekly observations of Colgate's market share.

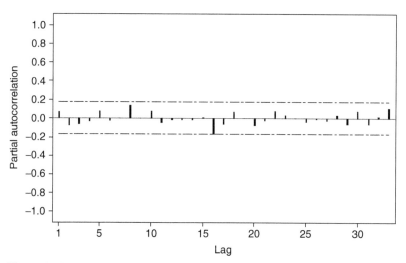

Figure 8.50 PACF of the residuals after fitting an IMA(1, 1) model of the first 134 weekly observations of Colgate's market share.

after that the ACF cuts off. Similarly, the PACF for ∇z_{2t} in Figure 8.54 lingers on for larger lags, all indicating that the appropriate model for ∇z_{2t} is a first order moving average model or equivalently that for z_{2t} is an IMA(1, 1) model. The estimation summary for such a model is shown in Table 8.14 and indicates a good fit. Furthermore, the ACF of the residuals in Figure 8.55 and the PACF of the residuals in Figure 8.56 both indicate a good fit. Thus both time series z_{1t} and z_{2t} seem to follow IMA(1, 1) processes before the intervention.

TABLE 8.13   Estimates and Summary Statistics for the IMA(1, 1) Model Fitted to the Colgate Market Share Data

```
Estimates of Parameters: Colgate Data

Type     Coef     SE Coef           T         P
MA 1    0.8095      0.0512      15.81     0.000

Differencing: 1 regular difference
Number of observations: Original series 134, after differencing 133
Residuals:    SS = 0.319019 (backforecasts excluded)
              MS = 0.002417 DF = 132

Modified Box-Pierce (Ljung-Box) Chi-Square statistic


Lag                 12     24     36     48
Chi-Square         6.9   15.1   26.2   32.6
DF                  11     23     35     47
P-Value          0.808  0.890  0.859  0.945
```

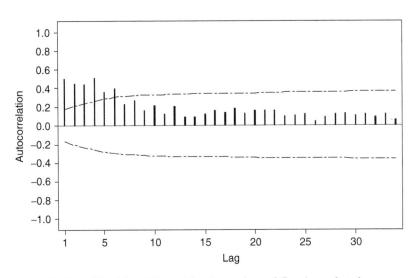

Figure 8.51   ACF of first 134 weekly observations of Crest's market share.

8.10.3.4   *Graphical Investigation of the Intervention*   We now take a closer look at the actual event. What occurred around the time of the ADA endorsement of Crest? We need to know that to be able to model the event. To answer that question, we provide a number of different graphs, first some overall plots of all the data and then some close-ups right before and after the endorsement to get insights into the details—details that often get drowned in time series plots of all the data.

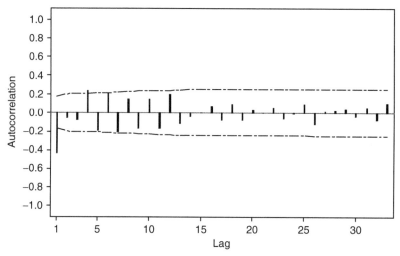

Figure 8.52   ACF of first difference of the first 134 weekly observations of Crest's market share.

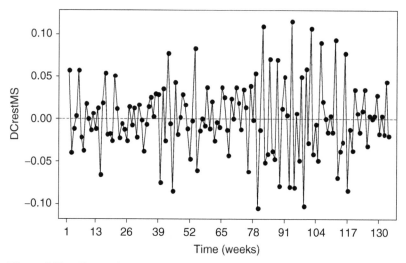

Figure 8.53   Time series plot of first difference of the first 134 weekly observations of Crest's market share. DCrestMS, Difference of Crest market share.

Figure 8.57 shows a plot of all the Crest market share data. We clearly see that the market share rapidly increased around observation 135. Next in Figure 8.58, we show a box plot of the Crest's market share data before and after the intervention. This type of plot provides only a very rough impression of the effect of the intervention.

In Figures 8.59 and 8.60, the close-up time series plots of the original data as well as the first differences are shown. In the former, the impact of the

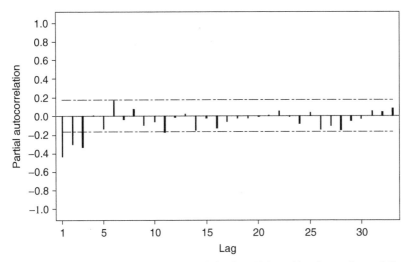

Figure 8.54   PACF of first difference of the first 134 weekly observations of Crest's market share.

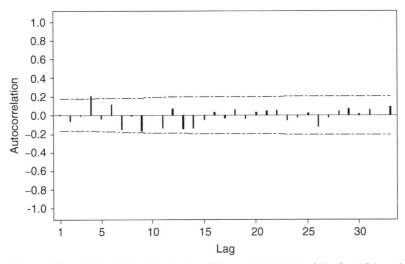

Figure 8.55   ACF of the residuals after fitting an IMA(1, 1) of the first 134 weekly observations of Crest's market share.

intervention is quite clear. The plot of the first differences in Figure 8.60 as expected shows an erratic behavior right around the intervention.

8.10.3.5   Intervention Models for the Market Share Data   In Section 8.10.2, we used the intervention model of the form $(1 - B)z_t = \delta P_t(T) + (1 - \theta B)a_t$ for a simulated nonstationary data. This model can also be used for the market share data since both series exhibit similar behavior as the data we had in

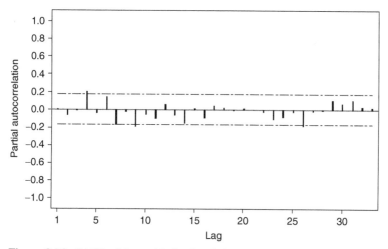

Figure 8.56 PACF of the residuals after fitting an IMA(1, 1) of the first 134 weekly observations of Crest's market share.

TABLE 8.14 Estimation Summary After Fitting an IMA(1, 1) to the Crest's Market Share Data (First 134 Observations)

```
Estimates of Parameters: IMA(1,1) Model for Crest

Type      Coef   SE Coef     T     P
MA   1  0.6975  0.0624   11.18  0.000

Differencing: 1 regular difference
Number of observations: Original series 134, after differencing 133
Residuals: SS = 0.164163 (backforecasts excluded)
          MS = 0.001244 DF = 132
Modified Box-Pierce (Ljung-Box) Chi-Square statistic

Lag               12      24      36      48
Chi-Square      19.7    28.9    35.7    47.2
DF                11      23      35      47
P-Value        0.049   0.185   0.434   0.463
```

Section 8.10.2. However, Wichern and Jones (1977), who originally analyzed this data, suggest the following model

$$(1 - B)z_{j,t} = \delta_{j1}X_{1,t} + \delta_{j2}X_{2,t} + (1 - \theta_j B)a_{jt}, \quad j = 1, 2 \tag{8.56}$$

where

$$X_{1,t} = P_t(135) = \begin{cases} 1 & t = 135 \\ 0 & t \neq 135 \end{cases} \tag{8.57}$$

$$X_{2,t} = P_t(136) = \begin{cases} 1 & t = 136 \\ 0 & t \neq 136 \end{cases} \tag{8.58}$$

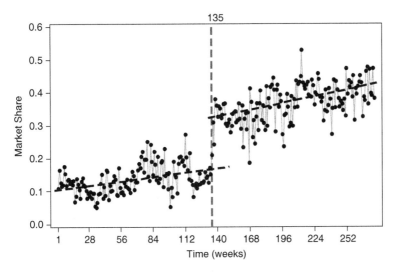

Figure 8.57   Time series plot of all the Crest market share data.

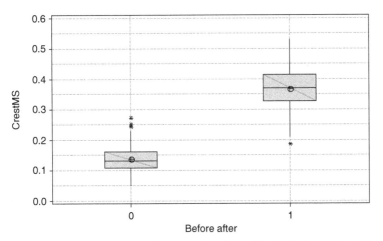

Figure 8.58   Box plot of the Crest market share (CrestMS) data before (0) and after (1) the intervention.

They provide the justification of having two pulses in the model instead of one as: "Other ways to allow for the intervention effect" were considered. However, an initial examination of the differenced data indicated that the market share adjustment to the intervention was essentially accomplished over two consecutive weeks and that there was no *obvious* relationship between the two single week adjustments. Consequently, we decided to incorporate two indicator variables with multiplicative coefficients, rather than, say, a single indicator variable with more complicated dynamics to capture the intervention effect. Therefore,

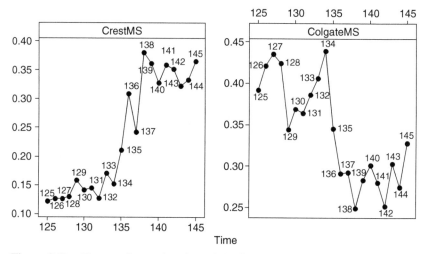

Figure 8.59 Close-up time series plots of the Crest market share (CrestMS) and the Colgate market share (ColgateMS) data from week 125 (10 weeks before) to week 145 (10 weeks after) the intervention.

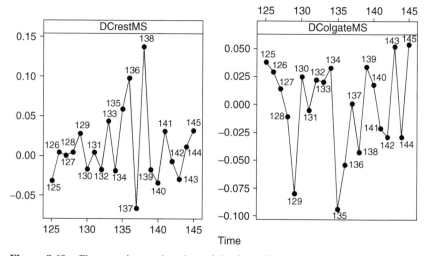

Figure 8.60 Close-up time series plots of the first difference of the Crest and the Colgate market share (DCrestMS and DColgateMS, respectively) data from week 125 (10 weeks before) to week 145 (10 weeks after) the intervention.

we also proceed with model in Equation (8.56) for both Colgate's and Crest's market share data.

Colgate's Market Share The parameter estimates of the model for Colgate's market share is given in Table 8.15. These estimates match the ones obtained by Wichern and Jones. The parameter estimates for δ_{11} and δ_{12} are

TABLE 8.15   The Parameter Estimates of the Intervention Model for Colgate's Market Share

Parameters	Estimate	Standard error	t-Ratio	p-Value
δ_{11}	-0.052	0.046	-1.12	0.264
δ_{12}	-0.061	0.046	-1.31	0.191
θ_1	0.811	0.034	24.06	0.000

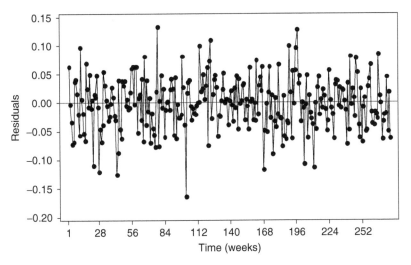

Figure 8.61   Residuals of the intervention model for Colgate's market share data.

small compared to their standard errors (hence giving in absolute value small t-ratios). Therefore, the necessity of having both of these pulses in the model rather than just one can be questioned. We, however, as suggested by Wichern and Jones (1977) proceed with the current model and our diagnostic checks.

In checking the residuals for the intervention model, our attention should in particular be focused on those right before and after the intervention, in this case around time $t = 134$. We see from Figure 8.61 that in this case, the model seems to fit quite well. This is further confirmed by looking at the ACF of the residuals in Figure 8.62. In Figure 8.63, we plotted the residuals versus the fitted values and in Figure 8.64, we show a time series plot of the fitted values superimposed on the observed values. All of these plots seem to confirm that the model fits the data quite well.

Crest's Market Share   The estimates for the intervention model to the Crest data are given in Table 8.16. We again get a reasonably good fit. We see from the time series plot in Figure 8.65 that there are a few large residuals right around the intervention, which is most likely because the intervention model fits the Crest data less well. This is more clearly seen in the close-up residual plot

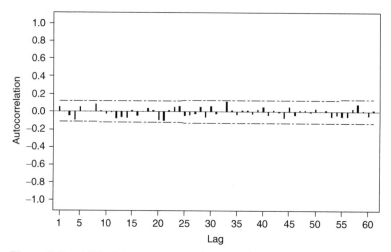

Figure 8.62   ACF of the residuals of the intervention model for Colgate's market share data.

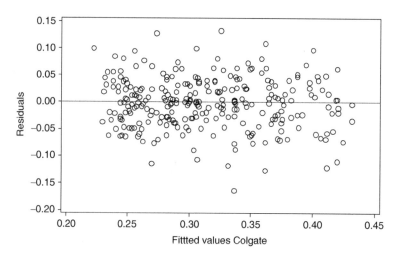

Figure 8.63   Residuals versus fitted values for Colgate intervention modeling.

shown in Figure 8.66 where the residuals 137 and 138 are somewhat larger than the rest. This should not be surprising since the indicator variables model the changes in level at observations 135 and 136 but from Figure 8.66 we can see that observations 137 and 138 show some additional variability.

Owing to our concerns about the residuals right after the intervention for the model with two pulses, we considered other alternatives that did not yield better results. Thus, we settled with the first model. For comparison purposes,

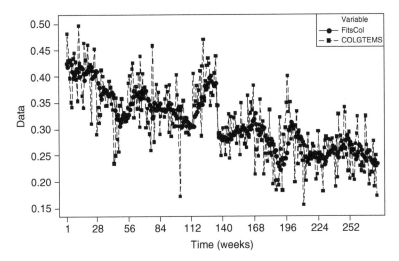

Figure 8.64 Observed and fitted values for Colgate intervention modeling.

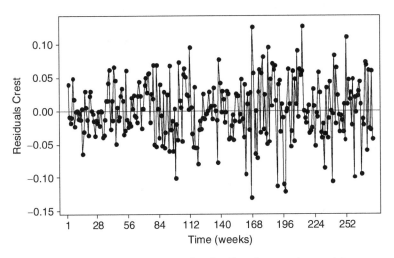

Figure 8.65 Residuals before and after for Crest intervention model.

consider the models:

Colgate market share:

$$z_{1,t} - z_{1,t-1} = -0.052X_{1,t} - 0.061X_{2,t} + a_{1,t} - 0.811a_{1,t-1}$$

Crest market share:

$$z_{2,t} - z_{2,t-1} = 0.066X_{1,t} + 0.112X_{2,t} + a_{2,t} - 0.780a_{2,t-1}$$

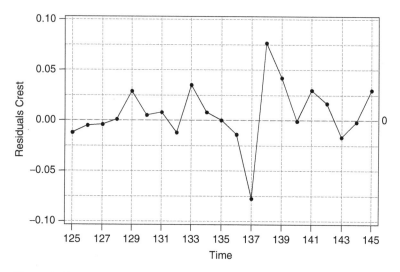

Figure 8.66   Close-up plot of the residuals before and after for Crest intervention model.

TABLE 8.16   The Parameter Estimates of the Intervention Model for Crest's Market Share

Parameters	Estimate	Standard error	t-Ratio	p-Value
δ_{21}	0.066	0.043	1.51	0.132
δ_{22}	0.112	0.043	2.59	0.010
θ_2	0.780	0.037	21.17	0.000

We can calculate the gain (or loss) of market share as the following. Consider, for example, Colgate's market share at weeks 135 and 136.

$$z_{1,135} = z_{1,134} - 0.052X_{1,135} - 0.061X_{2,135} + a_{1,135} - 0.811a_{1,134}$$
$$= z_{1,134} - 0.052(1) - 0.061(0) + a_{1,135} - 0.811a_{1,134}$$
$$= z_{1,134} - 0.052 + a_{1,135} - 0.811a_{1,134}$$

and

$$z_{1,136} = z_{1,135} - 0.052X_{1,136} - 0.061X_{2,136} + a_{1,136} - 0.811a_{1,135}$$
$$= z_{1,135} - 0.052(0) - 0.061(1) + a_{1,136} - 0.811a_{1,135}$$
$$= z_{1,135} - 0.061 + a_{1,136} - 0.811a_{1,135}$$
$$= (z_{1,134} - 0.052 + a_{1,135} - 0.811a_{1,134}) - 0.061 + a_{1,136} - 0.811a_{1,135}$$
$$= z_{1,134} - (0.052 + 0.061) + a_{1,136} + 0.189a_{1,135} - 0.811a_{1,134}$$

In fact, we can show that after observation 136, that is, $t \geq 136$, we have

$$z_{1,t} = z_{1,134} - (0.052 + 0.061) + a_{1,t}$$
$$+ 0.189(a_{1,t-1} + \cdots + a_{1,135}) - 0.811a_{1,134} \qquad (8.59)$$

Equation 8.59 implies that the market share of Colgate suffered a drop of 11.3% ($0.052 + 0.061 = 0.113$) after ADA's endorsement of its rival Crest. The surprising aspect of this is that the drop happened very quickly within 2 weeks after the intervention. Meanwhile, Crest enjoyed a gain of 17.8% market share. Wichern and Jones attributed the discrepancy of these two numbers to the fact that Crest might have taken some of market share of the brands other than Colgate. We believe that this case study provides an excellent example of the use of intervention analysis to study the impact of an intervention at a known time. In this example, the intervention resulted in permanent changes in market shares of two rival products.

EXERCISES

8.1 For the impulse response functions below, determine the corresponding degrees of the transfer function as well as the possible delay.

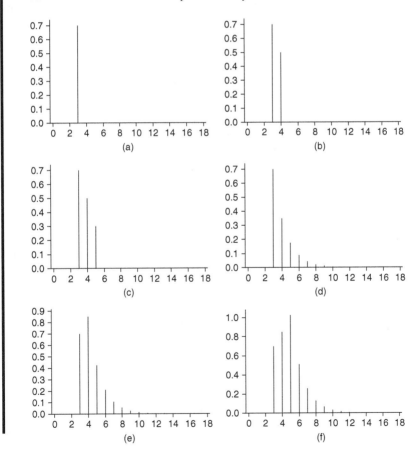

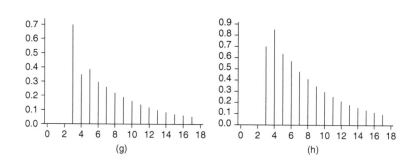

(g)   (h)

8.2 Table B.8 shows the monthly residential sales and retail price for electricity in the United States from January 1990 to July 2010. First consider only the data from January 1990 to December 2009. Develop a transfer function–noise model for the residential sales as the output and retail price of electricity as the input.

8.3 Using the model in Exercise 8.2, forecast the residential electricity sales in the United States for the first 7 months of 2010. Compare your results to the ones from Exercise 5.3a.

8.4 Table B.15 contains quarterly unemployment and GDP data for the United Kingdom (series P of BJR). Fit an appropriate transfer function–noise model with unemployment as the output and GDP as the input.

8.5 In Exercise 3.14, we introduced a dataset on the number of paper checks processed by a branch of a national bank that started accepting electronic checks over the Internet in January 2006. In that exercise, we considered only the first half of the dataset, that is, before the intervention. Now consider the entire dataset in Table B.3 and fit an intervention model to study the impact of the bank's decision.

CHAPTER 9

ADDITIONAL TOPICS

9.1 SPURIOUS RELATIONSHIPS

In Chapter 8, we showed that the functional relationship between an input and an output, that is, $y = f(x)$ is sometimes more complex than a simple linear regression relationship, $y_t = a + bx_t + \varepsilon_t$, implying that a change in x at time t is immediately followed by a proportional change in y. Sometimes the response may be delayed and distributed over time.

Another problem with analyzing process data is that we may sometimes find relationships where none exist. This is called *spurious regression*. This problem is easy to deal with when the relationship does not make obvious sense. The well-known example of the correlation between number of storks and the population of Oldenburg, Germany, shown in Figure 9.1, comes to mind; see Box *et al.* (2005, p. 8). However, in other cases, it can be difficult to know whether the relationship is real or not. In this section, we demonstrate the problem of spurious regression for autocorrelated data using a simulated example.

9.1.1 Review of Basic Regression Theory and Assumptions

For the following discussion, it is important to recall that when performing a regression analysis we make certain assumptions. Some of these are important and others are less so. In the simplest case, suppose we are fitting a straight line $y_t = a + bx_t + \varepsilon_t$ to a set of data $(x_t, y_t), t = 1, \ldots, n$ where ε_t is the error term. Typically, we use least squares estimation to find the best estimates of the intercept $\hat{a}$ and the slope $\hat{b}$ such that the sum of squares $\sum_{t=1}^{n} (y_t - [a + bx_t])^2$ is minimized. Minimizing the sum of squares can be found by calculus, and yields the following two equations.

$$\hat{b} = \frac{\sum_{t=1}^{n} (x_t - \bar{x})(y_t - \bar{y})}{\sum_{t=1}^{n} (x_t - \bar{x})^2}, \quad \hat{a} = \bar{y} - \hat{b}\bar{x} \tag{9.1}$$

where $\bar{x} = n^{-1} \sum_{t=1}^{n} x_t$ and $\bar{y} = n^{-1} \sum_{t=1}^{n} y_t$; see Montgomery *et al.* (2006) and Draper and Smith (1998) for more on regression. In the derivation of the estimates

---

Time Series Analysis and Forecasting by Example, First Edition. Søren Bisgaard and Murat Kulahci.
© 2011 John Wiley & Sons, Inc. Published 2011 by John Wiley & Sons, Inc.

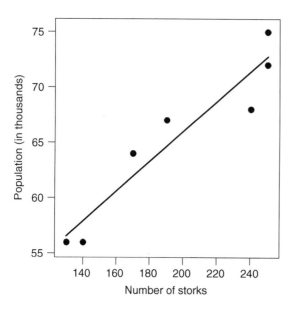

Figure 9.1 A plot of the population of Oldenburg at the end of each year against the number of storks observed in that year, 1930–1936. Adapted from Box *et al.* (2005).

$\hat{a}$ and $\hat{b}$, we make the following assumptions: (i) the errors ε_t are uncorrelated, (ii) the errors have zero mean, $E\{\varepsilon_t\} = 0$, and (iii) the variance of the errors is constant, $V\{\varepsilon_t\} = \sigma^2$, for all t. For the derivation of the least squares estimates, $\hat{a}$ and $\hat{b}$, we need not make any assumption about the errors being normally distributed. However, if we proceed to use a t-test and compute p-values to test if the slope b appears to be different from zero, we also need to assume (iv) that the errors are normally distributed, $\varepsilon_t \sim N(0, \sigma^2)$. Of these, assumption (iv) about normality is not particularly important. The t-test performs fairly well and is quite robust to non-normal errors. The most critical and often overlooked assumption is that of uncorrelated errors; see Box *et al.* (2005, p. 118). This is of particular interest when we study process data sampled over time because such data is most likely positively autocorrelated. We now show how this may cause problems.

9.1.2 What is Spurious Relationship?

Suppose there is no relationship between the input and the output. In that case, we obviously would like a regression analysis of a sample of data from a process to show that there is indeed no relationship. However, if the data is nonstationary and autocorrelated, we may encounter what is called a spurious relationship. A very famous example of this was used by the Cambridge statistician G. Udny Yule in a paper in 1926 aptly entitled "Why do we sometime get Nonsense Correlation between Time Series?" He plotted the annual standardized mortality per 1000 versus the annual proportion of marriages per 1000 in the Church of England of all marriages from 1866 to 1911. In doing so, he found a correlation of 0.9512. The plot is shown in Figure 9.2. However, if we plot the

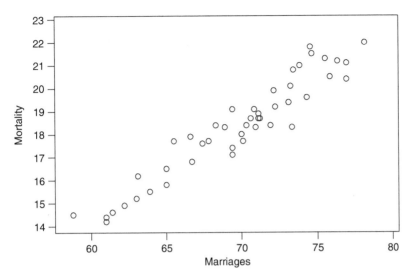

Figure 9.2   Mortality versus marriages from 1866 to 1911.

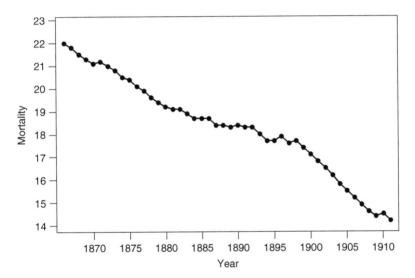

Figure 9.3   Standardized mortality per 1000.

data as time series as in Figures 9.3 and 9.4, we see that both are trending downward.

For our further discussion of the spurious correlation phenomenon, we use simulated data. A simple nonstationary autocorrelated process is the random walk if $x_t = x_{t-1} + a_t$, where a_t's are assumed to be independent normally distributed random errors, $a_t \sim N(0, \sigma_a^2)$. A more general form of the random walk is $x_t = x_{t-1} + a_t - \theta a_{t-1}$, where θ is a constant, $-1 < \theta < 1$. This process can also be

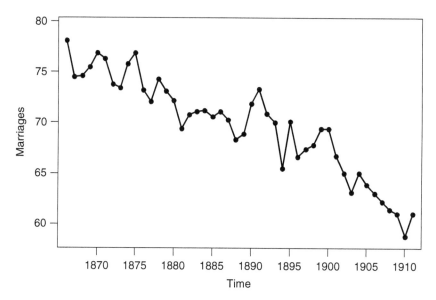

Figure 9.4 Marriages per 1000.

written as $x_t - x_{t-1} = a_t - \theta a_{t-1}$. In this formulation, we see that the difference $\nabla x_t = x_t - x_{t-1}$ follows a first order moving average process, $a_t - \theta a_{t-1}$, which is an IMA(1, 1) process.

Now suppose both x_t and y_t follow IMA(1, 1) processes

$$x_t = x_{t-1} + a_t - \theta a_{t-1} \tag{9.2}$$

$$y_t = y_{t-1} + b_t - \theta_2 b_{t-1} \tag{9.3}$$

where a_t and b_t are mutually independent random normally distributed errors $a_t \sim N(0, \sigma_a^2)$ and $b_t \sim N(0, \sigma_b^2)$. Thus, x_t and y_t are not related to each other. For the simulation experiment, we generated two sets of 100 independent normally distributed error series $\{a_t\}$ and $\{b_t\}$ with zero means and $\sigma_a^2 = 1$ and $\sigma_b^2 = 1$. To start the simulation, we let $x_0 = 0, y_0 = 0, a_0 = 0$, and $b_0 = 0$. We then used a spreadsheet to compute the series $\{x_t\}$ and $\{y_t\}$ from (9.2) and (9.3) both with $\theta_1 = \theta_2 = -0.7$. The data is given in Appendix A and shown in Figure 9.5 as time series plots.

From Figure 9.5, we notice that the two data series clearly look nonstationary and autocorrelated; each series takes large swings up and down, but successive observations are closely related. Because we simulated the data, we know the two series are not related to each other. Nevertheless, we now proceed with fitting a linear regression equation.

Figure 9.6 shows a scatter plot of y_t versus x_t for the 100 simulated observations with a superimposed regression line. We see that there appears to be a negative relationship between x_t and y_t. To understand how this could happen, we look at Figure 9.5. In the later part of the time series plot, approximately after observation 40, y_t trends downward while x_t continues to trend upward. Thus, it

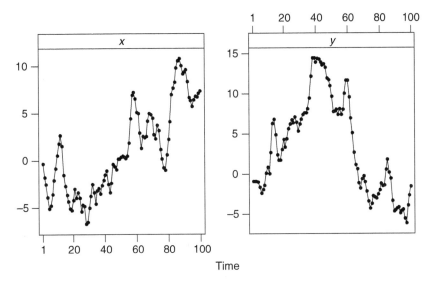

Figure 9.5 Time series plots of two simulated IMA(1, 1) processes, x_t and y_t.

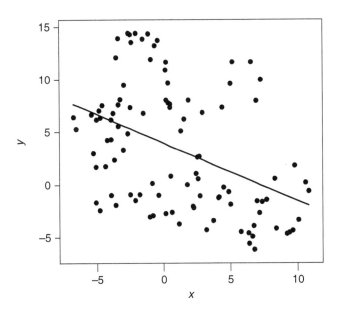

Figure 9.6 Scatter plot of the simulated process data with the regression line.

appears as if the two series overall are negatively correlated even though the data is a random walk. If we fit a regression equation, the best fit is $y_t = 3.95 - 0.55x_t$ with a t-value of -4.74, a p-value of 0.000 for the slope estimate, and an $R^2 = 18.7\%$. In other words, the R^2 is modest, but the regression relationship nevertheless looks quite respectable, despite the fact that x_t and y_t are unrelated.

How can we avoid being tricked into thinking that there is a relationship when none exists? The best answer is clearly to conduct a randomized factorial experiment. However, it is often difficult to convince management to make changes to a process; the current conditions are perhaps not great, but management may fear that a change will make things worse. Thus, when we are left with the option of using existing process data, there is no fail-safe cure for the problem of spurious relations. However, we should always check the assumptions by a thorough residual analysis. Figure 9.7 shows the standard type of plots of residuals—a normal plot, a plot of the residuals versus the predicted values, a histogram, and a time series plot of the residuals. The plot that clearly reveals any abnormal behavior is the time series plot of the residuals. Instead of a random scatter, this plot exhibits serial correlation. This is a telltale sign of violation of the assumption of independent errors. In these situations, it is sometimes recommended to perform a Durbin–Watson test; see MJK. However, we prefer to compute the autocorrelation function (ACF) of the residuals. This approach is more general and easier to interpret. The estimated ACF shown in Figure 9.8 also shows a serious violation of the independence assumption.

Thus, the general lesson learned from this example is that if we analyze process data to see whether there exists an input variable that can be used to control an output variable, we need to be aware of the possibility of spurious relationships. One way to check for this possibility is to carefully analyze the residuals. If they show signs of autocorrelation, the apparent relationship may be spurious.

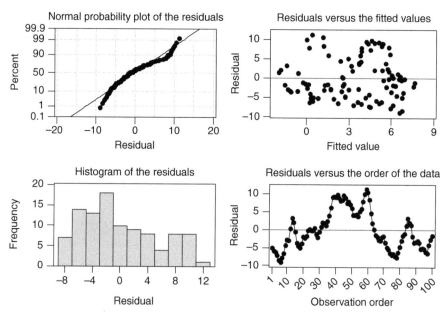

Figure 9.7 Diagnostic residual plots after fitting a straight line to the simulated process data.

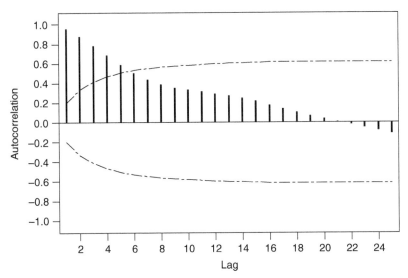

Figure 9.8 Estimated autocorrelation plot of the residuals and 95% confidence limits after fitting a straight line to the process data.

9.1.3 Identifying Spurious Relationships

Let us now consider a simple check for spurious relationships. Interestingly, this method was discussed a long time ago by William S. Gosset, also known as *Student*, the famous brewer, pioneer statistician, and inventor of the t-test. In an early contribution to the time series literature published in 1914, he explained that if two time series are truly related through a linear relationship $y_t = a + bx_t + \varepsilon_t$, then the same relationship should also apply one time unit earlier for $y_{t-1} = a + bx_{t-1} + \varepsilon_{t-1}$. Thus, if we subtract the two, we see that the same relationship holds for the differences $y_t - y_{t-1} = b(x_t - x_{t-1}) + \varepsilon_t - \varepsilon_{t-1}$ or $\nabla y_t = b \nabla x_t + \nabla \varepsilon_t$. In other words, Gosset suggested that if there is a linear relationship, then the differenced data $(\nabla x_t, \nabla y_t)$ should exhibit a similar proportional relationship with the same slope b.

Thus, if we are given a set of process data $\{x_t\}$ and $\{y_t\}$ that indicates a possible relationship, but is a nonstationary time series, then it is a good idea to check the first differences to see if they also exhibit a proportional relationship. If they do, the relationship is likely real. If not, the relationship could be an artifact of the nonstationary behavior of the process data $\{x_t\}$ and $\{y_t\}$.

For the present data, Figure 9.9 shows the relationship between $\nabla x_t = x_t - x_{t-1}$ and $\nabla y_t = y_t - y_{t-1}$. The regression equation now is $\nabla y_t = -0.018 + 0.14 \nabla x_t$ with a t-value for the slope of 1.32, a p-value of 0.19, and $R^2 = 1.8\%$. Thus, there no longer appears to be a proportional relationship. Note that it could be argued that we should have fitted a line without an intercept, but this would not have made much of a difference.

For good measure, we also computed the residuals and the autocorrelation of the residuals after fitting a regression line to the differences, which are shown

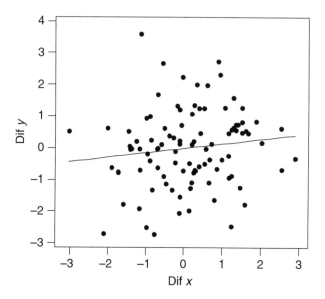

Figure 9.9   First differences of the simulated process data.

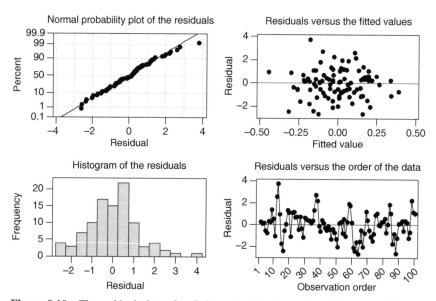

Figure 9.10   The residual plots after fitting a straight line to the differences of the simulated data.

in Figures 9.10 and 9.11. This time we see from Figure 9.10 that the residuals are quite well behaved. There is a hint of positive autocorrelation in the time series plots of the residuals. This is further confirmed by Figure 9.11 that shows a significant autocorrelation at lag 1. Given that the data was generated using an IMA(1, 1) process, this should not be a surprise as it indicates an MA(1) process

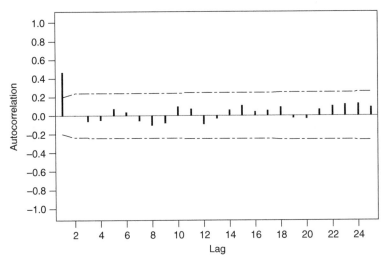

Figure 9.11   ACF of the residuals after fitting a straight line to the differences of the simulated data.

for the residuals. We hasten to say that this autocorrelation also invalidates the regression based on the differences just as it did for the previous regression based on the original data. As a rough check for possible spurious regression, Gosset's trick is useful. However, the method of prewhitening discussed in Chapter 8 is far better. If the relationship between two time series involves a delay, then Gosset's method may be misleading. Thus, we can consider Gosset's method a rough preliminary prewhitening method.

Analyzing process data for possible relationships is a quite common task. However, one should always be aware of the many possible pitfalls. Box *et al.* (2005, pp. 397–407) provide a comprehensive overview of the many different types of problems that may occur in industrial practice when analyzing happenstance data. In Chapter 8, we discussed the problem of delayed relationships and showed how prewhitening can be used to study transfer functions. In this section, we have discussed the problem of autocorrelation in the data causing spurious relationships.

Spurious relationship is an issue that has been studied extensively in the econometrics literature. Good references are Box and Newbold (1971), Granger and Newbold (1974), and Granger and Newbold (1986). Unfortunately, the regression and econometrics literature does not emphasize this as much as we think it should. In the next section, we discuss the issue of autocorrelation in regression analysis in general.

9.2   AUTOCORRELATION IN REGRESSION

In this section, we discuss a notorious example from economics to show how regression analysis can be seriously misleading if the assumption of independence of the errors is violated. As discussed in the previous section as well as in Chapter 8, the issue of autocorrelation is of importance when we use regression to estimate input–output relationships from data sampled over time that likely may be autocorrelated. The data we use was originally discussed by Coen *et al.* (1969). It is of historical and pedagogical significance because it generated quite a controversy when it was first published; the discussion following the paper is still interesting to read. A particularly incisive analysis of this example was subsequently provided by Box and Newbold (1971). The reason we like to use this example is that it is such a poignant illustration of the problem of autocorrelated errors that we think it ought to be known by anyone using regression analysis using time series data. We first provide an analysis similar to that provided by Coen *et al.* (1969) followed by the reanalysis similar to that provided by Box and Newbold (1971), but with more details, accompanying graphics and commentary. The relevant data, edited for easier use, is provided in Appendix A.

9.2.1 Example 1: Economic Forecasting by Coen *et al.*

The purpose of Coen *et al.*'s (CGK, 1969) article was to provide an illustration of the use of lagged regression relationships for economic forecasting. Their main point was to demonstrate that changes to the output of an economic system may be the result of changes in the inputs that occurred several time periods before. Delay is of course also frequently the case with industrial processes. CGK's primary example was a supposed lagged relationship between the *Financial Times* ordinary share index (Y_t) and two inputs, X_{1t}—the UK car production and X_{2t}—the *Financial Times* commodity index. The 51 observations used were quarterly observations starting with the second quarter of 1954 until the end of 1964. Those three time series are plotted in Figure 9.12. It is evident from these plots that the three time series are autocorrelated and most likely nonstationary. Note that in Appendix A we have also provided data for the two input series stretching six and seven time periods back prior to the second quarter of 1954 to accommodate the lagged relationships to be discussed below.

9.2.2 Cross Correlation Analysis

To determine the appropriate lags, CGK used a number of approaches including the cross correlation function (CCF) we discussed in Chapter 8. Now CGK recognized that if we try to compute cross correlations from nonstationary time series, we are likely to get nonsensical answers. Strictly speaking, cross correlations are not defined for nonstationary processes. The mean of a nonstationary process, used in this computation, is not defined because it keeps changing. Indeed, if we compute the cross correlation anyway, the underlying trend is going to dominate the estimated cross correlations. For illustration, Figure 9.13 shows the cross correlation function computed (incorrectly) on the basis of the original data $X_{1,t}$ and Y_t. We see that the cross correlations are very large for both positive and negative lags.

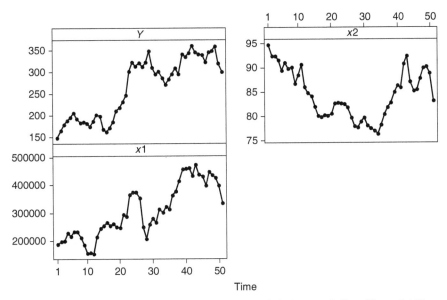

Figure 9.12 Time series plots, quarter by quarter, of the "output," $Y_t = $ *Financial Times* ordinary share index, the two "inputs," $X_{1t} = $ United Kingdom car production, and $X_{2t} = $ *Financial Times* commodity index from the second quarter of 1954 to the end of 1964.

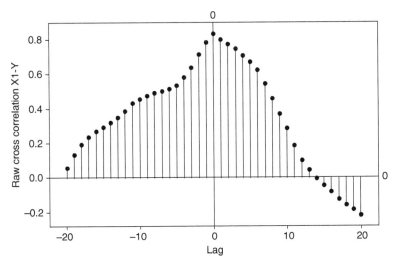

Figure 9.13 Cross correlation function computed (incorrectly) from two nonstationary processes.

To deal with the nonstationarity, CGK proposed to detrend the data by fitting a linear regression line to the data and use the residuals as the "new" data. This is an old trick but not a safe approach. As we have discussed in Chapter 8, a better method is to prewhiten the data. In the present case, if we use

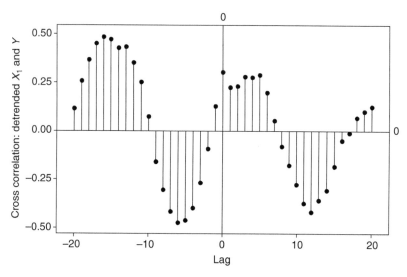

Figure 9.14 Cross correlation between a detrended $X_{1,t}$ and a detrended Y_t.

the detrending method suggested by CGK, we get a cross correlation diagram as shown in Figure 9.14.

The cross correlation between the detrended $X_{1,t}$ and Y_t and a similar one for the relationship between the detrended $X_{2,t}$ and Y_t suggested to CGK that $X_{1,t}$ should be lagged by six time units and $X_{2,t}$ by seven time units. Thus, they proceeded to fit the following regression model to the (non-detrended) data:

$$Y_t = \alpha + \beta_1 X_{1,t-6} + \beta_2 X_{2,t-7} + \eta_t \qquad (9.4)$$

where they assumed that the error term η_t was independent and identically distributed noise. As the result of their analysis, they obtained the following regression equation:

$$Y_t = 653 + \underset{[14.11]}{0.000475\, X_{1,t-6}} - \underset{[-9.88]}{6.127\, X_{2,t-7}} + \eta_t \qquad (9.5)$$

with a very respectable R^2 of 90.2%. The numbers shown below the estimated regression coefficients are the t-values indicating that the coefficients are very significant. These t-values are, of course, highly dependent on the assumed independent error structure. Further, the R^2 is deceptively large because of the trend in the data.

Box and Newbold (1971) in their reanalysis of this example commented that "it must surely be rare that the noise structure for economic models can be so represented." Thus, we proceed to check the residuals from the CGK model. In particular, Figure 9.15 shows a plot of the autocorrelations, which indicates a pronounced autocorrelation at lag 1. Indeed, we ought to be suspicious of the CGK analysis based on the assumption of independent errors η_t.

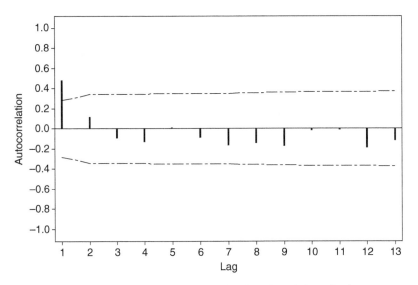

Figure 9.15   ACF of the residuals after fitting Equation (9.4) to the data.

9.2.3   An Alternative Analysis: Box–Newbold's Approach

Box and Newbold (1971) in their reanalysis of the CGK data suggested that a more plausible error structure would be one where the η_t's were assumed to be dependent and possibly represented by a nonstationary time series model. Indeed, it is often possible to model dependent noise by an integrated moving average (IMA) process given by

$$\eta_t = \eta_{t-1} + a_t - \theta a_{t-1} \tag{9.6}$$

where the a_t's are white noise and θ is a constant $-1 < \theta \le 1$. By rearranging the terms, this model for the errors can also be written as

$$\eta_t - \eta_{t-1} = \nabla \eta_t = a_t - \theta a_{t-1} \tag{9.7}$$

This model, can incidentally, also be written by recursive substitution as

$$\eta_t = (1 - \theta) \sum_{j=1}^{\infty} a_{t-j} + a_t \tag{9.8}$$

which shows that it is a noisy random walk. Yet another way to write this model is as an infinite autoregressive process

$$\eta_t = (1 - \theta)(\eta_{t-1} + \theta \eta_{t-1} + \theta^2 \eta_{t-2} + \ldots) + a_t$$

which we recognize as an exponentially weighted moving average of the past errors. Note that if $\theta = 0$, this process is a pure random walk, and if $\theta = 1$, this model is the independent noise assumed by CGK.

The beauty of using this model for the errors is that we are not required to make a strong assumption about independence. Instead, using the integrated moving average process, the data, via the estimated coefficient θ, will tell us what is the most appropriate assumption about the error structure. In other words, if the estimated value of θ turns out to be close to 1, we would have confirmed CGK's assumption. However, if it turns out that $\hat{\theta}$ is less than 1, then the data is likely an autocorrelated noisy random walk and standard least squares regression is not appropriate.

Box and Newbold (1971) proposed to substitute the more realistic integrated moving average error structure into the CGK model as shown in Equation (9.4). Furthermore, instead of detrending the data by linear regression, they suggested it would make more sense to allow for the trend by adding a time dependent regression term in the model. This trick is used to avoid picking up spurious relationships. In other words, Box and Newbold's version of the CGK model was

$$Y_t = \alpha + \beta_0 t + \beta_1 X_{1,t-6} + \beta_2 X_{2,t-7} + \eta_t \tag{9.9}$$

However, Box and Newbold argued that when the time series involved is nonstationary, it is often better to difference the data to achieve stationarity, rather than removing the trend with a linear regression line; see BJR, p. 89. Indeed, by taking first differences of the model in Equation (9.9), the relationship becomes

$$Y_t - Y_{t-1} = \alpha - \alpha + \beta_0[t - (t-1)] + \beta_1(X_{1,t-6} - X_{1,t-7})$$
$$+ \beta_2(X_{2,t-7} - X_{2,t-8}) + (\eta_t - \eta_{t-1}) \tag{9.10}$$

If we now substitute in the terms for the first differences, $y_t = \nabla Y_t = Y_t - Y_{t-1}, x_{1,t} = \nabla X_{1,t} = X_{1,t} - X_{1,t-1}, x_{2,t} = \nabla X_{2,t} = X_{2,t} - X_{2,t-1}$, and $\nabla \eta_t = \eta_t - \eta_{t-1} = a_t - \theta a_{t-1}$, we get

$$y_t = \beta_0 + \beta_1 x_{1,t-6} + \beta_2 x_{2,t-7} + a_t - \theta a_{t-1} \tag{9.11}$$

Notice that the differencing leaves the slopes, the quantities of primary interest, untouched. However, differencing is another (better) way to avoid the problem of picking up spurious regression relationships caused by the nonstationary nature of the time series.

The model in Equation (9.11) is unfortunately not a linear regression equation because of the more complicated error term. Therefore, it cannot be estimated by a regular regression algorithm; it requires a more sophisticated procedure such as iterative nonlinear least squares or a maximum likelihood algorithm. We find that the maximum likelihood estimates and the associated standard errors (in parentheses underneath the estimates) of the model terms in Equation (9.11) are

$$\hat{\beta}_0 = \underset{(2.62)}{1.78}, \quad \hat{\beta}_1 = \underset{(0.00009)}{0.00016},$$
$$\hat{\beta}_2 = \underset{(1.15)}{-1.16}, \quad \hat{\theta} = \underset{(0.05)}{-0.05} \tag{9.12}$$

Considering the estimates relative to their standard errors, none of the terms seem significant. However CGK in their analysis reported a very significant relationship between the *Financial Times* share index, UK car production, and

Financial Times commodity index. The Box and Newbold model with a more realistic assumption about the errors indicated that there might not be a relationship at all. Indeed, Box and Newbold in their analysis argued that the relationship more likely than not was spurious.

The Box and Newbold analysis also showed that the moving average coefficient $\hat{\theta}$ is essentially zero. This implies, as indicated above, that the error η_t is essentially a random walk $\eta_t = \eta_{t-1} + a_t$. This is very different from assuming that the errors are uncorrelated as assumed in a regular least squares regression analysis. In other words, what we assume about the errors makes a great difference in regression analysis. If the errors are autocorrelated, this example provides a vivid illustration of how misleading the t-values can be. Indeed, we may think that there is a relationship when in fact there is none.

To finalize our analysis we show in Figure 9.16, a plot of the ACF of the residuals after fitting the Box and Newbold model. We see that there is no longer any appreciable autocorrelation in the residuals, indicating that the model is plausible. Other useful diagnostic tools are plots of the residuals versus the predictors and the cross correlations between the residuals and the prewhitened predictors.

We should further mention that in the Box and Newbold paper, they tried a number of other time series models for the error including a first and a second order autoregressive model. However, in every case, they obtained very similar results.

In the engineering, managerial, financial, and economic context, we are often faced with the problem of trying to assess whether there exists a relationship between several inputs and an output of a process. However, data observed over time is likely to be autocorrelated. When that is the case, we need to be very careful.

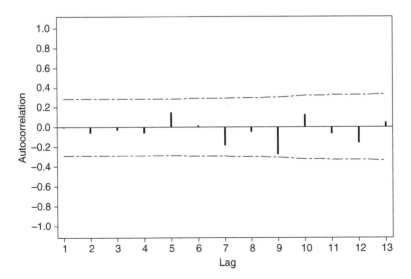

Figure 9.16 ACF of the residuals for the Box and Newbold model.

A major reason for spurious relationships between time series is that two unrelated time series that are internally autocorrelated sometimes by chance can produce very large cross correlations. This may occur if Y_t is highly correlated with its previous values Y_{t-1}, Y_{t-2}, and so on, and if X_t is highly correlated with its previous values X_{t-1}, X_{t-2}, and so on. This may then produce what appears as a high correlation between Y_t and X_t. However, if we calculated the partial correlation between Y_t and X_t taking into account the correlation between Y_t and its previous value Y_{t-1} and X_t and its previous value X_{t-1}, then there might not be much left. In other words, the effect of X_t on Y_t may already be contained in the past history of Y_t and no additional information is provided by including X_t in the model. It is for this reason that the safest approach to assessing the relationship between the input and the output of a process when the data is autocorrelated is to use prewhitening as demonstrated in Chapter 8.

The usual model assessment tools such as t-test in linear regression are often based on the assumptions that the errors are uncorrelated and normally distributed. We recommend that you verify these assumptions via residual analysis. Most standard statistical software packages emphasize testing the normality assumption with normal probability plots and histograms of the residuals. However, most standard statistical data analysis techniques such as analysis of variance and regression analysis are robust to slight deviations from normality. Violation of the assumption of uncorrelated errors is however a much more serious issue. As illustrated in this section, violation of this assumption can easily lead to erroneous conclusions.

9.3 PROCESS REGIME CHANGES

In most of the examples in this book, we have assumed that the underlying mechanism or model that generated a time series stayed constant. However, in practice that may very well not be the case. For example, when new tax laws are introduced, the economy may likely change because managers of firms start to change their behaviors in response to the changed laws. The same may be the case for industrial processes. Such fundamental changes to the underlying structure of the process that generates a time series is called a *regime change*. In this chapter, we will provide an example from process engineering that illustrates some simple graphical ways to analyze such data. The principles illustrated via this example are universal and applicable to any application area. A more mathematical approach can be found, for example, in Hamilton (1994).

In Sections 9.1 and 9.2 as well as in Chapter 8, we discussed how autocorrelation and delays may make it difficult to analyze and interpret input–output relationships. In particular, we have demonstrated that a simple scatter plot of contemporaneous observations of y and x may sometimes provide a false or misleading impression of the true relationship between the input and the output.

Besides the reasons previously discussed, another reason input–output relations may be obscured is what we will call regime changes.

A regime is defined as a regular behavior of a process. Thus, a regime change is a drastic change to the regular behavior of a process. A regime change may, for example, occur when process operators make significant changes to how a process is managed and operated. We will in particular study the graphical patterns typically experienced when a process in steady state is subject to drastic operational changes. These distinct graphical patterns in scatter plots are important to recognize and understand.

9.3.1 Example 2: Temperature Readings from a Ceramic Furnace

To explain the phenomenon of a regime change, we use an example from a large ceramic furnace. To produce good products, it is important that the temperature profile across the length of the furnace maintains a particular convex shape. However, because of the large size of the furnace, its exposure to the environment, and the relatively weak controls, the temperature profile may change over time. To monitor the temperature profile, a number of thermocouples are placed along the length of the furnace. Data from adjacent thermocouples will typically be positively cross correlated. Because of the large mass of the furnace and the relatively high sampling rate, the data will also be positively autocorrelated. Indeed, if we consider two thermocouples located in relatively close proximity to each other, the temperatures will, under normal circumstances, follow each other closely.

Sometimes the process may drift off and the desired profile may change. As a result, the process will produce low yield. When that happens, the process engineers will try to bring the yield back up by making adjustments to how the furnace is managed. For example, they may change the distribution of fuel to the burners, change the flue damper positions, or rearrange the cooling airflow to achieve a different temperature profile in the furnace in the hope these changes will improve the yield. When the changes are introduced, the process will typically react by going through an initial transition period lasting several time units and then, if left alone, settle down to a new steady state or regime.

While these changes are taking place, the process engineers will intensely study a large number of temperature readings from the process to make sure the operational changes produce the intended changes to the temperature profile. In this section, we will study only two adjacent temperature measurements made with two thermocouples located relatively close to each other. These two time series will illustrate the main point of what typically happens during a regime change.

Temperature observations from the two adjacent thermocouples, x_5 and x_6, from 139 consecutive time periods are listed in Appendix A. For proprietary reasons, we have made a (for our purposes) nonessential scale change. In Figure 9.17, we have made a simple scatter plot of the data. Because the temperature scale resolution is $1°$, the temperature data is recorded as whole numbers. Thus, several observations in Figure 9.17 coincide. In Figure 9.18, we have added jitter (small

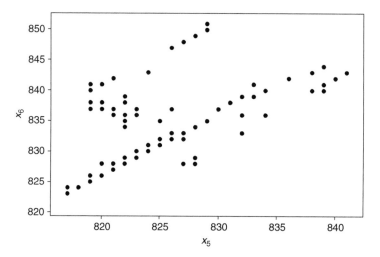

Figure 9.17   A scatter plot of the temperature readings (No jitter).

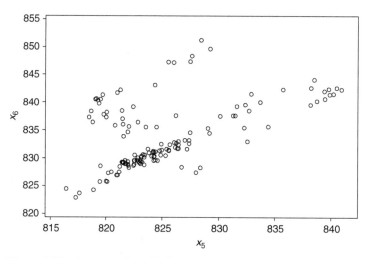

Figure 9.18   A scatter plot with jitter added and using open-circle plotting symbols to better show the data concentration.

random perturbations to the data points) and used open-circle plotting symbols so that it is easier to see the actual concentration of data points.

Both Figures 9.17 and 9.18 display a rather peculiar pattern. Normally, one would expect that two highly correlated observations would scatter as a skinny, upward sloping ellipse. However, in Figures 9.17 and 9.18, we see several clusters. Indeed, we can see what looks like three parallel positively correlated scatter ellipses with additional stray points.

To gain a better understanding of what is happening, we display the data as time series plots in Figure 19.9. From this graph, we see that the two time

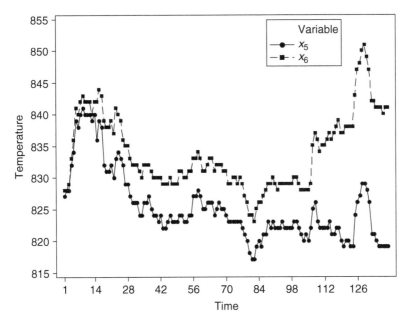

Figure 9.19   Time series plot of the two temperatures.

series initially follow each other closely, but then begin to diverge around observation 15. Note that even after they begin to diverge, the time series seem to continue to resemble each other in terms of their up and down movements.

On what we learned from Figure 9.19, it makes sense to take a look at the difference $d_t = x_{6,t} - x_{5,t}$ between the temperatures. The difference, d_t, is shown as a time series plot in Figure 9.20. From this plot, we gain a better insight into what is really happening. From observation 1 to about observation 14, we see a more or less constant difference. This is followed by a short transitional period from observation 15 to about 26. This transitional period appears to follow the pattern of a second order dynamic system when disturbed; first, the output rises, then overshoots and then comes down a bit, and eventually stabilizes at a new equilibrium. After about observation 27, we see a more or less constant difference of about $6-7°$ lasting to about observation 106. At about observation 107, we see again a drastic transition that lasts to about observation 124, followed by a more stable difference of $22°$. Thus, we will call the initial phase from observation 1 to observation 14 regime 1, followed by transition period 1 lasting to observation 26. After observation 27 follows a long steady period that lasts until observation 106. We call this regime 2. Regime 2 is then followed by transition period 2 lasting from observation 107 until observation 124, after which the process again enters a steady period from observations 125 to 139, called regime 3.

To further verify these regime and transition periods, we used brushing, a powerful computer tool available in many statistical software packages including Minitab for directly highlighting groups of data points on multiple interrelated graphical panels. This technique is not easily demonstrated in a static paper,

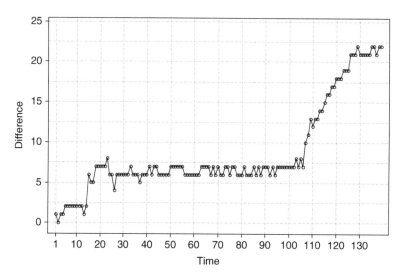

Figure 9.20 A time series plot of the difference between x_6 and x_5.

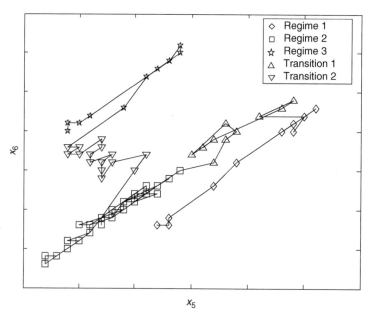

Figure 9.21 Scatter plot of the two temperature observations with different plotting symbols used for the different phases and the points connected in time order.

but it allows us to brush specific points or groups of points with the computer mouse and then move from plot to plot to look for patterns. To give a flavor of this, we used different plotting symbols for each of the three regimes and the two transition phases in the scatter plot in Figure 9.21. We also connected the

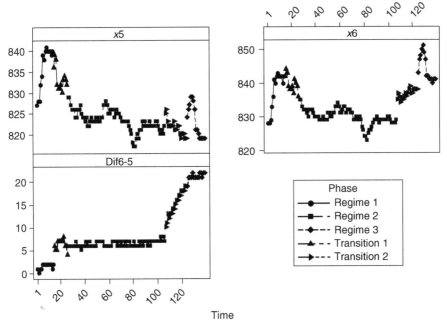

Figure 9.22   Time series plots of the two temperatures and their difference, using different plotting symbols for the three regimes and the two transition periods.

individual observations in time order to emphasize the peculiar pattern in this data. For further illustration, we provide time series plots of the two time series and their difference shown in Figure 9.22.

The picture that has emerged from this graphical analysis, and most clearly seen in Figure 9.21, is that during each of the three steady state regimes, the data is highly linearly related. The main difference between the three regimes is one of different intercepts. Thus, if we were to model the relationship between x_5 and x_6 for the steady state regimes, we could write a model as

$$x_{6,t} = a_j + b_j x_{5,t} + \varepsilon_{j,t}, j = 1, 2, 3 \qquad (9.13)$$

where a_j is an intercept that changes from regime to regime, b_j a slope also changing from regime to regime, and $\varepsilon_{j,t}$ an error also changing from regime to regime and possibly autocorrelated. Indeed, as can be seen from Figure 9.21 and crudely verified with regression, the slopes for each regime are almost the same and close to 1, but the intercept changes from regime to regime. In the econometrics literature, such linear relationships between multiple time series are called *co-movement* or *co-integration*; see for example, Peña *et al.* (2001).

The relationships during the transition periods are less easily modeled. However, it is these transitions that give rise to the irregular patterns in the scatter plots where points are seen to be wandering off and hence obscure the

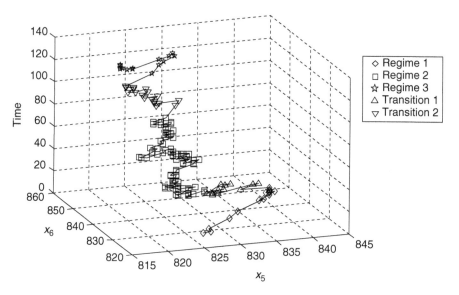

Figure 9.23 A three-dimensional plot of the two temperatures with time.

overall impression of a linear relationship during each regime as observed most clearly in Figure 9.21.

To further illustrate the correlated time series aspect of the data, we have shown in Figure 9.23 the two temperature observations, x_5 and x_6, in a three-dimensional scatter plot with time as the third coordinate. The data points are connected in time order. From this graph, we clearly see how the two temperatures follow each other over time.

We have again used different plotting symbols for each of the different phases. However, this type of graphical representation is most useful when viewed with different color plotting symbols and in a modern statistical computing environment where the three-dimensional plot can be spun around, the data viewed from different angles, and patterns can be explored with the brushing feature.

Gaining understanding of process behavior and exploring the relationship between process variables are important perquisite for process engineering. In some cases, the relationships are simple linear relations and easily verified through plots and regression analysis. In others, the relationship may be more complex and obscured by process dynamics and time series effects. Indeed, as illustrated with the furnace example above, the relationship may change over time. This phenomenon of regime change is not much discussed in the process engineering literature. However, references can be found in the econometrics literature where similar regime changes may happen, for example, when governments change economic policies; see for example, Hamilton (1994).

Regime changes do occasionally occur in the engineering context. The telltale sign of a regime change, as described in this section, is most easily seen in scatter plots. They will typically exhibit a set of parallel skinny ellipses with stray points signifying the transition periods between regimes. The kind of graphical analysis demonstrated in this section is useful in the early diagnostic phase of analyzing processes suspected of having undergone regime changes. If the purpose of the study is to learn more about the behavior of a process in a steady state regime, then one may proceed to analyze more formally particular subsets of the data from individual regimes using regular time series analysis methodology. However, if the purpose is to study the transition periods and formally model those, then one would need to resort to transfer functions and intervention analysis as discussed in Chapter 8.

9.4   ANALYSIS OF MULTIPLE TIME SERIES

In Chapters 3–7, we considered univariate time series. That is, we were only concerned with a single time series. In Chapter 8, however, we introduced the exogenous input variable, x_t. By introducing such a variable, our intention was to improve the understanding of the output of interest by exploring and modeling the dynamic relationship between the input and the output variables. Our conjecture was that this new model would capture more about the behavior of the output than a univariate model. There was one caveat though. The relationship that we envisioned to exist between the input and the output was unidirectional. That is, while the input is expected to impact the behavior of the output, a reverse relationship is deemed nonexistent. However, what if this assumption is not correct? What if there is a feedback coming from the output toward the input as well? If such a feedback exists, a model that considers and explicitly includes that dual relationship can greatly improve our understanding of both variables and have considerable impact on our forecasts for those variables. In general, we can of course have more than two variables at a time and a general model structure is needed to properly study all possible feedback relationships that may exist among these variables. For that, we will discuss the multivariate extension of ARIMA models.

We will first introduce the vector time series $Z_t = (z_{1,t}, z_{2,t}, \ldots, z_{k,t})$, which consists of k univariate time series. Consequently, the mean vector μ simply consists of the mean values of the individual series. We can, in the multivariate case, talk about not only the autocorrelation and partial autocorrelation function (PACF) of the univariate series but also their cross covariance at different lags. Then as in the univariate case, we declare Z_t to be weakly stationary if the mean vector and the cross covariance function are not dependent on time. More formally, if (i) $E(Z_t) = E(Z_{t+l}) = \mu$, constant for all l, (ii) $\text{Cov}(Z_t) = E[(Z_t - \mu)(Z_t - \mu)'] = \Gamma(0)$, and (iii) $\text{Cov}(Z_t, Z_{t+l}) = \Gamma(l)$ depends only on l, then Z_t will be deemed (weakly) stationary. We can then show that for a stationary Z_t, the cross correlation function is given as

$$\rho(l) = V^{-1/2}\Gamma(l)V^{-1/2} \tag{9.14}$$

where **V** is a diagonal matrix with the variances of the individual time series, $\gamma_{ii}(0)$, that is,

$$
\mathbf{V} = \text{diag}\{\gamma_{11}(0), \gamma_{22}(0), \dots, \gamma_{kk}(0)\}
$$

$$
= \begin{bmatrix} \gamma_{11}(0) & 0 & \cdots & 0 \\ 0 & \gamma_{22}(0) & \cdots & 0 \\ \vdots & \vdots & \ddots & \vdots \\ 0 & 0 & \cdots & \gamma_{kk}(0) \end{bmatrix} \tag{9.15}
$$

9.4.1   Models for Stationary Multivariate Time Series

Similar to the univariate model, the stationary multivariate time series can be modeled as

$$
\mathbf{\Phi}(B)Z_t = \mathbf{\delta} + \mathbf{\Theta}(B)\mathbf{a}_t \tag{9.16}
$$

where $\mathbf{\Phi}(B) = \mathbf{I} - \mathbf{\Phi}_1 B - \mathbf{\Phi}_2 B^2 - \cdots - \mathbf{\Phi}_p B^p$, $\mathbf{\Theta}(B) = \mathbf{I} - \mathbf{\Theta}_1 B - \mathbf{\Theta}_2 B^2 - \cdots - \mathbf{\Theta}_q B^q$, and $\mathbf{a}_t$ represents the sequence of random shocks with $E(\mathbf{a}_t) = \mathbf{0}$ and $\text{Cov}(\mathbf{a}_t) = \mathbf{\Sigma}$. Since the random vectors are white noise, we have $\mathbf{\Gamma_a}(l) = 0$ for all $l \neq 0$. The concepts of causality, stationarity, and invertibility that we discussed for univariate ARMA models are also valid for the multivariate models with a twist since $\mathbf{\Phi}(B)$ and $\mathbf{\Theta}(B)$ are now matrices. Therefore, we will call a multivariate ARMA model in Equation (9.16) (causal) stationary if the roots of

$$
\det[\mathbf{\Phi}(B)] = \det[\mathbf{I} - \mathbf{\Phi}_1 B - \mathbf{\Phi}_2 B^2 - \cdots - \mathbf{\Phi}_p B^p] = 0 \tag{9.17}
$$

are all greater than 1 in absolute value. Note that "det" stands for determinant. Similarly, the model in Equation (9.16) is invertible if the roots of

$$
\det[\mathbf{\Theta}(B)] = \det[\mathbf{I} - \mathbf{\Theta}_1 B - \mathbf{\Theta}_2 B^2 - \cdots - \mathbf{\Theta}_q B^q] = 0 \tag{9.18}
$$

are all greater than 1 in absolute value.

Consider as an example a multivariate ARMA(1, 1) model with three variables

$$
\mathbf{\Phi}(B) = \mathbf{I} - \mathbf{\Phi}_1 B
$$

$$
= \begin{bmatrix} 1 & 0 & 0 \\ 0 & 1 & 0 \\ 0 & 0 & 1 \end{bmatrix} - \begin{bmatrix} \phi_{11} & \phi_{12} & \phi_{13} \\ \phi_{21} & \phi_{22} & \phi_{23} \\ \phi_{31} & \phi_{32} & \phi_{33} \end{bmatrix} B \tag{9.19}
$$

and

$$
\mathbf{\Theta}(B) = \mathbf{I} - \mathbf{\Theta}_1 B
$$

$$
= \begin{bmatrix} 1 & 0 & 0 \\ 0 & 1 & 0 \\ 0 & 0 & 1 \end{bmatrix} - \begin{bmatrix} \theta_{11} & \theta_{12} & \theta_{13} \\ \theta_{21} & \theta_{22} & \theta_{23} \\ \theta_{31} & \theta_{32} & \theta_{33} \end{bmatrix} B \tag{9.20}
$$

Then we have

$$
\left[\begin{bmatrix} 1 & 0 & 0 \\ 0 & 1 & 0 \\ 0 & 0 & 1 \end{bmatrix} - \begin{bmatrix} \phi_{11} & \phi_{12} & \phi_{13} \\ \phi_{21} & \phi_{22} & \phi_{23} \\ \phi_{31} & \phi_{32} & \phi_{33} \end{bmatrix} B \right] \begin{bmatrix} z_{1,t} \\ z_{2,t} \\ z_{3,t} \end{bmatrix}
$$
$$
= \begin{bmatrix} \delta_1 \\ \delta_2 \\ \delta_3 \end{bmatrix} + \left[\begin{bmatrix} 1 & 0 & 0 \\ 0 & 1 & 0 \\ 0 & 0 & 1 \end{bmatrix} - \begin{bmatrix} \theta_{11} & \theta_{12} & \theta_{13} \\ \theta_{21} & \theta_{22} & \theta_{23} \\ \theta_{31} & \theta_{32} & \theta_{33} \end{bmatrix} B \right] \begin{bmatrix} a_{1,t} \\ a_{2,t} \\ a_{3,t} \end{bmatrix} \quad (9.21)
$$

or

$$
\begin{aligned}
z_{1,t} &= \delta_1 + \phi_{11} z_{1,t-1} + \phi_{12} z_{2,t-1} + \phi_{13} z_{3,t-1} + a_{1,t} - \theta_{11} a_{1,t-1} \\
&\quad - \theta_{12} a_{2,t-1} - \theta_{13} a_{3,t-1} \\
z_{2,t} &= \delta_2 + \phi_{21} z_{1,t-1} + \phi_{22} z_{2,t-1} + \phi_{23} z_{3,t-1} + a_{2,t} - \theta_{21} a_{1,t-1} \\
&\quad - \theta_{22} a_{2,t-1} - \theta_{23} a_{3,t-1}. \\
z_{3,t} &= \delta_3 + \phi_{31} z_{1,t-1} + \phi_{32} z_{2,t-1} + \phi_{33} z_{3,t-1} + a_{3,t} - \theta_{31} a_{1,t-1} \\
&\quad - \theta_{32} a_{2,t-1} - \theta_{33} a_{3,t-1}
\end{aligned} \quad (9.22)
$$

As we can see in Equation (9.22), in the models of each univariate time series we have the past values not only of the same series but also of the other series. Moreover, through the MA component, in the individual models there are contributions from all of the three error terms. This makes the model identification in multivariate ARMA models quite difficult, particularly for large k. Nonetheless, similar principles based on ACF and PACF can be applied as we will discuss further in our example below.

9.4.2 Models for Nonstationary Multivariate Time Series

We discussed differencing in great detail as a tool for transforming nonstationary time series to achieve stationarity. The same applies in the multivariate case. That is, stationarity for multivariate nonstationary time series can also be achieved by an appropriate degree of differencing. Of course, the key is to find that appropriate degree of differencing. In general, we can write the ARIMA model for multivariate time series as

$$
\mathbf{\Phi}(B)\mathbf{D}(B)Z_t = \mathbf{\Theta}(B)\mathbf{a}_t \quad (9.23)
$$

where

$$
\mathbf{D}(B) = \mathrm{Diag}\left\{ (1-B)^{d_1}, (1-B)^{d_2}, \ldots, (1-B)^{d_k} \right\}
$$
$$
= \begin{bmatrix} (1-B)^{d_1} & 0 & \cdots & 0 \\ 0 & (1-B)^{d_2} & \cdots & 0 \\ \vdots & \vdots & \ddots & \vdots \\ 0 & 0 & \cdots & (1-B)^{d_k} \end{bmatrix} \quad (9.24)
$$

The model in Equation (9.23) is flexible enough to allow for various degrees of differencing for each time series. However, identification of these various degrees is very complex and simultaneous differencing of the time series can cause various complications in modeling (Box and Tiao, 1977; Reinsel, 1993). When differencing all time series simultaneously, the major concern is *co-integration*. This is usually of great interest for economic time series. The nonstationary time series of order d, that is, the dth differences of these time series are stationary, are said to be co-integrating if there exists a particular linear combination of these time series that is nonstationary of order b where b is less than d. In most cases, $d = 1$ so we can simplify the above statement as follows: The nonstationary time series of order 1 is said to be co-integrating if there exists a particular linear combination of these time series that is stationary. The practical implication is that if the time series are co-integrating, that would mean that a common trend is affecting all these nonstationary series. This will cause these time series to exhibit similar long-term behavior. For a more general definition and detail, see BJR, pp. 570–573; BD, pp. 254–255; and Engle and Granger (1991).

9.4.3 Model Selection Criteria for Multivariate Time Series Models

In Chapter 6, we discussed model selection criteria such as Akaike's information criterion (AIC), AICC, and Bayesian information criterion (BIC) for univariate time series. They can also be used with proper modification for multivariate data. From Reinsel (1993), for a multivariate ARMA(p, q) model for k time series with sample size of N, we have AIC given by

$$AIC \approx N \ln(|\hat{\Sigma}|) + 2r \tag{9.25}$$

where $|\hat{\Sigma}|$ is the determinant of the estimate of the residual covariance and r is the number of parameters estimated in the ARMA model. Similarly, the *BIC* is

$$BIC \approx N \ln(|\hat{\Sigma}|) + 2r \ln(N) \tag{9.26}$$

These criteria are fairly straightforward to calculate but as in the univariate case, we recommend using care and caution when using these criteria and to avoid relying solely on one of these criteria. The model selection procedure should involve not only these criteria but also a thorough analysis of the residuals.

Tiao and Box (1981) provide a test statistic for determining the appropriate order for a multivariate AR(p) process akin to PACF in univariate time series. More specifically, the test statistic is a likelihood ratio to test whether $\boldsymbol{\Phi}_m = 0$ and given as

$$M_m = -\left(n - mk - \frac{1}{2}\right) \ln\left(\frac{|\mathbf{S}_m|}{|\mathbf{S}_{m-1}|}\right) \tag{9.27}$$

where $n = N - m$ for a model without a constant term or otherwise $n = N - m - 1$ and $\mathbf{S}_m$ the residual sum of squares of the AR(m) model calculated as $\mathbf{S}_m = \sum_{t=m+1}^{N} \hat{\mathbf{a}}_t \hat{\mathbf{a}}_t'$. If $\boldsymbol{\Phi}_m = 0$, the M_m statistic in Equation (9.27) approximately follows a chi-square distribution with degrees of freedom k^2, $\chi_{k^2}^2$. In other words, large values of M_m suggest that $\boldsymbol{\Phi}_m$ is not equal to 0 and hence we move on to test for $\boldsymbol{\Phi}_{m+1} = 0$.

9.4.4 Example 3: Box and Jenkins' Gas Furnace

In this section, we will revisit the gas furnace example we discussed in Chapter 8. In that example, the percentage of CO_2 in outlet gas, y_t, was modeled using a transfer function–noise model with the input gas rate as the exogenous variable, x_t. The final transfer function–noise model for the output together with the univariate model used for x_t were

$$x_t = \phi_1 x_{t-1} + \phi_2 x_{t-2} + \phi_3 x_{t-3} + \alpha_t \tag{9.28}$$

and

$$y_t = \frac{\omega_0 - \omega_1 B - \omega_2 B^2}{1 - \delta_1 B} x_{t-3} + \frac{1}{(1 - \phi_1 B - \phi_2 B^2)} a_t \tag{9.29}$$

We will now analyze the gas furnace data pretending that the unidirectional relationship between the input and the output is not known *a priori*. Hence, we will attempt to analyze two time series $z_{1,t}$ (for x_t) and $z_{2,t}$ (for y_t) with a possible reciprocating feedback relationship. But before we start the analysis, we rewrite the Equations (9.28) and (9.29) as

$$z_{1,t} = \phi_{111} z_{1,t-1} + \phi_{112} z_{1,t-2} + \phi_{113} z_{1,t-3} + a_{1,t} \tag{9.30}$$

and

$$(1 - \delta_1 B)(1 - \phi_1 B - \phi_2 B^2) z_{2,t}$$
$$= (1 - \phi_1 B - \phi_2 B^2)(\omega_0 - \omega_1 B - \omega_2 B^2) z_{1,t-3} + (1 - \delta_1 B) a_{2,t}$$
$$\Rightarrow z_{2,t} = \phi_{213} z_{1,t-3} + \phi_{214} z_{1,t-4} + \phi_{215} z_{1,t-5} + \phi_{216} z_{1,t-6} + \phi_{217} z_{1,t-7}$$
$$+ \phi_{221} z_{2,t-1} + \phi_{222} z_{2,t-2} + \phi_{223} z_{2,t-3} + a_{2,t} - \theta_{221} a_{2,t-1} \tag{9.31}$$

Now we proceed with our analysis. First, we plot the time series. Figures 9.24 and 9.25 give demeaned $z_{1,t}$ obtained by subtracting -0.057 from each observation and demeaned $z_{2,t}$ obtained by subtracting 53.5 from each observation, respectively. From these figures, we do not observe any severe deviations from stationarity. Hence, we continue with the ACF and cross correlation function of the time series given in Figures 9.26–9.28. Since we only have two time series of interest, this kind of analysis is still feasible. However, in more general situations with large k, it will be quite cumbersome to go through these plots to identify any existing patterns. Instead, Tiao and Box (1981) propose a more practical approach by replacing the numerical values with $(+, -, \cdot)$ signs to indicate whether the numerical value is greater than $2/\sqrt{N}$, less than $-2/\sqrt{N}$, or between these two values, respectively. It should be noted that these limits are based on the assumption that the standard deviation of the terms in the cross correlation matrix, $\hat{\rho}_{ij}$ is $1/\sqrt{N}$ as we discussed in the previous chapters. However, the standard deviation of the $\hat{\rho}_{ij}$ can be greater than $1/\sqrt{N}$ particularly for highly autocorrelated series. In that case, since the limits $\pm 2/\sqrt{N}$ will be tighter than what they should be, some $\hat{\rho}_{ij}$'s will be considered significant when they actually are not. This may lead to models with more parameters than needed. Therefore, as usual, these limits

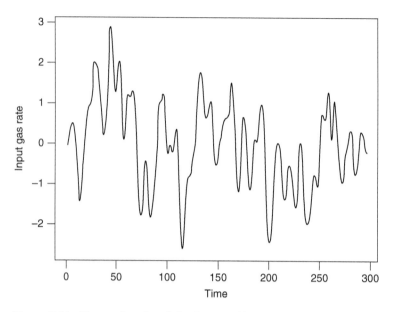

Figure 9.24 Time series plot of the demeaned input gas rate, $z_{1,t}$.

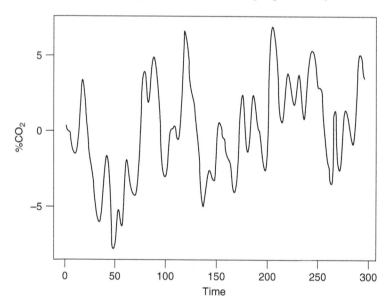

Figure 9.25 Time series plot of the demeaned $\%CO_2$, $z_{2,t}$.

should be applied with caution keeping in mind that the conclusions drawn thereof are only tentative.

Table 9.1 shows the cross correlation matrices for various lags. Note that the diagonal elements of these matrices are the ACFs of the individual time series whereas the off-diagonal elements are the cross correlation between the

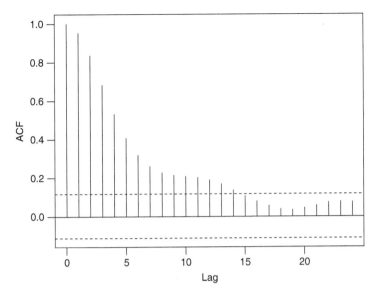

Figure 9.26   ACF of the input gas rate, $z_{1,t}$.

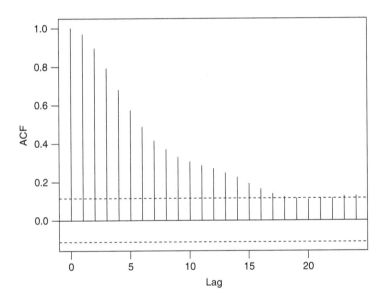

Figure 9.27   ACF of the % CO_2, $z_{2,t}$.

two time series. Even for two variables only, the interpretation of these matrices is not that straightforward, which makes finding a pattern quite difficult. Instead, in Table 9.2, we provide two possible representations of the cross correlation matrices in terms of the $(+, -, \cdot)$ signs as we discussed earlier. A certain pattern in the behavior of $\hat{\rho}_{ij}$'s starts to emerge. It seems like the $\hat{\rho}_{ij}$'s are decaying slowly while they remain significant even for large lags. This certainly is not

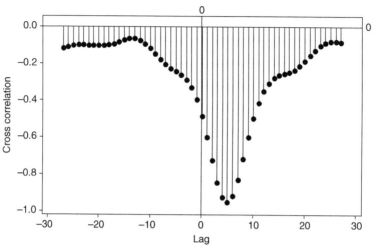

Figure 9.28   The estimated cross correlation function between the input gas feed rate and the output CO_2.

the behavior of a pure MA process. At this point, we decide to consider pure AR models with which the pattern of $\hat{\rho}_{ij}$'s match. It should be noted that as in the univariate models, an ARMA process may also have an exponential decay pattern in ACF; hence, a similar argument can be made for the pattern we see in Table 9.2. Depending on the order of the AR model, we determine at the end of our analysis, that an ARMA model may indeed provide an alternative. We will discuss this issue at the end of our analysis.

Statistical software packages that can fit a multivariate ARMA model are somewhat limited. For this example, we use the *vars* package in *R*, a popular open source statistical software with which multivariate AR model of various orders can be fit fairly easily. We then calculate the M_m statistics and AIC values that are given in Table 9.3.

Considering that the upper 5% point for a chi-square distribution with 4 degrees of freedom is 9.5, M_m statistics suggests that an AR(6) model is appropriate. AIC, however, suggests an AR(4) model. We nonetheless proceed with the AR(6) model and decide on whether Φ_5 and Φ_6 are indeed necessary once we fit the model.

Table 9.4 provides the parameter estimates of the AR(6) model. Underneath each estimate, its standard error is provided in parentheses. Using the ratio of the estimate to the standard error being less than 2 in absolute value as a rough indicator for significance, we identify the nonsignificant terms as indicated by dashed boxes in Table 9.4. We then drop those terms from the model, that is, set them equal to zero and reestimate the remaining parameters.

The reduced model for which the nonsignificant terms are set to zero is given in Table 9.5. There are several interesting features in the representation of the model given in Table 9.5. First of all the terms corresponding to the lagged $z_{2,t}$ terms for the model of $z_{1,t}$, that is, the $(1,2)$ elements of each $\hat{\Phi}$ matrix, are

TABLE 9.1 The Cross Correlation Matrices for Different Lags for the Gas Furnace Data

Lag 1
$$\begin{bmatrix} 0.95 & -0.60 \\ -0.39 & 0.97 \end{bmatrix}$$

Lag 2
$$\begin{bmatrix} 0.83 & -0.72 \\ -0.33 & 0.90 \end{bmatrix}$$

Lag 3
$$\begin{bmatrix} 0.68 & -0.84 \\ -0.29 & 0.79 \end{bmatrix}$$

Lag 4
$$\begin{bmatrix} 0.53 & -0.92 \\ -0.26 & 0.68 \end{bmatrix}$$

Lag 5
$$\begin{bmatrix} 0.41 & -0.95 \\ -0.24 & 0.57 \end{bmatrix}$$

Lag 6
$$\begin{bmatrix} 0.32 & -0.92 \\ -0.83 & 0.48 \end{bmatrix}$$

Lag 7
$$\begin{bmatrix} 0.26 & -0.83 \\ -0.21 & 0.42 \end{bmatrix}$$

Lag 8
$$\begin{bmatrix} 0.23 & -0.72 \\ -0.18 & 0.37 \end{bmatrix}$$

Lag 9
$$\begin{bmatrix} 0.21 & -0.60 \\ -0.15 & 0.33 \end{bmatrix}$$

Lag 10
$$\begin{bmatrix} 0.21 & -0.50 \\ -0.12 & 0.31 \end{bmatrix}$$

Lag 11
$$\begin{bmatrix} 0.20 & -0.41 \\ -0.09 & 0.31 \end{bmatrix}$$

Lag 12
$$\begin{bmatrix} 0.19 & -0.35 \\ -0.08 & 0.27 \end{bmatrix}$$

TABLE 9.2 Two Possible Representations of the Cross Correlation Matrices in Terms of the $(+, -, \cdot)$ Signs

$$
\begin{array}{cccccc}
Lag\ 1 & Lag\ 2 & Lag\ 3 & Lag\ 4 & Lag\ 5 & Lag\ 6 \\
\begin{bmatrix} + & - \\ - & + \end{bmatrix} &
\begin{bmatrix} + & - \\ - & + \end{bmatrix} &
\begin{bmatrix} + & - \\ - & + \end{bmatrix} &
\begin{bmatrix} + & - \\ - & + \end{bmatrix} &
\begin{bmatrix} + & - \\ - & + \end{bmatrix} &
\begin{bmatrix} + & - \\ - & + \end{bmatrix}
\end{array}
$$

$$
\begin{array}{cccccc}
Lag\ 7 & Lag\ 8 & Lag\ 9 & Lag\ 10 & Lag\ 11 & Lag\ 12 \\
\begin{bmatrix} + & - \\ - & + \end{bmatrix} &
\begin{bmatrix} + & - \\ - & + \end{bmatrix} &
\begin{bmatrix} + & - \\ - & + \end{bmatrix} &
\begin{bmatrix} + & - \\ - & + \end{bmatrix} &
\begin{bmatrix} + & - \\ \cdot & + \end{bmatrix} &
\begin{bmatrix} + & - \\ \cdot & + \end{bmatrix}
\end{array}
$$

$$
\begin{array}{cc}
 & \quad z_{1,t} \qquad\qquad\qquad z_{2,t} \\
\begin{array}{c} z_{1,t} \\ z_{2,t} \end{array} &
\begin{bmatrix}
+{+}{+}{+}{+}{+}{+}{+}{+}{+}{+}{+} & -{-}{-}{-}{-}{-}{-}{-}{-}{-}{-}{-} \\
-{-}{-}{-}{-}{-}{-}{-}{-}{-}\cdot\ \cdot & +{+}{+}{+}{+}{+}{+}{+}{+}{+}{+}{+}
\end{bmatrix}
\end{array}
$$

TABLE 9.3 M_m and AIC Statistics for the Furnace Data

Order	1	2	3	4	5	6
M_m	**1714.37**	**690.16**	**32.90**	**23.08**	6.02	**13.39**
AIC	−1065.97	−1749.12	−1762.71	**−1766.29**	−1752.22	−1745.89
Order	7	8	9	10	11	12
M_m	1.82	8.56	4.00	1.46	4.69	2.88
AIC	−1727.34	−1715.94	−1699.65	−1680.66	−1665.19	−1647.67

TABLE 9.4 The Parameter Estimates and their Standard Error (in parentheses) for the AR(6) Model for the Furnace Data with Nonsignificant Terms Indicated with Dashed Boxes

$\hat{\Phi}_1$

$$
\begin{bmatrix}
1.931 & -0.051 \\
(0.059) & (0.047) \\
0.063 & 1.546 \\
(0.076) & (0.060)
\end{bmatrix}
$$

$\hat{\Phi}_2$

$$
\begin{bmatrix}
-1.204 & 0.100 \\
(0.129) & (0.086) \\
-0.134 & -0.594 \\
(0.165) & (0.110)
\end{bmatrix}
$$

$\hat{\Phi}_3$

$$
\begin{bmatrix}
0.170 & -0.080 \\
(0.147) & (0.900) \\
-0.441 & -0.171 \\
(0.188) & (0.115)
\end{bmatrix}
$$

$\hat{\Phi}_4$

$$
\begin{bmatrix}
-0.160 & 0.027 \\
0.149 & (0.090) \\
0.152 & 0.132 \\
(0.190) & (0.115)
\end{bmatrix}
$$

$\hat{\Phi}_5$

$$
\begin{bmatrix}
0.380 & -0.042 \\
(0.140) & (0.079) \\
-0.119 & 0.057 \\
(0.179) & (0.101)
\end{bmatrix}
$$

$\hat{\Phi}_6$

$$
\begin{bmatrix}
-0.214 & 0.031 \\
(0.086) & (0.034) \\
0.248 & -0.042 \\
(0.110) & (0.043)
\end{bmatrix}
$$

all set to zero indicating that there is no feedback from $z_{2,t}$ to $z_{1,t}$. Hence, it is appropriate to model it using a univariate model as we did in Chapter 8. Moreover, the order of that univariate model is indicated to be 6. However, if we look at the parameter estimates in relation to their standard errors, the estimates

TABLE 9.5   The Reduced Model for the Furnace Data

$$
\begin{bmatrix} \hat{\Phi}_1 \\ \begin{array}{cc} 1.894 & \cdot \\ (0.045) & \\ \cdot & 1.446 \\ & (0.033) \end{array} \end{bmatrix}
\begin{bmatrix} \hat{\Phi}_2 \\ \begin{array}{cc} -1.075 & \cdot \\ (0.057) & \\ \cdot & -0.515 \\ & (0.026) \end{array} \end{bmatrix}
\begin{bmatrix} \hat{\Phi}_3 \\ \begin{array}{cc} \cdot & \cdot \\ & \\ -0.509 & \cdot \\ (0.026) & \end{array} \end{bmatrix}
$$

$$
\begin{bmatrix} \hat{\Phi}_4 \\ \begin{array}{cc} \cdot & \cdot \\ & \\ \cdot & \cdot \\ & \end{array} \end{bmatrix}
\begin{bmatrix} \hat{\Phi}_5 \\ \begin{array}{cc} 0.243 & \cdot \\ (0.057) & \\ \cdot & \cdot \\ & \end{array} \end{bmatrix}
\begin{bmatrix} \hat{\Phi}_6 \\ \begin{array}{cc} -0.114 & \cdot \\ (0.045) & \\ 0.300 & \cdot \\ (0.044) & \end{array} \end{bmatrix}
$$

for AR(1) and AR(2) terms are 20–40 times their standard errors, whereas the estimates for AR(5) and AR(6) terms are only 4 and 2 times their standard errors, respectively. This suggests that a univariate AR(2) model for $z_{1,t}$ may be enough. For $z_{2,t}$, we see lagged $z_{1,t}$ terms in the model. This means that there is a feedback from $z_{1,t}$ to $z_{2,t}$. However, that feedback is not immediate since the first $z_{1,t}$ term that is significant appears in $\hat{\Phi}_3$. Hence, the first feedback occurs between $z_{2,t}$ and $z_{1,t-3}$. This suggests a delay of three time units as concluded in the transfer function we fit in Chapter 8. Thus, the outcome of this analysis matches quite well with the model we came up with in the previous chapter. The models for the times series are then

$$
\hat{z}_{1,t} = 1.894\hat{z}_{1,t-1} - 1.075\hat{z}_{1,t-2} + 0.243\hat{z}_{1,t-5} - 0.114\hat{z}_{1,t-6}
$$
$$
\hat{z}_{2,t} = -0.509\hat{z}_{1,t-3} + 0.3\hat{z}_{1,t-6} + 1.446\hat{z}_{2,t-1} - 0.515\hat{z}_{2,t-2} \tag{9.32}
$$

As for the residual analysis, in Table 9.6 we provide the cross correlations of the residuals using $(+, -, \cdot)$ signs as discussed earlier to facilitate the interpretation.

In Table 9.6, we see that there are only three significant correlations left in the residuals. A closer look reveals that all of these three estimates are barely significant compared to the $\pm 2/\sqrt{N}$ limits. Therefore, we believe it is safe to assume that there is no auto- and cross correlation left in the residuals from the model in Equation (9.32) and that the model is indeed appropriate.

Following the study in Tiao and Box (1981), we also provide the parameter estimates in successive AR fits from order 1 to 6 in Table 9.7. To ease the interpretation, we once again use the $(+, -, \cdot)$ notation.

It is interesting to note that for AR(1) and AR(2) models, the unidirectional relationship between the input and the output is not clear. But once we reach order 3, the univariate input model becomes clear. Coincidentally, the delay we find between the input and the output in the transfer function–noise model from Chapter 8 was also 3. Hence, once we properly identify the univariate model for the input as well as the delay between the input and the output, that is, for order 4 and higher, the unidirectional feedback relationship is also identified through the multivariate AR model.

TABLE 9.6 The Cross Correlations Using $(+,-,\cdot)$ Signs of the Residuals from the Reduced AR(6) Model in Equation (9.32)

$$
\begin{array}{c}
 \\
\hat{a}_{1,t} \\
\hat{a}_{2,t}
\end{array}
\begin{array}{c}
\hat{a}_{1,t} \qquad\qquad \hat{a}_{2,t} \\
\left[
\begin{array}{cc}
\cdots\cdots\cdots+\cdot & \cdots\cdots\cdots\cdots \\
\cdots\cdots\cdots\cdots & +\cdots-\cdots\cdots
\end{array}
\right]
\end{array}
$$

TABLE 9.7 The Pattern of Parameter Estimates of Multivariate AR Fits from Orders 1 to 6

Order	$\hat{\Phi}_1$	$\hat{\Phi}_2$	$\hat{\Phi}_3$	$\hat{\Phi}_4$	$\hat{\Phi}_5$	$\hat{\Phi}_6$	$\hat{\Sigma}$
1	$\begin{bmatrix}+ & +\\ - & +\end{bmatrix}$						$\begin{bmatrix}0.100 & 0.089\\ & 0.338\end{bmatrix}$
2	$\begin{bmatrix}+ & -\\ + & +\end{bmatrix}$	$\begin{bmatrix}- & +\\ - & -\end{bmatrix}$					$\begin{bmatrix}0.037 & -0.004\\ & 0.068\end{bmatrix}$
3	$\begin{bmatrix}+ & \cdot\\ \cdot & +\end{bmatrix}$	$\begin{bmatrix}- & \cdot\\ - & -\end{bmatrix}$	$\begin{bmatrix}+ & \cdot\\ \cdot & +\end{bmatrix}$				$\begin{bmatrix}0.036 & -0.002\\ & 0.063\end{bmatrix}$
4	$\begin{bmatrix}+ & \cdot\\ \cdot & +\end{bmatrix}$	$\begin{bmatrix}- & \cdot\\ \cdot & -\end{bmatrix}$	$\begin{bmatrix}\cdot & \cdot\\ - & \cdot\end{bmatrix}$	$\begin{bmatrix}\cdot & \cdot\\ + & +\end{bmatrix}$			$\begin{bmatrix}0.036 & -0.003\\ & 0.059\end{bmatrix}$
5	$\begin{bmatrix}+ & \cdot\\ \cdot & +\end{bmatrix}$	$\begin{bmatrix}- & \cdot\\ \cdot & -\end{bmatrix}$	$\begin{bmatrix}\cdot & \cdot\\ - & \cdot\end{bmatrix}$	$\begin{bmatrix}\cdot & \cdot\\ \cdot & \cdot\end{bmatrix}$	$\begin{bmatrix}\cdot & \cdot\\ \cdot & \cdot\end{bmatrix}$		$\begin{bmatrix}0.036 & -0.004\\ & 0.059\end{bmatrix}$
6	$\begin{bmatrix}+ & \cdot\\ \cdot & +\end{bmatrix}$	$\begin{bmatrix}- & \cdot\\ \cdot & -\end{bmatrix}$	$\begin{bmatrix}\cdot & \cdot\\ - & \cdot\end{bmatrix}$	$\begin{bmatrix}\cdot & \cdot\\ \cdot & \cdot\end{bmatrix}$	$\begin{bmatrix}+ & \cdot\\ \cdot & \cdot\end{bmatrix}$	$\begin{bmatrix}- & \cdot\\ + & \cdot\end{bmatrix}$	$\begin{bmatrix}0.036 & -0.002\\ & 0.058\end{bmatrix}$

If the simplification were not possible through dropping insignificant terms and the analysis yielded an AR model of large order, then we would also consider the possibility of adding MA terms to simplify the model structure as we did in the univariate cases. However, the seemingly high order of AR was primarily due to the 3 periods of delay between the input and the output.

This example shows that the transfer function–noise models in Chapter 8 represent a special case of multivariate ARIMA models. But if there exists *a priori* process knowledge that suggests a unidirectional relationship, it may be simpler and easier to consider the transfer function–noise model, rather than the multivariate ARIMA model. Nevertheless, for more general cases, multivariate ARIMA models offer a flexible class of models to be used for cross- and autocorrelated multivariate data.

9.5 STRUCTURAL ANALYSIS OF MULTIPLE TIME SERIES

In some cases, it is possible to gain a better understanding of the general structure of the multivariate data by considering certain linear combinations of the time series rather than the time series themselves. These linear combinations are, however, not haphazardly constructed. We consider below two different approaches offering different structural analysis of the data.

9.5.1   Principal Component Analysis of Multiple Time Series

In engineering, management, finance, and economics, it is increasingly common that a large number of time series are observed simultaneously over time. In that case, the observations will most likely be correlated with each other (contemporaneously) and in time (temporarily). To gain understanding of the behavior of a system, especially when troubleshooting a process, graphical and statistical tools appropriate for such multivariate situations can be helpful. The purpose of this section is to demonstrate a few techniques, mostly graphical, that are useful when dealing with such situations. We focus primarily on contemporaneous correlation. Indeed, one of our goals is to demonstrate principal component analysis (PCA), a method akin to a Pareto analysis. Rather than taking a typical mathematical approach, we will focus on the geometry of PCA to enhance intuition.

9.5.1.1   *Example 4: Monitoring an Industrial Process*   Our discussion will be based on an example of monitoring the temperature of an industrial process at five different locations, $\mathbf{x}_t = (x_{1t}, x_{2t}, \ldots, x'_{5t})$. However, the approach used could be applied to gain understanding of the structure of any multivariate data set, regardless of the application. The 50 consecutive observations are provided in Appendix A. For proprietary reasons, the data has been rescaled. The analysis that follows can be performed with most standard statistical software packages.

First, we need to get an overview of the structure of the data. Figure 9.29 provides five panels of time series plots. None of the series looks particularly stable which, of course, is one of the motivations for studying this process. From the time series plots, it is sometimes possible to see relationships. For example, x_1 and x_2 seem to peak at about the same time around observation 30.

Figure 9.30 provides a matrix plot of the five time series. Notice that in Figure 9.30, we have followed the recommendation of Cleveland (1985) of using open circles as plotting symbol, so that the concentration of multiple observations in the same location is better appreciated.

From Figure 9.30, we see that the temperatures from the five adjacent locations are correlated with each other. Given the physical nature of the process, this is not surprising. If the temperature changes in any part of the process, it will also likely change in neighboring locations. The actual correlation coefficients are provided in Table 9.8. Both from Figure 9.31 and Table 9.8, we see that neighboring observations tend to be more positively correlated than observations further apart. This is particularly notable for the relationship between x_1 and x_2, and between x_3 and x_4, and the correlation between x_3 and x_5 is also relatively strong.

When studying multivariate process data, the dynamic brushing feature now available in many statistical packages is indispensable. Unfortunately, this feature is not easily demonstrated in a static book format. But suppose we brushed

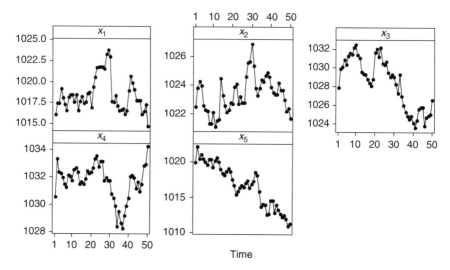

Figure 9.29 Time series plot of the five temperatures.

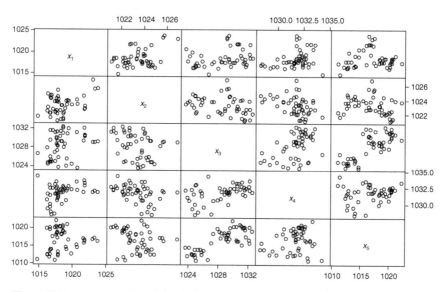

Figure 9.30 A matrix plot of the 5 × 50 consecutive temperature observations.

observations 21–30 corresponding to the sudden process upset or elevated temperatures seen in the time series plot of x_1 in Figure 9.29. On the computer screen, when brushed these 10 observations would be highlighted with a different color. In the matrix plot in Figure 9.30, we have instead indicated these observations with square plotting symbols. We see, for example, that these observations, when looked at jointly between x_1 and x_2, are lying on the extreme upper edge of the data cluster. Further brushing of individual or clusters of points can help the analyst in gaining insight into the physical meaning of the patterns seen in the data.

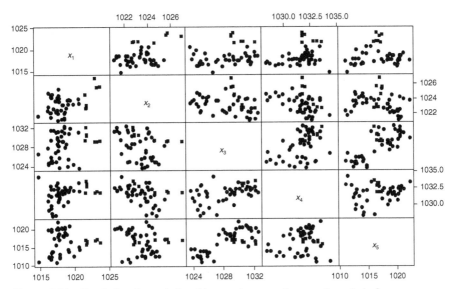

Figure 9.31 Emulating dynamic brushing we have used squared symbols for observations 21–30 correspond to the process upset (i.e., high temperature values) for x_1 as seen in Figure 9.29.

TABLE 9.8 The Correlation between the Five Temperatures with the Associated p-Values in Parentheses

	x_2	x_3	x_4	x_5
x_1	0.371 (0.008)	0.302 (0.033)	0.334 (0.018)	0.091 (0.532)
x_2		−0.303 (0.032)	−0.407 (0.003)	−0.278 (0.050)
x_3			0.458 (0.001)	0.770 (0.000)
x_4				0.194 (0.178)

For the manufacturing process to function properly, the temperatures at different locations are supposed to form a concave profile. To visualize this profile, it is customary to just plot the averages. Instead, we use five box plots with a line connecting the averages as in Figure 9.32. This method shows the profile and provides useful information about the temperature variation at each location. For example, we see that x_3 and x_5 exhibit more variation than the other readings. We see from Figure 9.29 that this is because these two time series are trending downward.

So far, we have demonstrated a number of simple graphical tools that can be used with great benefit to study the behavior of a process and help the process engineer better understand its behavior. Such graphs are indispensable for troubleshooting processes and should always be part of the initial study of a process. However, we now turn to a multivariate technique that is also sometimes helpful but not so commonly used.

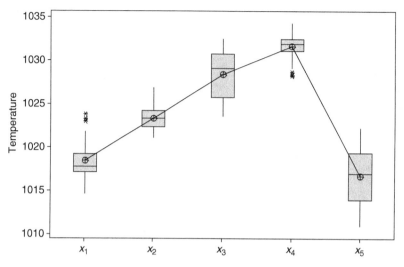

Figure 9.32 Box plot of the temperature readings.

9.5.1.2 PCA in Two and Three Dimensions

When trying to uncover the structure of a large multidimensional data set, PCA is often a useful technique. Almost any text on multivariate statistics, and in particular Johnson and Wichern (2002), provides technical details about PCA. Admittedly, the technique may at first appear a bit intimidating. But we will focus on the intuitively appealing geometry of PCA.

The sample correlations r_{ij} displayed in Table 9.8 can be written as a symmetric 5×5 matrix $\mathbf{R} = \{r_{ij}\}$. These sample correlations r_{ij}'s are computed as $r_{ij} = \text{Cov}\{x_i, x_j\}/(s_{x_i} s_{x_j})$, where $\text{Cov}\{x_i, x_j\} = \hat{\sigma}_{ij} = n^{-1} \sum (x_i - \bar{x}_i)(x_j - \bar{x}_j)$ is the sample covariance between x_i and x_j, and s_{x_i} and s_{x_j} are the usual sample standard deviations. PCA can be performed using either correlations or covariances. Since all five temperatures in our example are measured on the same scale, we prefer to use the (nonscaled) covariances for our PCA. Thus, we can create another symmetric matrix similar to Table 9.8 for the sample covariances $\hat{\sigma}_{ij}$ that in matrix notation is given by

$$\hat{\Sigma} = \{\hat{\sigma}_{ij}\} = \begin{pmatrix} \hat{\sigma}_{11} & \hat{\sigma}_{12} & \hat{\sigma}_{13} & \hat{\sigma}_{14} & \hat{\sigma}_{15} \\ & \hat{\sigma}_{22} & \hat{\sigma}_{23} & \hat{\sigma}_{24} & \hat{\sigma}_{25} \\ & & \hat{\sigma}_{33} & \hat{\sigma}_{34} & \hat{\sigma}_{35} \\ & & & \hat{\sigma}_{44} & \hat{\sigma}_{45} \\ & & & & \hat{\sigma}_{55} \end{pmatrix} \tag{9.33}$$

where the diagonal elements $\hat{\sigma}_{ii}$ are the sample variances s_{ii}^2.

From a symmetric matrix, we can find the eigenvectors and eigenvalues. The eigenvectors are the *principal axes* and they define a new rotated coordinate system. The first eigenvector associated with the largest eigenvalue defines the direction in which most of the variation in the data occurs.

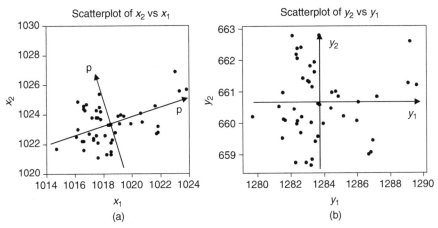

Figure 9.33   (a) 2D scatter plot of the 50 simultaneous temperature readings x_1 and x_2 with the principal axes $\mathbf{p}_1$, and $\mathbf{p}_2$ superimposed and (b) scatter plot of the rotated data.

To see this geometrically, consider for the moment only x_1 and x_2. Those two temperatures are plotted in Figure 9.33a and show a scatter that more or less forms an ellipse with axes slightly rotated counterclockwise. The first principal axis (eigenvector) is the vector $\mathbf{y}_1$ and the second principal axis (eigenvector) is the vector $\mathbf{y}_2$ pointing in a direction rotated 90° relative to the first principal axis. The *principal components*, also called *scores*, are the original data projected down onto the new rotated set of coordinate axes. Figure 9.33b shows the data plotted in the rotated set of coordinates (and somewhat rescaled). Ignoring the rescaling, it can visually be seen that the data in Figure 9.33b exhibits the same pattern as in Figure 9.33a.

To gain further insight, consider the three-dimensional equivalent of what we just looked at. Specifically, we will bring in one more temperature reading, x_3. Figure 9.34 is a three-dimensional scatter plot of x_1, x_2, and x_3. We have again superimposed the principal axes (eigenvectors) $\mathbf{p}_1, \mathbf{p}_2$, and $\mathbf{p}_3$. Each vector is pointing in a direction orthogonal (90°) to the others. As before, the first principal axis, $\mathbf{p}_1$, points in the direction of the scatter that exhibits the most variability. The second principal axis, $\mathbf{p}_2$, points in the direction of second most variability, and the third and last principal axis, $\mathbf{p}_3$, points in the direction of least variability. As in the two-dimensional example, the principal components are the orthogonal projections of the original data down on the rotated coordinate axes (vectors).

From Figure 9.34, we see that the three temperature observations form a three-dimensional ellipsoid. With some effort, it is also possible to see that the variability in the direction of the last principal axis $\mathbf{p}_3$ is relatively small. Indeed, try to visualize a plane spanned by the two vectors $\mathbf{p}_1$ and $\mathbf{p}_2$. Most of the scatter lies on that plane. Thus, because of their mutual correlations, the three different temperatures essentially vary in a two-dimensional subspace (plane) spanned by these two vectors. The projection of the data down on that plane (but rescaled) is shown in Figure 9.35. Thus, in the present case, we can explain 93% of the

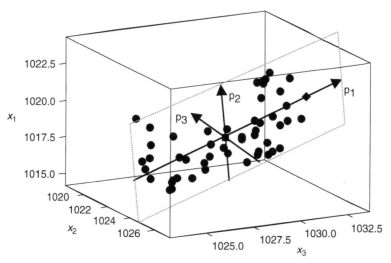

Figure 9.34   3D scatter plot of the 50 simultaneous temperature readings x_1, x_2, and x_3 with the principal axes $\mathbf{p}_1, \mathbf{p}_2$, and $\mathbf{p}_3$ superimposed.

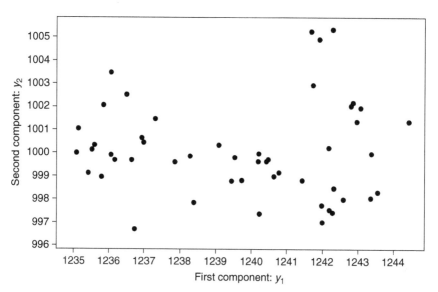

Figure 9.35   Scatter plot of x_1, x_2, and x_3 projected down onto the first two principal axes.

total variability as taking place in a two-dimensional subspace spanned by the two principal vectors $\mathbf{p}_1$ and $\mathbf{p}_2$. It is this ability to identify redundancy, much like a Pareto analysis, that is the key to appreciating the power of PCA.

9.5.1.3   *PCA of the Five Temperatures*   We are now ready to apply PCA to the full five-dimensional data set. Table 9.9 summarizes a typical computer

TABLE 9.9   Typical Information Provided from a PCA

Eigenvalues	15.316	4.841	2.439	1.094	0.490
Proportion variability	0.633	0.200	0.101	0.045	0.020
Cumulative variability	0.633	0.834	0.934	0.980	1.000
Variable:	p_1	p_2	p_3	p_4	p_5
x_1	0.139	−0.878	0.193	0.247	0.334
x_2	−0.102	−0.266	0.535	−0.337	−0.720
x_3	0.650	−0.151	−0.342	−0.661	0.019
x_4	0.136	−0.207	−0.594	0.470	−0.604
x_5	0.728	0.303	0.455	0.409	−0.069

output. The eigenvalues provide information about how much of the variability is associated with the variability along each of the five principal axes. Note that the sum of the eigenvalues equals the total variability in the data set. Thus, the first eigenvalue is 15.316 and accounts for 0.633 or 63.3% of the variability. Similarly, the second eigenvalue is 4.841 and accounts for 0.200 or 20% of the variability. The third eigenvalue is 2.439 and accounts for 10.1% of the variability. Combined, the first three eigenvalues account for 93.4% of all the variability. Thus, the data set is essentially three-dimensional.

Once we have performed a PCA, we need to interpret the results. Unfortunately, this is not always easy. However, we provide a few suggestions of how this can be done. Table 9.9 provides the values of the five eigenvectors $p_i, i = 1, \ldots, 5$. For example, the first eigenvector implies that the first principal component y_1 is given by the linear equation

$$y_1 = p'_1 x = 0.139x_1 - 0.102x_2 + 0.650x_3 + 0.136x_4 + 0.728x_5 \qquad (9.34)$$

It is important to assess how large the coefficients of the eigenvectors $p_i, i = 1, \ldots, 5$ are and hence how large an influence the individual x's have on each of the principal components. To do so, we have plotted the eigenvector weights in Figure 9.36. We see that the first principal component appears to be primarily driven by x_3 and x_5 and that the coefficients for x_3 and x_5 are about the same size. Thus, the first principal component is more or less proportional to the average of x_3 and x_5, $(x_3 + x_5)/2$. Similarly, the second principal component is mostly driven by x_1 (the negative sign is not relevant as we can multiply the vector by -1) and the third principal component is approximately the contrast $(x_2 + x_5)/2 - (x_3 + x_4)/2$. The remaining principal components represent only 6.6% of the variability, and therefore are mostly noise and not all that interesting.

These interpretations can be verified by plotting the principal components against the suspected relations as in Figure 9.37 to see how well they correlate. In the present case, the correlations are high (i.e., $\text{corr}\{y_1, c_1\} = 0.996$, $\text{corr}\{y_2, c_2\} = 0.938$,   and $\text{corr}\{y_3, c_3\} = 0.937$ where $c_1 = (x_3 + x_5)/2, c_2 = -x_1$,   and $c_3 = (x_2 + x_5)/2 - (x_3 + x_4)/2$,   so   we conclude that our interpretations match the principal components well.

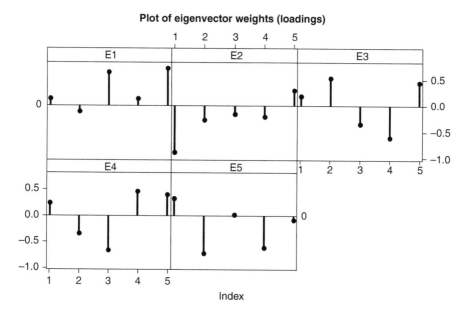

Figure 9.36   A plot of the coordinates of the five eigenvectors.

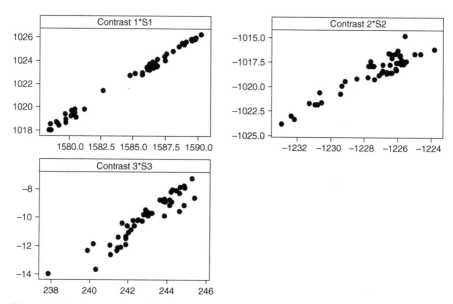

Figure 9.37   Scatter plot of the three first principal components and the simplified interpretations $c_1 = (x_3 + x_5)/2, c_2 = -x_1$, and $c_3 = (x_2 + x_5)/2 - (x_3 + x_4)/2$.

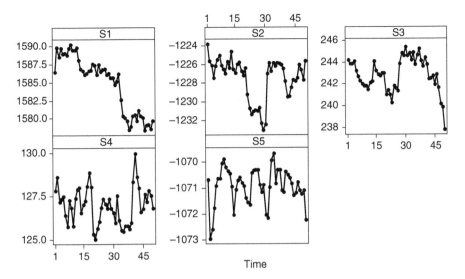

Figure 9.38   Time series plots of the five principal components.

To gain further insight, we plot the principal components as time series plots as in Figure 9.38 and compare with Figure 9.29. We see that x_3 and x_5, the main drivers for the first principal component, show a significant trend, are quite similar in appearance, and look similar to the first principal component in Figure 9.38. Thus, the first principal component, representing most of the variability, represents the downward trend that is most visible and very similar in x_3 and x_5. It is now a question for the process engineer to explain why x_3 and x_5 exhibit a downward trend over time and why these two temperatures are so much more unstable than the other temperatures. One explanation may be related to the control strategy. The control engineers keep x_4 under close control. However, the immediately adjacent temperatures x_3 and x_5, located on either side of x_4, are less well controlled. Since x_3 and x_5 are about the same distance from x_4 they exhibit very similar, and in this case, unstable behavior.

The time series plot of the second principal component looks much like the mirror image of x_1. Thus, the second largest source of variability appears to be the process upset that occurred between observations 20 and 30 and is reflected in several of the other time series but most visible for x_1. Again we need to question the process engineer about what may have caused this sudden increase in x_1. One possible explanation is that this temperature is in close proximity to the input side of the process and hence susceptible to process inhomogeneities in the raw material.

The third principal component is approximately a contrast $c_3 = (x_2 + x_5)/2 - (x_3 + x_4)/2$. From the time series plot of the third principal component, y_3, we see that it is relatively stable until observation 40 but then trends downward. When looking at the time series plot of the original data in Figure 9.29, we see that both x_2 and x_5 also seem to nosedive from observation 40 and onward. However, x_3 and x_4 both trend to go up. Thus, these temperatures trend in this later phase to move in opposite direction. Again this may be due to the engineers' control strategy. They most likely tried to reestablish control after the sudden drop in x_4 between observations 30 and 40 but ended up overshooting. The temperature x_3 followed along but x_2 and x_5 continued to nosedive despite the control action.

In practical applications, we typically have many more dimensions than what we can conveniently visualize with customary plotting tools. In our example, we had five simultaneous temperature readings and hence observations that vary in a five-dimensional space. But that is actually not a very large number. As sensor and computer technology continue to improve, we will increasingly be confronted with ever higher dimensional data sets. This will challenge our ability to study and control the behavior of processes. For such data, PCA is a tool that can help us in reducing the complexity.

9.5.2   Canonical Analysis of Multiple Time Series

Box and Tiao (1977) propose a canonical transformation of multiple time series generated by a stationary autoregressive process. The idea is to obtain k linear combinations of the k time series so that these combinations can be ordered from least predictable to most. The former shows the characteristics of a white noise series, whereas the latter approaches nonstationarity. First, consider the following model for a multivariate stationary (demeaned) process.

$$Z_t = \hat{Z}_{t-1}(1) + \mathbf{a}_t \tag{9.35}$$

where $\hat{Z}_{t-1}(1)$ is the 1-step-ahead prediction of Z_t, $\mathbf{a}_t$ is the vector of independent, normally distributed errors with zero mean and covariance matrix, $\boldsymbol{\Sigma}$, and independent of $\hat{Z}_{t-1}(1)$. We can then write the covariance matrix of $\mathbf{a}_t$ as

$$\boldsymbol{\Gamma}_Z(0) = \boldsymbol{\Gamma}_{\hat{Z}}(0) + \boldsymbol{\Sigma} \tag{9.36}$$

Box and Tiao (1977) then show that the k eigenvalues, $\lambda_1, \ldots, \lambda_k$ of $\boldsymbol{\Gamma}_Z(0)^{-1}\boldsymbol{\Gamma}_{\hat{Z}}(0)$ provide a measure of predictability of the linear combinations constructed using the corresponding k eigenvectors $\mathbf{e}_1, \ldots, \mathbf{e}_k$. If the eigenvalues $\lambda_1, \ldots, \lambda_k$ are ordered in an ascending order, then the linear combination obtained by $\mathbf{e}_1$ gives the least predictable linear combination with nearly white noise behavior with λ_1 approaching 0. On the other hand, the linear combination obtained by $\mathbf{e}_k$ gives the most predictable linear combination with nearly nonstationary behavior as λ_k approaches 1. If we then let

$$\mathbf{M} = \begin{bmatrix} \mathbf{e}_1' \\ \mathbf{e}_2' \\ \vdots \\ \mathbf{e}_k' \end{bmatrix} \tag{9.37}$$

be the $k \times k$ matrix consisting of the eigenvectors in rows, then Equation (9.35) using the transformed data becomes

$$Y_t = \hat{Y}_{t-1}(1) + \mathbf{b}_t \qquad (9.38)$$

where $Y_t = \mathbf{M}Z_t$, $\hat{Y}_{t-1}(1) = \mathbf{M}\hat{Z}_{t-1}(1)$ and $\mathbf{b}_t = \mathbf{M}\mathbf{a}_t$. It can be shown that the covariance matrices of $Y_t, \hat{Y}_{t-1}, \mathbf{b}_t$ are all diagonal.

   Now to better understand the implications of such a transformation, we will consider the "US hog series" example originally used by Quenouille (1957) and reanalyzed by Box and Tiao (1977). Quenouille (1957) provides 82 yearly observations from 1967 to 1948 of five variables; number of hogs ($H_{1,t}$), price of hogs ($H_{2,t}$), price of corn ($H_{3,t}$), supply of corn ($H_{4,t}$), and farm wage rate ($H_{5,t}$). These variables are scaled using log transformation to be comparable. The plots of the scaled observations are given in Figure 9.39. Similar to Quenouille (1957), Box and Tiao (1977) also proceed with a multivariate AR(1) model. To improve the fit, Box and Tiao suggest moving the variable number of hogs and farming wage rate backward by one period. The data matrix becomes $Z_t^* = [H_{1,t} \quad H_{2,t+1} \quad H_{3,t} \quad H_{4,t} \quad H_{5,t+1}]$ and is provided in Appendix A. We also demeaned the data using $\hat{\mu} = [698.9 \quad 894.9 \quad 771.4 \quad 1328.1 \quad 995.6]$. Then $Z_t = Z_t^* - \hat{\mu}$ is modeled as

$$Z_t = \mathbf{\Phi} Z_{t-1} + \mathbf{a}_t \qquad (9.39)$$

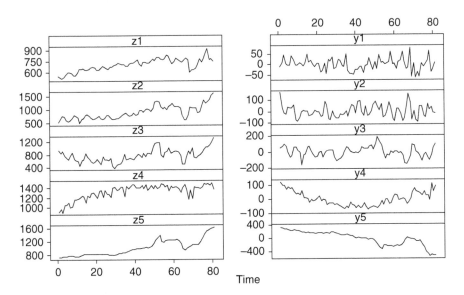

Figure 9.39   Time series plots of the original and transformed data.

The estimated parameters, together with the estimated covariance matrix for the data and the residuals calculated using R, are

$$\hat{\Phi} = \begin{bmatrix} 0.372 & -0.05 & 0.008 & 0.249 & 0.102 \\ -0.107 & 0.153 & 0.504 & 0.264 & 0.47 \\ -0.931 & -0.534 & 0.915 & 0.571 & 0.622 \\ 0.847 & 0.305 & -0.073 & 0.377 & -0.265 \\ -0.249 & -0.169 & 0.105 & 0.184 & 1.08 \end{bmatrix}$$

$$\hat{\Gamma}_z(0) = \begin{bmatrix} 6917.2 & 12680 & 6617.1 & 9652.4 & 15414 \\ 12680 & 62714 & 38318 & 20461 & 56405 \\ 6617.1 & 38318 & 37338 & 2665.5 & 35180 \\ 9652.4 & 20461 & 2665.5 & 21675 & 22199 \\ 15414 & 56405 & 35180 & 22199 & 57921 \end{bmatrix}$$

$$\hat{\Sigma} = \begin{bmatrix} 951.3 & -1038 & 67.3 & 285.6 & 238.1 \\ -1038 & 6354.6 & 3686.6 & -219.9 & 1942.9 \\ 67.3 & 3686.6 & 10906 & -5157 & 2375.7 \\ 285.6 & -219.9 & -5157 & 5690.6 & 142.9 \\ 238.1 & 1942.9 & 2375.7 & 142.9 & 1896.4 \end{bmatrix}$$

From these estimates, we calculate the following eigenvalues and eigenvectors of $\Gamma_Z(0)^{-1}\Gamma_{\hat{Z}}(0)$

$$\hat{\lambda} = \begin{bmatrix} 0.03 \\ 0.28 \\ 0.53 \\ 0.78 \\ 0.97 \end{bmatrix} \quad \text{and} \quad \hat{M} = \begin{bmatrix} 0.80 & 0.28 & -0.17 & -0.47 & -0.20 \\ -0.16 & -0.68 & 0.57 & 0.35 & 0.24 \\ -0.30 & 0.28 & 0.53 & 0.37 & -0.64 \\ -0.73 & -0.21 & -0.13 & -0.26 & 0.59 \\ -0.47 & -0.17 & 0.24 & 0.15 & -0.82 \end{bmatrix}$$

Note that each row of the $\hat{M}$ matrix is the eigenvector corresponding to the eigenvalue at the same row in $\hat{\lambda}$. The time series plots of the original and the transformed data as in $Y_t = MZ_t$ are given in Figure 9.39. The focus should be on the two extreme eigenvalues: the smallest and the largest. The smallest eigenvalue is very close to zero (0.03), indicating that the linear combination obtained using the first eigenvector (the first row of $\hat{M}$) is nearly white noise. As evident from the time series plot in Figure 9.39 and its ACF and PACF plots in Figure 9.40, the linear combination obtained using the first eigenvector, (the first row of $\hat{M}$), $y_{1,t}$ does not show any significant autocorrelation. This is quite striking considering, as noted by Box and Tiao (1977) as well, that it can be argued that almost all five original series exhibit some nonstationary behavior.

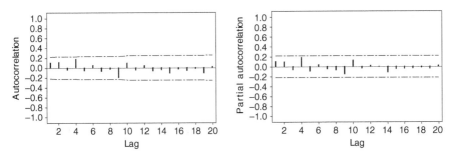

Figure 9.40   ACF and PACF of $y_{1,t}$ ($\hat{\lambda}_1 = 0.03$).

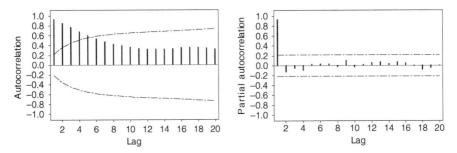

Figure 9.41   ACF and PACF of $y_{5,t}$ ($\hat{\lambda}_5 = 0.97$).

Yet a simple linear combination of these seemingly nonstationary series behaves quite stationary. This is quite closely related to the co-integration concept we discussed in Section 9.4. This shows that while all series may be nonstationary, there may in fact be a stationary contemporaneous relationship among them. On the other hand, the last transformed variable $y_{5,t}$ shows nearly nonstationary behavior. This was expected as the corresponding eigenvalue is very close to 1 (0.97). The slowly decaying ACF and the only significant first lag estimate for PACF being very close to 1 given in Figure 9.41 are also testimonials for nearly nonstationary behavior of $y_{5,t}$. Further practical interpretation of these results as well as discussion on other issues such as the singularity of the covariance matrices can be found in Box and Tiao (1977). It should be noted that our results match in general with the analysis reported by Box and Tiao except for the last eigenvalue, which was $\hat{\lambda}_e^{BT} = 0.8868$. This value is still pretty close to 1, suggesting that the corresponding linear combination should still be nearly nonstationary.

The analyses provided in this section show that an appropriate transformation of the data based on the eigenvalues and eigenvectors of $\mathbf{\Gamma}_Z(0)$ as in PCA or of $\mathbf{\Gamma}_Z(0)^{-1}\mathbf{\Gamma}_{\hat{Z}}(0)$ as in the canonical analysis can help us understand the structure of the data and its underlying dynamics. We therefore strongly recommend the use of such techniques for multivariate time series data in conjunction with other modeling efforts as provided in Section 9.4.

EXERCISES

9.1 Fit a multiple time series model to the data in Exercise 8.2 and compare the model you obtain to the transfer function–noise model from Exercise 8.2.

9.2 Fit a multiple time series model to the data in Exercise 8.4 and compare the model you obtain to the transfer function–noise model from Exercise 8.4.

9.3 Fit a multiple time series model to the temperature data in Section 9.5.1 given in Appendix A.

9.4 Fit a multiple time series model to the "US Hog series" data given in Appendix A.

9.5 Table B.16 contains the monthly crude oil production of OPEC nations from January 1990 to July 2010. Perform a PCA on this data and comment on your results.

DATASETS USED IN THE EXAMPLES

Time Series Analysis and Forecasting by Example, First Edition. Søren Bisgaard and Murat Kulahci.
© 2011 John Wiley & Sons, Inc. Published 2011 by John Wiley & Sons, Inc.

311

TABLE A.1 Temperature Readings from a Ceramic Furnace

1578.71	1579.63
1578.79	1578.82
1579.38	1578.59
1579.36	1578.56
1579.83	1579.56
1580.13	1579.46
1578.95	1579.59
1579.18	1579.66
1579.52	1579.89
1579.72	1580.03
1580.11	1579.76
1580.41	1579.84
1580.77	1580.41
1580.05	1580.3
1579.53	1580.17
1579.00	1579.81
1579.12	1579.71
1579.13	1579.77
1579.39	1580.16
1579.73	1580.38
1580.12	1580.18
1580.23	1579.59
1580.25	1580.06
1579.8	1581.21
1579.72	1580.89
1579.49	1580.82
1579.22	1580.48
1579.03	1579.97
1579.76	1579.64
1580.19	1580.42
1580.17	1580.06
1580.22	1580.12
1580.44	1579.92
1580.71	1579.57
1579.91	1579.56
1579.48	1579.4
1579.82	1578.9
1580.34	1578.5
1580.56	1579.3
1580.05	1579.93

80 observations (read down).

TABLE A.2 Chemical Process Temperature Readings

26.6	19.7	24.0	21.7	23.4
27.0	19.9	23.9	21.8	23.3
27.1	20.0	23.7	21.9	23.3
27.1	20.1	23.6	22.2	23.3
27.1	20.2	23.5	22.5	23.4
27.1	20.3	23.5	22.8	23.4
26.9	20.6	23.5	23.1	23.3
26.8	21.6	23.5	23.4	23.2
26.7	21.9	23.5	23.8	23.3
26.4	21.7	23.7	24.1	23.3
26.0	21.3	23.8	24.6	23.2
25.8	21.2	23.8	24.9	23.1
25.6	21.4	23.9	24.9	22.9
25.2	21.7	23.9	25.1	22.8
25.0	22.2	23.8	25.0	22.6
24.6	23.0	23.7	25.0	22.4
24.2	23.8	23.6	25.0	22.2
24.0	24.6	23.4	25.0	21.8
23.7	25.1	23.2	24.9	21.3
23.4	25.6	23.0	24.8	20.8
23.1	25.8	22.8	24.7	20.2
22.9	26.1	22.6	24.6	19.7
22.8	26.3	22.4	24.5	19.3
22.7	26.3	22.0	24.5	19.1
22.6	26.2	21.6	24.5	19.0
22.4	26.0	21.3	24.5	18.8
22.2	25.8	21.2	24.5	—
22.0	25.6	21.2	24.5	—
21.8	25.4	21.1	24.5	—
21.4	25.2	21.0	24.4	—
20.9	24.9	20.9	24.4	—
20.3	24.7	21.0	24.2	—
19.7	24.5	21.0	24.2	—
19.4	24.4	21.1	24.1	—
19.3	24.4	21.2	24.1	—
19.2	24.4	21.1	24.0	—
19.1	24.4	20.9	24.0	—
19.0	24.4	20.8	24.0	—
18.9	24.3	20.8	23.9	—
18.9	24.4	20.8	23.8	—
19.2	24.4	20.8	23.8	—
19.3	24.4	20.9	23.7	—
19.3	24.4	20.8	23.7	—
19.4	24.4	20.8	23.6	—
19.5	24.5	20.7	23.7	—
19.6	24.5	20.7	23.6	—
19.6	24.4	20.8	23.6	—
19.6	24.3	20.9	23.6	—
19.6	24.2	21.2	23.5	—
19.6	24.2	21.4	23.5	—

Source: BJR series C (read down).

TABLE A.3 Chemical Process Concentration Readings

17.0	17.6	16.8	16.9	17.1
16.6	17.5	16.7	17.1	17.1
16.3	16.5	16.4	16.8	17.1
16.1	17.8	16.5	17.0	17.4
17.1	17.3	16.4	17.2	17.2
16.9	17.3	16.6	17.3	16.9
16.8	17.1	16.5	17.2	16.9
17.4	17.4	16.7	17.3	17.0
17.1	16.9	16.4	17.2	16.7
17.0	17.3	16.4	17.2	16.9
16.7	17.6	16.2	17.5	17.3
17.4	16.9	16.4	16.9	17.8
17.2	16.7	16.3	16.9	17.8
17.4	16.8	16.4	16.9	17.6
17.4	16.8	17.0	17.0	17.5
17.0	17.2	16.9	16.5	17.0
17.3	16.8	17.1	16.7	16.9
17.2	17.6	17.1	16.8	17.1
17.4	17.2	16.7	16.7	17.2
16.8	16.6	16.9	16.7	17.4
17.1	17.1	16.5	16.6	17.5
17.4	16.9	17.2	16.5	17.9
17.4	16.6	16.4	17.0	17.0
17.5	18.0	17.0	16.7	17.0
17.4	17.2	17.0	16.7	17.0
17.6	17.3	16.7	16.9	17.2
17.4	17.0	16.2	17.4	17.3
17.3	16.9	16.6	17.1	17.4
17.0	17.3	16.9	17.0	17.4
17.8	16.8	16.5	16.8	17.0
17.5	17.3	16.6	17.2	18.0
18.1	17.4	16.6	17.2	18.2
17.5	17.7	17.0	17.4	17.6
17.4	16.8	17.1	17.2	17.8
17.4	16.9	17.1	16.9	17.7
17.1	17.0	16.7	16.8	17.2
17.6	16.9	16.8	17.0	17.4
17.7	17.0	16.3	17.4	—
17.4	16.6	16.6	17.2	—
17.8	16.7	16.8	17.2	—

Source: BJR series A (read down).

TABLE A.4 International Airline Passengers

Year	January	February	March	April	May	June	July	August	September	October	November	December
1949	112	118	132	129	121	135	148	148	136	119	104	118
1950	115	126	141	135	125	149	170	170	158	133	114	140
1951	145	150	178	163	172	178	199	199	184	162	146	166
1952	171	180	193	181	183	218	230	242	209	191	172	194
1953	196	196	236	235	229	243	264	272	237	211	180	201
1954	204	188	235	227	234	264	302	293	259	229	203	229
1955	242	233	267	269	270	315	364	347	312	274	237	278
1956	284	277	317	313	318	374	413	405	355	306	271	306
1957	315	301	356	348	355	422	465	467	404	347	305	336
1958	340	318	362	348	363	435	491	505	404	359	310	337
1959	360	342	406	396	420	472	548	559	463	407	362	405
1960	417	391	419	461	472	535	622	606	508	461	390	432

Source: BJR series G.

TABLE A.5 Company X's Sales Data

Year	Month	Sales	Year	Month	Sales	Year	Month	Sales
1965	January	154	1968	January	346	1971	January	628
1965	February	96	1968	February	261	1971	February	308
1965	March	73	1968	March	224	1971	March	324
1965	April	49	1968	April	141	1971	April	248
1965	May	36	1968	May	148	1971	May	272
1965	June	59	1968	June	145			
1965	July	95	1968	July	223			
1965	August	169	1968	August	272			
1965	September	210	1968	September	445			
1965	October	278	1968	October	560			
1965	November	298	1968	November	612			
1965	December	245	1968	December	467			
1966	January	200	1969	January	518			
1966	February	118	1969	February	404			
1966	March	90	1969	March	300			
1966	April	79	1969	April	210			
1966	May	78	1969	May	196			
1966	June	91	1969	June	186			
1966	July	167	1969	July	247			
1966	August	169	1969	August	343			
1966	September	289	1969	September	464			
1966	October	347	1969	October	680			
1966	November	375	1969	November	711			
1966	December	203	1969	December	610			
1967	January	223	1970	January	613			
1967	February	104	1970	February	392			
1967	March	107	1970	March	273			
1967	April	85	1970	April	322			
1967	May	75	1970	May	189			
1967	June	99	1970	June	257			
1967	July	135	1970	July	324			
1967	August	211	1970	August	404			
1967	September	335	1970	September	677			
1967	October	460	1970	October	858			
1967	November	488	1970	November	895			
1967	December	326	1970	December	664			

Source: Chatfield and Prothero (1973).

TABLE A.6   Internet Users Data

88	91	147	140	142	172	112	89	121	193
84	99	149	134	150	172	104	88	135	204
85	104	143	131	159	174	102	85	145	208
85	112	132	131	167	174	99	86	149	210
84	126	131	129	170	169	99	89	156	215
85	138	139	126	171	165	95	91	165	222
83	146	147	126	172	156	88	91	171	228
85	151	150	132	172	142	84	94	175	226
88	150	148	137	174	131	84	101	177	222
89	148	145	140	175	121	87	110	182	220

Source: Makridakis *et al*.(2003).

TABLE A.7   Historical Sea Level (mm) Data in Copenhagen, Denmark

Year	Sea Level	Year	Sea Level	Year	Sea Level	Year	Sea Level
1889	6939	1919	6936	1949	7035	1979	6967
1890	6977	1920	6943	1950	6988	1980	6968
1891	6933	1921	7022	1951	6911	1981	7061
1892	6949	1922	6992	1952	6999	1982	7012
1893	6986	1923	7006	1953	7001	1983	—
1894	6950	1924	6958	1954	6949	1984	6966
1895	6966	1925	7027	1955	7011	1985	6994
1896	6952	1926	6991	1956	6996	1986	6997
1897	6952	1927	7008	1957	7009	1987	6962
1898	7009	1928	6980	1958	6976	1988	7014
1899	7051	1929	6973	1959	6948	1989	7072
1900	6947	1930	6956	1960	6912	1990	7060
1901	6927	1931	6992	1961	7043	1991	—
1902	6950	1932	6999	1962	7004	1992	7066
1903	7021	1933	6946	1963	6930	1993	6976
1904	6943	1934	6979	1964	6972	1994	6994
1905	6976	1935	6997	1965	6972	1995	7033
1906	6982	1936	6970	1966	6981	1996	6926
1907	6960	1937	6924	1967	7065	1997	7019
1908	6921	1938	7024	1968	6973	1998	7057
1909	6964	1939	6924	1969	6955	1999	—
1910	6962	1940	6940	1970	6981	2000	7038
1911	6993	1941	6921	1971	7000	2001	—
1912	7000	1942	6951	1972	6921	2002	7056
1913	7030	1943	7028	1973	7003	2003	7006
1914	7011	1944	6994	1974	6960	2004	7092
1915	6945	1945	7027	1975	6992	2005	7043
1916	6994	1946	6999	1976	6954	2006	7002
1917	6956	1947	6913	1977	6975	—	—
1918	6951	1948	7016	1978	6995	—	—

Source: www.psmsl.org.

TABLE A.8   Gas Furnace Data

X_t	Y_t	X_t	Y_t	X_t	Y_t	X_t	Y_t	X_t	Y_t	X_t	Y_t
-0.109	53.8	1.608	46.9	-0.288	51.0	-0.049	53.2	-2.473	55.6	0.185	56.3
0.000	53.6	1.905	47.8	-0.153	51.8	0.060	53.9	-2.330	58.0	0.662	56.4
0.178	53.5	2.023	48.2	-0.109	52.4	0.161	54.1	-2.053	59.5	0.709	56.4
0.339	53.5	1.815	48.3	-0.187	53.0	0.301	54.0	-1.739	60.0	0.605	56.0
0.373	53.4	0.535	47.9	-0.255	53.4	0.517	53.6	-1.261	60.4	0.501	55.2
0.441	53.1	0.122	47.2	-0.229	53.6	0.566	53.2	-0.569	60.5	0.603	54.0
0.461	52.7	0.009	47.2	-0.007	53.7	0.560	53.0	-0.137	60.2	0.943	53.0
0.348	52.4	0.164	48.1	0.254	53.8	0.573	52.8	-0.024	59.7	1.223	52.0
0.127	52.2	0.671	49.4	0.330	53.8	0.592	52.3	-0.050	59.0	1.249	51.6
-0.180	52.0	1.019	50.6	0.102	53.8	0.671	51.9	-0.135	57.6	0.824	51.6
-0.588	52.0	1.146	51.5	-0.423	53.3	0.933	51.6	-0.276	56.4	0.102	51.1
-1.055	52.4	1.155	51.6	-1.139	53.0	1.337	51.6	-0.534	55.2	0.025	50.4
-1.421	53.0	1.112	51.2	-2.275	52.9	1.460	51.4	-0.871	54.5	0.382	50.0
-1.520	54.0	1.121	50.5	-2.594	53.4	1.353	51.2	-1.243	54.1	0.922	50.0
-1.302	54.9	1.223	50.1	-2.716	54.6	0.772	50.7	-1.439	54.1	1.032	52.0
-0.814	56.0	1.257	49.8	-2.510	56.4	0.218	50.0	-1.422	54.4	0.866	54.0
-0.475	56.8	1.157	49.6	-1.790	58.0	-0.237	49.4	-1.175	55.5	0.527	55.1
-0.193	56.8	0.913	49.4	-1.346	59.4	-0.714	49.3	-0.813	56.2	0.093	54.5
0.088	56.4	0.620	49.3	-1.081	60.2	-1.099	49.7	-0.634	57.0	-0.458	52.8
0.435	55.7	0.255	49.2	-0.910	60.0	-1.269	50.6	-0.582	57.3	-0.748	51.4
0.771	55.0	-0.280	49.3	-0.876	59.4	-1.175	51.8	-0.625	57.4	-0.947	50.8
0.866	54.3	-1.080	49.7	-0.885	58.4	-0.676	53.0	-0.713	57.0	-1.029	51.2
0.875	53.2	-1.551	50.3	-0.800	57.6	0.033	54.0	-0.848	56.4	-0.928	52.0
0.891	52.3	-1.799	51.3	-0.544	56.9	0.556	55.3	-1.039	55.9	-0.645	52.8
0.987	51.6	-1.825	52.8	-0.416	56.4	0.643	55.9	-1.346	55.5	-0.424	53.8
1.263	51.2	-1.456	54.4	-0.271	56.0	0.484	55.9	-1.628	55.3	-0.276	54.5
1.775	50.8	-0.944	56.0	0.000	55.7	0.109	54.6	-1.619	55.2	-0.158	54.9
1.976	50.5	-0.570	56.9	0.403	55.3	-0.310	53.5	-1.149	55.4	-0.033	54.9
1.934	50.0	-0.431	57.5	0.841	55.0	-0.697	52.4	-0.488	56.0	0.102	54.8
1.866	49.2	-0.577	57.3	1.285	54.4	-1.047	52.1	-0.160	56.5	0.251	54.4
1.832	48.4	-0.960	56.6	1.607	53.7	-1.218	52.3	-0.007	57.1	0.280	53.7
1.767	47.9	-1.616	56.0	1.746	52.8	-1.183	53.0	-0.092	57.3	0.000	53.3
1.608	47.6	-1.875	55.4	1.683	51.6	-0.873	53.8	-0.620	56.8	-0.493	52.8
1.265	47.5	-1.891	55.4	1.485	50.6	-0.336	54.6	-1.086	55.6	-0.759	52.6
0.790	47.5	-1.746	56.4	0.993	49.4	0.063	55.4	-1.525	55.0	-0.824	52.6
0.360	47.6	-1.474	57.2	0.648	48.8	0.084	55.9	-1.858	54.1	-0.740	53.0
0.115	48.1	-1.201	58.0	0.577	48.5	0.000	55.9	-2.029	54.3	-0.528	54.3
0.088	49.0	-0.927	58.4	0.577	48.7	0.001	55.2	-2.024	55.3	-0.204	56.0
0.331	50.0	-0.524	58.4	0.632	49.2	0.209	54.4	-1.961	56.4	0.034	57.0
0.645	51.1	0.040	58.1	0.747	49.8	0.556	53.7	-1.952	57.2	0.204	58.0
0.960	51.8	0.788	57.7	0.900	50.4	0.782	53.6	-1.794	57.8	0.253	58.6
1.409	51.9	0.943	57.0	0.993	50.7	0.858	53.6	-1.302	58.3	0.195	58.5
2.670	51.7	0.930	56.0	0.968	50.9	0.918	53.2	-1.030	58.6	0.131	58.3
2.834	51.2	1.006	54.7	0.790	50.7	0.862	52.5	-0.918	58.8	0.017	57.8
2.812	50.0	1.137	53.2	0.399	50.5	0.416	52.0	-0.798	58.8	-0.182	57.3
2.483	48.3	1.198	52.1	-0.161	50.4	-0.336	51.4	-0.867	58.6	-0.262	57.0
1.929	47.0	1.054	51.6	-0.553	50.2	-0.959	51.0	-1.047	58.0	—	—
1.485	45.8	0.595	51.0	-0.603	50.4	-1.813	50.9	-1.123	57.4	—	—
1.214	45.6	-0.080	50.5	-0.424	51.2	-2.378	52.4	-0.876	57.0	—	—
1.239	46.0	-0.314	50.4	-0.194	52.3	-2.499	53.5	-0.395	56.4	—	—

Source: BJR series J.

TABLE A.9   Sales with Leading Indicator

Sales	Leading Indicator	Sales	Leading Indicator	Sales	Leading Indicator
200.1	10.01	220.0	10.77	249.4	12.90
199.5	10.07	218.7	10.88	249.0	13.12
199.4	10.32	217.0	10.49	249.9	12.47
198.9	9.75	215.9	10.50	250.5	12.47
199.0	10.33	215.8	11.00	251.5	12.94
200.2	10.13	214.1	10.98	249.0	13.10
198.6	10.36	212.3	10.61	247.6	12.91
200.0	10.32	213.9	10.48	248.8	13.39
200.3	10.13	214.6	10.53	250.4	13.13
201.2	10.16	213.6	11.07	250.7	13.34
201.6	10.58	212.1	10.61	253.0	13.34
201.5	10.62	211.4	10.86	253.7	13.14
201.5	10.86	213.1	10.34	255.0	13.49
203.5	11.20	212.9	10.78	256.2	13.87
204.9	10.74	213.3	10.80	256.0	13.39
207.1	10.56	211.5	10.33	257.4	13.59
210.5	10.48	212.3	10.44	260.4	13.27
210.5	10.77	213.0	10.50	260.0	13.70
209.8	11.33	211.0	10.75	261.3	13.20
208.8	10.96	210.7	10.40	260.4	13.32
209.5	11.16	210.1	10.40	261.6	13.15
213.2	11.70	211.4	10.34	260.8	13.30
213.7	11.39	210.0	10.55	259.8	12.94
215.1	11.42	209.7	10.46	259.0	13.29
218.7	11.94	208.8	10.82	258.9	13.26
219.8	11.24	208.8	10.91	257.4	13.08
220.5	11.59	208.8	10.87	257.7	13.24
223.8	10.96	210.6	10.67	257.9	13.31
222.8	11.40	211.9	11.11	257.4	13.52
223.8	11.02	212.8	10.88	257.3	13.02
221.7	11.01	212.5	11.28	257.6	13.25
222.3	11.23	214.8	11.27	258.9	13.12
220.8	11.33	215.3	11.44	257.8	13.26
219.4	10.83	217.5	11.52	257.7	13.11
220.1	10.84	218.8	12.10	257.2	13.30
220.6	11.14	220.7	11.83	257.5	13.06
218.9	10.38	222.2	12.62	256.8	13.32
217.8	10.90	226.7	12.41	257.5	13.10
217.7	11.05	228.4	12.43	257.0	13.27
215.0	11.11	233.2	12.73	257.6	13.64
215.3	11.01	235.7	13.01	257.3	13.58
215.9	11.22	237.1	12.74	257.5	13.87
216.7	11.21	240.6	12.73	259.6	13.53
216.7	11.91	243.8	12.76	261.1	13.41
217.7	11.69	245.3	12.92	262.9	13.25
218.7	10.93	246.0	12.64	263.3	13.50
222.9	10.99	246.3	12.79	262.8	13.58
224.9	11.01	247.7	13.05	261.8	13.51
222.2	10.84	247.6	12.69	262.2	13.77
220.7	10.76	247.8	13.01	262.7	13.40

Source: BJR series M (read down).

TABLE A.10 Crest/Colgate Market Share

Year	Week	Crest	Colgate	Year	Week	Crest	Colgate	Year	Week	Crest	Colgate
1958	1	0.108	0.424	1958	43	0.165	0.235	1959	33	0.221	0.325
1958	2	0.166	0.482	1958	44	0.160	0.250	1959	34	0.183	0.337
1958	3	0.126	0.428	1958	45	0.075	0.360	1959	35	0.136	0.338
1958	4	0.115	0.397	1958	46	0.118	0.282	1959	36	0.206	0.323
1958	5	0.119	0.352	1958	47	0.100	0.257	1959	37	0.127	0.357
1958	6	0.176	0.342	1958	48	0.102	0.345	1959	38	0.139	0.381
1958	7	0.155	0.434	1958	49	0.131	0.344	1959	39	0.189	0.371
1958	8	0.118	0.445	1958	50	0.148	0.358	1959	40	0.194	0.294
1958	9	0.136	0.428	1958	51	0.137	0.322	1959	41	0.114	0.384
1958	10	0.137	0.395	1958	52	0.090	0.332	1959	42	0.229	0.286
1958	11	0.124	0.354	1959	1	0.088	0.315	1959	43	0.148	0.335
1958	12	0.131	0.497	1959	2	0.172	0.316	1959	44	0.155	0.310
1958	13	0.120	0.425	1959	3	0.111	0.341	1959	45	0.106	0.304
1958	14	0.133	0.401	1959	4	0.097	0.387	1959	46	0.156	0.305
1958	15	0.067	0.363	1959	5	0.098	0.402	1959	47	0.053	0.403
1958	16	0.086	0.341	1959	6	0.090	0.347	1959	48	0.112	0.365
1958	17	0.140	0.464	1959	7	0.127	0.414	1959	49	0.084	0.305
1958	18	0.122	0.431	1959	8	0.116	0.426	1959	50	0.191	0.172
1958	19	0.105	0.405	1959	9	0.137	0.322	1959	51	0.149	0.321
1958	20	0.079	0.460	1959	10	0.111	0.372	1959	52	0.143	0.343
1958	21	0.130	0.410	1959	11	0.107	0.381	1960	1	0.094	0.354
1958	22	0.142	0.423	1959	12	0.097	0.339	1960	2	0.184	0.316
1958	23	0.120	0.310	1959	13	0.134	0.405	1960	3	0.205	0.292
1958	24	0.115	0.413	1959	14	0.160	0.304	1960	4	0.206	0.305
1958	25	0.103	0.411	1959	15	0.147	0.439	1960	5	0.191	0.294
1958	26	0.078	0.452	1959	16	0.104	0.336	1960	6	0.195	0.289
1958	27	0.093	0.405	1959	17	0.128	0.405	1960	7	0.179	0.301
1958	28	0.086	0.290	1959	18	0.128	0.359	1960	8	0.272	0.304
1958	29	0.099	0.342	1959	19	0.165	0.379	1960	9	0.203	0.306
1958	30	0.078	0.311	1959	20	0.184	0.303	1960	10	0.165	0.405
1958	31	0.095	0.327	1959	21	0.172	0.340	1960	11	0.138	0.344
1958	32	0.094	0.413	1959	22	0.207	0.312	1960	12	0.216	0.353
1958	33	0.056	0.400	1959	23	0.221	0.291	1960	13	0.132	0.383
1958	34	0.050	0.380	1959	24	0.159	0.259	1960	14	0.120	0.349
1958	35	0.065	0.371	1959	25	0.198	0.342	1960	15	0.083	0.374
1958	36	0.091	0.344	1959	26	0.197	0.458	1960	16	0.118	0.411
1958	37	0.094	0.345	1959	27	0.251	0.275	1960	17	0.125	0.287
1958	38	0.124	0.363	1959	28	0.146	0.340	1960	18	0.109	0.420
1958	39	0.153	0.392	1959	29	0.133	0.385	1960	19	0.119	0.470
1958	40	0.078	0.379	1959	30	0.243	0.338	1960	20	0.154	0.354
1958	41	0.114	0.349	1959	31	0.192	0.370	1960	21	0.122	0.392
1958	42	0.088	0.337	1959	32	0.150	0.290	1960	22	0.126	0.421

TABLE A.10   (*Continued*)

Year	Week	Crest	Colgate	Year	Week	Crest	Colgate	Year	Week	Crest	Colgate
1960	23	0.126	0.435	1961	13	0.373	0.260	1962	3	0.432	0.155
1960	24	0.130	0.424	1961	14	0.265	0.298	1962	4	0.450	0.247
1960	25	0.158	0.344	1961	15	0.316	0.248	1962	5	0.530	0.201
1960	26	0.141	0.369	1961	16	0.245	0.308	1962	6	0.431	0.266
1960	27	0.145	0.364	1961	17	0.328	0.356	1962	7	0.420	0.290
1960	28	0.127	0.386	1961	18	0.368	0.278	1962	8	0.411	0.231
1960	29	0.171	0.406	1961	19	0.287	0.314	1962	9	0.423	0.255
1960	30	0.152	0.439	1961	20	0.369	0.214	1962	10	0.433	0.242
1960	31	0.211	0.345	1961	21	0.406	0.253	1962	11	0.393	0.271
1960	32	0.309	0.291	1961	22	0.316	0.287	1962	12	0.389	0.266
1960	33	0.242	0.292	1961	23	0.362	0.238	1962	13	0.387	0.244
1960	34	0.380	0.249	1961	24	0.308	0.253	1962	14	0.439	0.204
1960	35	0.362	0.283	1961	25	0.286	0.336	1962	15	0.421	0.213
1960	36	0.328	0.301	1961	26	0.420	0.255	1962	16	0.363	0.295
1960	37	0.359	0.280	1961	27	0.299	0.249	1962	17	0.401	0.254
1960	38	0.352	0.251	1961	28	0.383	0.195	1962	18	0.394	0.242
1960	39	0.322	0.303	1961	29	0.354	0.269	1962	19	0.459	0.228
1960	40	0.333	0.274	1961	30	0.418	0.201	1962	20	0.441	0.181
1960	41	0.365	0.328	1961	31	0.425	0.184	1962	21	0.388	0.264
1960	42	0.367	0.244	1961	32	0.445	0.203	1962	22	0.373	0.277
1960	43	0.305	0.323	1961	33	0.408	0.193	1962	23	0.385	0.284
1960	44	0.298	0.288	1961	34	0.282	0.322	1962	24	0.314	0.248
1960	45	0.307	0.293	1961	35	0.410	0.261	1962	25	0.347	0.280
1960	46	0.318	0.321	1961	36	0.425	0.183	1962	26	0.408	0.249
1960	47	0.280	0.330	1961	37	0.358	0.289	1962	27	0.341	0.279
1960	48	0.298	0.273	1961	38	0.393	0.243	1962	28	0.361	0.282
1960	49	0.336	0.304	1961	39	0.375	0.302	1962	29	0.414	0.267
1960	50	0.339	0.292	1961	40	0.273	0.350	1962	30	0.380	0.252
1960	51	0.344	0.251	1961	41	0.237	0.401	1962	31	0.274	0.190
1960	52	0.310	0.350	1961	42	0.331	0.332	1962	32	0.352	0.284
1961	1	0.317	0.302	1961	43	0.335	0.351	1962	33	0.439	0.207
1961	2	0.369	0.306	1961	44	0.395	0.280	1962	34	0.355	0.327
1961	3	0.320	0.272	1961	45	0.357	0.308	1962	35	0.435	0.259
1961	4	0.290	0.296	1961	46	0.296	0.299	1962	36	0.408	0.286
1961	5	0.361	0.265	1961	47	0.307	0.199	1962	37	0.383	0.275
1961	6	0.235	0.364	1961	48	0.390	0.283	1962	38	0.357	0.244
1961	7	0.320	0.284	1961	49	0.298	0.333	1962	39	0.374	0.341
1961	8	0.337	0.330	1961	50	0.381	0.233	1962	40	0.366	0.331
1961	9	0.289	0.351	1961	51	0.354	0.296	1962	41	0.346	0.250
1961	10	0.339	0.336	1961	52	0.436	0.267	1962	42	0.381	0.220
1961	11	0.187	0.383	1962	1	0.357	0.253	1962	43	0.329	0.293
1961	12	0.414	0.214	1962	2	0.427	0.239	1962	44	0.474	0.205

(*continued overleaf*)

TABLE A.10   (Continued)

Year	Week	Crest	Colgate	Year	Week	Crest	Colgate	Year	Week	Crest	Colgate
1962	45	0.397	0.254	1963	1	0.396	0.230	1963	9	0.466	0.194
1962	46	0.436	0.268	1963	2	0.420	0.220	1963	10	0.478	0.212
1962	47	0.417	0.211	1963	3	0.432	0.235	1963	11	0.365	0.218
1962	48	0.430	0.203	1963	4	0.453	0.228	1963	12	0.472	0.216
1962	49	0.388	0.271	1963	5	0.430	0.216	1963	13	0.399	0.276
1962	50	0.453	0.290	1963	6	0.327	0.324	1963	14	0.391	0.190
1962	51	0.316	0.323	1963	7	0.388	0.268	1963	15	0.473	0.249
1962	52	0.414	0.253	1963	8	0.377	0.257	1963	16	0.384	0.172

Source: Wichern and Jones (1977).

TABLE A.11   Simulated Process Data

−0.38	−1.82	−2.52	−3.94	−5.10	−4.80	−3.64	−2.14
−0.86	0.53	1.77	2.65	1.50	−1.54	−2.71	−3.70
−4.39	−5.11	−5.30	−4.24	−3.07	−3.97	−3.45	−3.97
−5.43	−4.77	−4.85	−6.76	−6.57	−5.07	−3.79	−2.55
−3.43	−4.61	−3.27	−3.01	−3.58	−2.65	−2.09	−1.56
−1.14	−2.53	−3.43	−2.45	−0.44	−0.67	−0.98	0.17
0.15	0.32	0.45	0.33	0.21	0.48	1.83	4.38
6.91	7.23	6.54	5.12	5.00	2.88	1.29	2.53
2.42	2.55	4.13	4.98	4.86	4.47	2.73	2.25
3.71	3.17	1.15	0.15	−0.82	−1.05	0.61	2.21
4.08	6.97	7.67	8.26	9.77	10.55	10.80	10.03
9.19	9.38	9.62	8.36	6.63	6.34	5.74	6.39
6.78	6.73	7.12	7.32	—	—	—	—

The simulated x_t data recorded row-wise.

−0.94	−0.89	−0.91	−0.97	−1.68	−2.40	−1.93	−1.45
0.13	0.87	0.00	2.72	6.31	6.83	4.90	2.38
1.75	1.71	3.04	4.30	3.37	4.37	5.61	6.22
6.74	6.38	7.09	6.47	5.37	6.19	6.83	7.40
7.64	7.60	8.15	9.50	12.16	14.48	14.44	13.93
14.39	14.36	13.96	13.59	13.76	13.29	11.96	11.72
10.99	9.72	7.74	7.87	8.09	7.38	8.11	7.42
8.03	10.02	11.71	11.71	9.64	6.93	5.14	2.66
1.11	0.62	−0.17	−1.84	−0.63	−0.25	−1.01	−2.15
−3.41	−4.31	−3.69	−2.75	−2.93	−3.05	−2.58	−2.05
−1.21	−1.54	−1.41	0.57	1.83	0.19	−0.59	−3.33
−4.66	−4.55	−4.35	−4.14	−4.93	−4.60	−4.49	−5.58
−6.17	−3.93	−2.68	−1.61	—	—	—	—

The simulated y_t data recorded row-wise.

TABLE A.12   Coen *et al.* (1969) Data

Observation	F. T. Index	F. T. Commodity Index	UK Car Production	Observation	F. T. Index	F. T. Commodity Index	UK Car Production
1*	148.5	96.21	121,874	27	312.9	80.42	293,062
2	165.4	93.74	126,260	28	323.7	82.67	285,809
3	178.5	91.37	145,248	29	349.3	82.78	366,265
4	187.3	86.31	160,370	30	310.4	82.61	374,241
5	195.9	84.98	163,648	31	295.8	82.47	375,764
6	205.2	86.46	178,195	32	301.2	81.86	354,411
7	191.5	90.04	187,197	33	285.8	79.7	249,527
8	183	94.74	195,916	34	271.7	77.89	206,165
9	184.5	92.43	199,253	35	283.6	77.61	258,410
10	181.1	92.41	227,616	36	295.7	78.9	279,342
11	173.7	91.65	215,363	37	309.3	79.72	264,824
12	185.4	89.38	231,728	38	295.7	78.08	312,983
13	201.8	91.05	231,767	39	342	77.54	300,932
14	198	89.89	211,211	40	335.1	76.99	323,424
15	168	90.16	185,200	41	344.4	76.25	312,780
16	161.6	86.78	152,404	42	360.9	78.13	363,336
17	170.2	88.45	156,163	43	346.5	80.38	378,275
18	184.5	90.69	151,567	44	340.6	81.78	414,457
19	211	86.03	213,683	45	340.3	82.81	459,158
20	218.3	84.85	244,543	46	323.3	84.99	460,397
21	231.7	84.07	253,111	47	345.6	86.31	462,279
22	247.4	81.96	266,580	48	349.3	85.95	434,255
23	301.9	80.03	253,543	49	359.7	90.73	475,890
24	323.8	79.8	261,675	50	320	92.42	439,365
25	314.1	80.19	249,407	51	299.9	87.18	413,666
26	321	80.13	246,248				

*The first observation for *Financial Times* index corresponds to the second quarter of 1954, whereas the first observations for *Financial Times* commodity index and for UK car production correspond to the third and fourth quarters of 1952, respectively.

TABLE A.13 Temperature Data from a Ceramic Furnace

Observation	$x5$	$x6$	Observation	$x5$	$x6$	Observation	$x5$	$x6$	Observation	$x5$	$x6$
1	827	828	36	826	832	71	823	829	106	822	829
2	828	828	37	827	832	72	823	829	107	825	835
3	828	829	38	825	831	73	823	830	108	826	837
4	832	833	39	824	830	74	823	830	109	823	836
5	834	836	40	824	830	75	823	829	110	822	834
6	839	841	41	823	830	76	823	830	111	822	835
7	838	840	42	824	830	77	822	829	112	822	835
8	840	842	43	822	829	78	821	827	113	822	836
9	841	843	44	822	829	79	820	826	114	823	837
10	840	842	45	823	829	80	818	824	115	821	836
11	840	842	46	824	830	81	817	824	116	821	837
12	840	842	47	823	829	82	817	823	117	822	838
13	839	840	48	823	829	83	819	825	118	822	839
14	840	842	49	823	829	84	820	826	119	820	837
15	836	842	50	824	831	85	819	826	120	819	837
16	839	844	51	824	831	86	821	827	121	820	838
17	838	843	52	823	830	87	821	828	122	820	838
18	832	839	53	823	830	88	823	829	123	819	838
19	831	838	54	824	831	89	822	829	124	819	838
20	831	838	55	824	831	90	823	830	125	824	843
21	832	839	56	827	833	91	822	829	126	826	847
22	830	837	57	827	833	92	822	828	127	827	848
23	833	841	58	828	834	93	822	829	128	829	850
24	834	840	59	827	833	94	823	829	129	829	851
25	833	839	60	825	831	95	822	829	130	828	849
26	832	836	61	825	831	96	822	829	131	826	847
27	829	835	62	826	832	97	822	829	132	821	842
28	829	835	63	826	833	98	822	829	133	821	842
29	827	833	64	826	833	99	823	830	134	820	841
30	826	832	65	824	831	100	823	830	135	819	841
31	826	832	66	825	832	101	822	829	136	819	841
32	826	832	67	826	832	102	821	828	137	819	840
33	824	831	68	825	832	103	820	828	138	819	841
34	824	830	69	825	831	104	821	828	139	819	841
35	826	832	70	824	831	105	820	828			

TABLE A.14   Temperature Readings from an Industrial Process

$x1$	$x2$	$x3$	$x4$	$x5$
1016.1	1022.5	1027.9	1030.6	1020.0
1017.5	1023.8	1029.9	1033.4	1022.2
1017.5	1024.3	1030.1	1032.4	1020.5
1019.2	1024.0	1030.9	1032.3	1021.1
1018.1	1022.6	1030.4	1032.0	1020.5
1017.3	1022.3	1031.3	1031.5	1020.1
1016.6	1022.2	1031.6	1031.3	1019.6
1018.2	1021.3	1031.5	1032.2	1020.4
1018.5	1021.3	1032.2	1032.1	1020.4
1018.5	1022.1	1032.5	1031.8	1019.3
1017.6	1021.1	1031.4	1032.6	1020.2
1018.5	1021.5	1031.1	1032.7	1020.7
1016.6	1021.6	1029.6	1032.4	1020.1
1017.7	1024.5	1029.4	1031.5	1019.0
1018.3	1023.3	1029.3	1031.7	1018.4
1017.5	1022.6	1028.8	1031.5	1018.2
1017.6	1022.1	1028.4	1031.9	1018.7
1018.6	1022.3	1028.1	1032.5	1019.1
1018.8	1022.8	1028.8	1032.3	1018.7
1017.0	1022.7	1031.7	1032.4	1017.6
1019.4	1023.9	1032.1	1032.7	1016.7
1020.6	1024.1	1031.2	1033.4	1015.4
1021.7	1022.7	1032.2	1033.6	1015.9
1021.8	1023.2	1030.5	1032.8	1016.4
1021.8	1022.8	1030.4	1033.2	1016.7
1021.8	1022.8	1030.7	1033.2	1016.5
1021.6	1024.6	1029.5	1031.8	1017.1
1023.3	1025.6	1029.1	1032.0	1017.4
1023.8	1025.7	1029.3	1031.8	1016.4
1023.0	1026.9	1029.1	1031.8	1016.9
1017.7	1025.4	1028.3	1030.8	1017.3
1017.6	1023.8	1027.3	1030.5	1018.6
1018.4	1023.3	1029.3	1029.9	1018.2
1017.2	1023.8	1027.0	1028.4	1015.9
1016.6	1024.4	1026.0	1029.5	1013.7
1016.7	1024.3	1025.3	1028.7	1014.1
1016.8	1024.7	1024.8	1028.3	1014.0
1016.2	1024.9	1024.5	1029.2	1012.6
1016.6	1024.6	1024.8	1029.9	1012.7
1019.0	1023.9	1024.1	1030.5	1014.6
1020.7	1023.5	1023.6	1032.1	1014.6
1019.9	1023.4	1024.4	1032.2	1012.9
1019.1	1023.4	1025.6	1032.0	1014.0
1017.8	1024.2	1025.8	1031.2	1013.3
1017.8	1023.7	1025.8	1031.7	1012.7
1017.8	1023.7	1023.8	1031.0	1012.3
1016.2	1023.0	1024.7	1031.5	1012.5
1016.5	1022.2	1024.9	1032.9	1012.0
1017.3	1022.4	1025.0	1033.0	1011.0
1014.7	1021.7	1026.6	1034.3	1011.3

TABLE A.15   US Hog Series

$z1$	$z2$	$z3$	$z4$	$z5$	$z1$	$z2$	$z3$	$z4$	$z5$
538	509	944	900	719	766	810	813	1409	982
522	663	841	964	716	720	957	790	1417	987
513	751	911	893	724	682	970	712	1455	991
529	739	768	1051	732	743	903	831	1394	1004
565	598	718	1057	740	743	995	742	1469	1013
594	556	634	1107	748	730	1022	847	1357	1004
600	594	735	1003	756	723	998	850	1402	1013
584	667	858	1025	748	753	928	830	1452	1053
554	776	673	1161	740	782	1073	1056	1385	1149
553	754	609	1170	732	760	1294	1163	1464	1248
595	689	604	1181	744	799	1346	1182	1388	1316
637	498	457	1194	756	808	1301	1180	1428	1384
641	643	612	1244	778	779	1134	805	1487	1190
647	681	642	1232	799	770	1024	714	1467	1179
634	778	849	1095	799	777	1090	865	1432	1228
629	829	733	1244	799	841	1013	911	1459	1238
638	751	672	1218	799	823	1119	1027	1347	1246
662	704	594	1289	801	746	1195	846	1447	1253
675	633	559	1313	803	717	1235	869	1406	1253
658	663	604	1251	806	744	1120	928	1418	1253
629	709	678	1205	806	791	1112	924	1426	1255
625	763	571	1352	806	771	1129	903	1401	1223
648	681	490	1361	810	746	1055	777	1318	1114
682	627	747	1218	813	739	787	507	1411	982
676	667	651	1368	810	773	624	500	1467	929
655	804	645	1278	806	793	612	716	1380	978
640	782	609	1279	771	768	800	911	1161	1013
668	707	705	1208	771	592	1104	816	1362	1045
678	653	453	1404	780	633	1075	1019	1178	1100
692	639	382	1427	789	634	1052	714	1422	1097
710	672	466	1359	799	649	1048	687	1406	1090
727	669	506	1371	820	699	891	754	1412	1100
712	729	525	1423	834	786	921	791	1390	1188
708	784	595	1425	848	735	1193	876	1424	1303
705	842	829	1234	863	782	1352	962	1487	1422
680	886	654	1443	884	869	1243	1050	1472	1498
682	784	673	1401	906	923	1314	1037	1490	1544
713	770	691	1429	928	774	1380	1104	1458	1582
726	783	660	1470	949	787	1556	1193	1507	1607
729	877	643	1482	960	754	1632	1334	1372	1629
752	777	754	1417	971					

Source: Quenouille (1957) (read down).

DATASETS USED IN THE EXERCISES

Table B.1 Beverage Amount (ml)

Table B.2 Pressure of the Steam Fed to a Distillation Column (bar)

Table B.3 Number of Paper Checks Processed in a Local Bank

Table B.4 Monthly Sea Levels in Los Angeles, California (mm)

Table B.5 Temperature Readings from a Chemical Process (°C)

Table B.6 Daily Average Exchange Rates between US Dollar and Euro

Table B.7 Monthly US Unemployment Rates

Table B.8 Monthly Residential Electricity Sales (MWh) and Average Residential Electricity Retail Price (c/kWh) in the United States

Table B.9 Monthly Outstanding Consumer Credits Provided by Commercial Banks in the United States (million USD)

Table B.10 100 Observations Simulated from an ARMA(1, 1) Process

Table B.11 Quarterly Rental Vacancy Rates in the United States

Table B.12 Wölfer Sunspot Numbers

Table B.13 Viscosity Readings from a Chemical Process

Table B.14 UK Midyear Population

Table B.15 Unemployment and GDP data for the United Kingdom

Table B.16 Monthly Crude Oil Production of OPEC Nations

Table B.17 Quarterly Dollar Sales of Marshall Field & Company ($1000)

Time Series Analysis and Forecasting by Example, First Edition. Søren Bisgaard and Murat Kulahci.
© 2011 John Wiley & Sons, Inc. Published 2011 by John Wiley & Sons, Inc.

TABLE B.1   Beverage Amount (ml)

250.1	249.6	249.1	249.8	251.3
249.1	250.3	249.3	249.2	252.4
250.0	250.7	250.2	249.6	251.1
249.6	248.8	250.7	252.0	249.6
250.1	249.5	250.7	249.0	249.5
249.8	247.9	251.0	249.4	249.1
250.2	248.2	250.6	249.8	251.7
250.6	250.0	250.0	249.9	251.7
252.2	250.1	250.8	248.9	251.7
250.2	249.2	249.9	248.8	250.3
250.2	249.6	248.4	248.2	250.4
249.5	248.7	248.7	250.3	250.6
249.3	248.6	250.5	251.1	250.1
248.2	248.5	249.8	250.3	249.5
248.0	248.6	250.6	251.1	252.7
248.4	248.7	249.6	251.0	252.8
248.8	249.7	246.6	250.6	251.4
249.2	250.3	246.9	251.7	250.9
250.0	250.7	247.5	249.6	251.6
250.1	249.5	247.3	250.6	251.2
248.3	248.0	249.0	251.6	250.4
248.2	250.6	249.9	249.2	250.1
248.8	250.1	249.5	250.3	249.3
249.8	250.2	249.5	250.6	250.1
250.4	251.4	249.0	250.7	251.4
249.2	250.5	249.5	250.5	250.0
250.2	250.9	251.2	251.9	248.3
249.1	252.5	252.8	251.7	248.7
248.4	252.7	251.6	250.9	246.5
248.8	250.8	250.3	251.6	248.8
249.5	249.8	251.1	250.9	247.7
250.4	249.4	251.3	252.0	247.3
251.5	250.2	251.7	252.5	248.4
251.5	249.7	251.2	250.4	249.5
253.5	252.0	251.1	252.7	250.6
252.9	250.6	250.3	251.3	250.4
249.2	250.0	251.1	251.3	251.7
248.8	248.7	250.5	250.1	250.2
249.0	248.5	251.6	250.9	250.5
249.6	249.5	253.8	250.7	249.3
250.0	249.3	251.8	250.9	250.1
248.2	249.7	249.1	250.8	249.8
248.7	251.2	248.3	252.5	248.1
250.6	250.0	249.4	253.1	247.6
251.6	248.0	250.9	252.3	248.4
249.4	249.0	250.1	252.9	249.6
247.3	248.4	250.3	252.4	250.3
248.8	249.7	251.6	251.5	249.9
249.3	248.7	250.4	250.8	250.2
249.6	247.5	251.4	251.2	251.1

250 Observations (read down).

TABLE B.2   Pressure of the Steam Fed to a Distillation Column (bar)

3.94	3.99	3.98
4.03	3.98	3.93
4.02	4.05	4.03
4.02	3.99	4.01
3.95	4.00	3.98
4.04	3.95	4.09
4.04	4.06	3.96
3.95	3.94	4.02
4.03	4.03	3.94
3.99	4.03	4.04
4.05	4.00	3.99
4.02	4.05	4.04
3.91	3.87	3.99
4.01	4.12	4.00
4.06	3.91	3.95
4.01	4.07	4.01
3.96	3.99	4.04
4.00	3.89	4.00
4.02	4.06	3.98
3.97	3.97	3.99
4.01	4.01	3.99
3.99	4.02	3.97
3.97	3.99	4.04
3.97	4.03	4.02
4.08	3.95	3.99
3.92	4.07	4.01
4.08	3.98	4.00
4.01	3.96	4.02
4.02	4.01	3.94
3.92	4.01	4.07
4.05	3.98	4.00
3.98	3.99	4.02
4.00	4.01	3.99
4.05	4.01	4.00
3.97	4.03	3.97
3.97	4.00	4.02
4.01	3.93	3.96
3.98	4.00	4.09
4.03	4.03	4.00
4.01	4.06	3.95

120 Observations (read down).

TABLE B.3 Number of Paper Checks Processed in a Local Bank

Year	Week	Number of Checks	Year	Week	Number of Checks	Year	Week	Number of Checks	Year	Week	Number of Checks
2004	1	476	2005	1	520	2006	1	451	2007	1	309
	2	448		2	519		2	405		2	303
	3	456		3	492		3	365		3	320
	4	435		4	483		4	300		4	335
	5	457		5	482		5	289		5	322
	6	482		6	504		6	321		6	317
	7	483		7	517		7	328		7	311
	8	487		8	509		8	301		8	336
	9	481		9	515		9	316		9	325
	10	471		10	512		10	322		10	327
	11	493		11	504		11	314		11	301
	12	490		12	514		12	308		12	297
	13	475		13	513		13	307		13	283
	14	514		14	502		14	307		14	284
	15	504		15	497		15	334		15	295
	16	483		16	539		16	326		16	289
	17	487		17	558		17	316		17	283
	18	482		18	527		18	278		18	272
	19	520		19	503		19	292		19	300
	20	500		20	494		20	317		20	313
	21	501		21	506		21	296		21	284
	22	516		22	510		22	319		22	330
	23	498		23	494		23	303		23	347
	24	527		24	468		24	296		24	327
	25	546		25	473		25	299		25	302
	26	527		26	459		26	304		26	319
	27	470		27	469		27	308		27	309
	28	489		28	463		28	330		28	316
	29	499		29	447		29	315		29	331
	30	474		30	485		30	316		30	340
	31	520		31	467		31	299		31	306
	32	488		32	495		32	340		32	315
	33	507		33	509		33	306		33	298
	34	510		34	502		34	299		34	296
	35	509		35	495		35	301		35	287
	36	512		36	525		36	287		36	259
	37	527		37	494		37	275		37	302
	38	528		38	523		38	279		38	249
	39	490		39	493		39	297		39	271
	40	517		40	534		40	296		40	288
	41	509		41	544		41	316		41	314
	42	476		42	504		42	313		42	311
	43	483		43	479		43	345		43	306
	44	496		44	483		44	345		44	304
	45	473		45	497		45	312		45	317
	46	483		46	505		46	305		46	306
	47	453		47	490		47	298		47	295
	48	473		48	494		48	304		48	256
	49	502		49	510		49	296		49	267
	50	517		50	486		50	294		50	271
	51	538		51	503		51	301		51	309
	52	557		52	483		52	330		52	315
	53	544										

TABLE B.4   Monthly Sea Levels in Los Angeles, California (mm)

Year	Mon	Val	Year	Mon	Val	Year	Mon	Val	Year	Mon	Val	Year	Mon	Val
1924	Jan	6993	1928	Jan	6993	1932	Jan	6962	1936	Jan	6938	1940	Jan	7035
	Feb	6990		Feb	6984		Feb	7014		Feb	6984		Feb	7017
	Mar	6978		Mar	6898		Mar	6932		Mar	6965		Mar	6965
	Apr	6910		Apr	6910		Apr	6968		Apr	6926		Apr	6944
	May	6947		May	6956		May	6993		May	6990		May	6981
	Jun	6959		Jun	7005		Jun	6971		Jun	7005		Jun	6993
	Jul	6971		Jul	7038		Jul	7038		Jul	7069		Jul	7038
	Aug	7002		Aug	7072		Aug	7096		Aug	7112		Aug	7087
	Sep	6990		Sep	7057		Sep	7084		Sep	7112		Sep	7133
	Oct	7014		Oct	7032		Oct	7051		Oct	7090		Oct	7099
	Nov	6968		Nov	7017		Nov	7002		Nov	7087		Nov	7087
	Dec	6993		Dec	6996		Dec	6959		Dec	7038		Dec	7151
1925	Jan	6962	1929	Jan	7002	1933	Jan	6968	1937	Jan	6978	1941	Jan	7072
	Feb	6886		Feb	6929		Feb	6901		Feb	7008		Feb	7145
	Mar	6968		Mar	6892		Mar	6920		Mar	6950		Mar	7115
	Apr	6892		Apr	6926		Apr	6917		Apr	6935		Apr	7008
	May	6959		May	6965		May	6874		May	7014		May	7014
	Jun	7014		Jun	7008		Jun	6938		Jun	7002		Jun	6993
	Jul	7048		Jul	7093		Jul	7005		Jul	7090		Jul	7069
	Aug	7048		Aug	7096		Aug	7014		Aug	7084		Aug	7099
	Sep	7048		Sep	7118		Sep	7017		Sep	7096		Sep	7157
	Oct	7118		Oct	7072		Oct	6956		Oct	7072		Oct	7185
	Nov	7063		Nov	7045		Nov	6962		Nov	7032		Nov	7115
	Dec	7104		Dec	6993		Dec	6941		Dec	7057		Dec	7115
1926	Jan	7084	1930	Jan	7029	1934	Jan	6953	1938	Jan	7029	1942	Jan	7042
	Feb	7008		Feb	7008		Feb	6914		Feb	6987		Feb	7032
	Mar	6996		Mar	6984		Mar	6962		Mar	6944		Mar	6996
	Apr	6965		Apr	6978		Apr	6959		Apr	6938		Apr	7020
	May	6941		May	6953		May	6978		May	6971		May	6984
	Jun	6956		Jun	7026		Jun	6984		Jun	6999		Jun	7051
	Jul	7026		Jul	7057		Jul	7048		Jul	7057		Jul	7081
	Aug	7057		Aug	7121		Aug	7035		Aug	7099		Aug	7084
	Sep	7026		Sep	7121		Sep	7096		Sep	7093		Sep	7112
	Oct	7038		Oct	7112		Oct	7026		Oct	7032		Oct	7102
	Nov	7002		Nov	7112		Nov	6987		Nov	7011		Nov	7057
	Dec	7002		Dec	7084		Dec	7002		Dec	7048		Dec	7032
1927	Jan	6953	1931	Jan	7023	1935	Jan	6959	1939	Jan	6993	1943	Jan	7075
	Feb	6953		Feb	7072		Feb	6968		Feb	6926		Feb	7017
	Mar	6886		Mar	6978		Mar	6923		Mar	6929		Mar	6965
	Apr	6910		Apr	6987		Apr	6920		Apr	6947		Apr	6959
	May	6932		May	6978		May	6923		May	6956		May	7023
	Jun	7002		Jun	7002		Jun	7002		Jun	6999		Jun	7075
	Jul	7017		Jul	7106		Jul	7008		Jul	7032		Jul	7096
	Aug	7029		Aug	7087		Aug	7078		Aug	7090		Aug	7102
	Sep	7048		Sep	7081		Sep	7051		Sep	7173		Sep	7115
	Oct	7054		Oct	7051		Oct	7042		Oct	7124		Oct	7102
	Nov	7035		Nov	7023		Nov	6996		Nov	7142		Nov	7048
	Dec	7038		Dec	6968		Dec	7023		Dec	7066		Dec	7069

(*continued overleaf*)

TABLE B.4 (*Continued*)

1944	Jan	7020	1948	Jan	7051	1952	Jan	7051	1956	Jan	6959	1960	Jan	7029
	Feb	7042		Feb	6990		Feb	6987		Feb	6962		Feb	6947
	Mar	7038		Mar	6953		Mar	6965		Mar	6938		Mar	6971
	Apr	6950		Apr	6947		Apr	6984		Apr	6965		Apr	6950
	May	6974		May	6978		May	6950		May	6981		May	6984
	Jun	7011		Jun	7008		Jun	6984		Jun	7051		Jun	7011
	Jul	7051		Jul	7060		Jul	7020		Jul	7066		Jul	7060
	Aug	7090		Aug	7054		Aug	7069		Aug	7084		Aug	7066
	Sep	7115		Sep	7090		Sep	7099		Sep	7066		Sep	7102
	Oct	7054		Oct	7017		Oct	7060		Oct	7038		Oct	7072
	Nov	7048		Nov	7020		Nov	7084		Nov	7048		Nov	7038
	Dec	7023		Dec	7008		Dec	7045		Dec	7020		Dec	7042
1945	Jan	7014	1949	Jan	7005	1953	Jan	7038	1957	Jan	7008	1961	Jan	7051
	Feb	6965		Feb	6999		Feb	6999		Feb	7014		Feb	6993
	Mar	6917		Mar	6910		Mar	6965		Mar	6956		Mar	6953
	Apr	6935		Apr	6935		Apr	6929		Apr	6959		Apr	6907
	May	6929		May	6959		May	6923		May	7014		May	6935
	Jun	7026		Jun	7029		Jun	6974		Jun	7081		Jun	6993
	Jul	7075		Jul	7023		Jul	7048		Jul	7130		Jul	7057
	Aug	7115		Aug	7054		Aug	7042		Aug	7151		Aug	7066
	Sep	7121		Sep	7072		Sep	7081		Sep	7102		Sep	7084
	Oct	7066		Oct	7023		Oct	7087		Oct	7121		Oct	7032
	Nov	7002		Nov	7020		Nov	7045		Nov	7118		Nov	7011
	Dec	7026		Dec	6984		Dec	7023		Dec	7136		Dec	7020
1946	Jan	6984	1950	Jan	6944	1954	Jan	7035	1958	Jan	7096	1962	Jan	7020
	Feb	6987		Feb	6941		Feb	7005		Feb	7063		Feb	7029
	Mar	6965		Mar	6914		Mar	6971		Mar	7048		Mar	6907
	Apr	6965		Apr	6947		Apr	6944		Apr	7020		Apr	6941
	May	6965		May	6950		May	7014		May	7014		May	6981
	Jun	7020		Jun	6993		Jun	7063		Jun	7026		Jun	6987
	Jul	7063		Jul	7081		Jul	7109		Jul	7084		Jul	7026
	Aug	7072		Aug	7038		Aug	7087		Aug	7142		Aug	7069
	Sep	7151		Sep	7057		Sep	7084		Sep	7173		Sep	7112
	Oct	7093		Oct	7051		Oct	7042		Oct	7185		Oct	7084
	Nov	7090		Nov	7011		Nov	7045		Nov	7118		Nov	6999
	Dec	7078		Dec	7032		Dec	7020		Dec	7102		Dec	7023
1947	Jan	7042	1951	Jan	6944	1955	Jan	7002	1959	Jan	7109	1963	Jan	7005
	Feb	7005		Feb	6959		Feb	6971		Feb	7109		Feb	7014
	Mar	6996		Mar	6993		Mar	6987		Mar	7048		Mar	6920
	Apr	6935		Apr	6938		Apr	6938		Apr	6990		Apr	6910
	May	7002		May	7002		May	6965		May	7005		May	6941
	Jun	7023		Jun	7054		Jun	7014		Jun	7057		Jun	7026
	Jul	7051		Jul	7112		Jul	7035		Jul	7087		Jul	7063
	Aug	7045		Aug	7109		Aug	7078		Aug	7124		Aug	7090
	Sep	7066		Sep	7142		Sep	7075		Sep	7151		Sep	7142
	Oct	7032		Oct	7130		Oct	6990		Oct	7173		Oct	7066
	Nov	7014		Nov	7115		Nov	6996		Nov	7069		Nov	7051
	Dec	7054		Dec	7054		Dec	6947		Dec	7118		Dec	7054

TABLE B.4   (*Continued*)

1964	Jan	7032	1968	Jan	7045	1972	Jan	7011	1976	Jan	6981
	Feb	6962		Feb	7032		Feb	7002		Feb	7005
	Mar	6923		Mar	6956		Mar	7002		Mar	6968
	Apr	6920		Apr	6974		Apr	6956		Apr	6953
	May	6901		May	6968		May	6965		May	6996
	Jun	6981		Jun	6971		Jun	7042		Jun	7054
	Jul	7017		Jul	7038		Jul	7054		Jul	7102
	Aug	7093		Aug	7069		Aug	7151		Aug	7066
	Sep	7084		Sep	7099		Sep	7139		Sep	7154
	Oct	7051		Oct	7075		Oct	7145		Oct	7112
	Nov	7002		Nov	7054		Nov	7124		Nov	7130
	Dec	7017		Dec	7017		Dec	7151		Dec	7121
1965	Jan	6990	1969	Jan	7087	1973	Jan	7060	1977	Jan	7072
	Feb	7042		Feb	7038		Feb	7066		Feb	7014
	Mar	6984		Mar	6947		Mar	6993		Mar	6956
	Apr	6990		Apr	6981		Apr	6920		Apr	6929
	May	6947		May	7008		May	6953		May	6956
	Jun	7029		Jun	7002		Jun	6984		Jun	6999
	Jul	7057		Jul	7054		Jul	7042		Jul	7051
	Aug	7133		Aug	7102		Aug	7069		Aug	7112
	Sep	7151		Sep	7115		Sep	7063		Sep	7099
	Oct	7106		Oct	7127		Oct	7054		Oct	7109
	Nov	7099		Nov	7099		Nov	7005		Nov	7096
	Dec	7136		Dec	7042		Dec	7011		Dec	7096
1966	Jan	7075	1970	Jan	6962	1974	Jan	7014	1978	Jan	7109
	Feb	7023		Feb	6990		Feb	6959		Feb	7081
	Mar	6996		Mar	7002		Mar	6990		Mar	7063
	Apr	6999		Apr	6923		Apr	6978		Apr	6959
	May	6999		May	6968		May	7005		May	7002
	Jun	7032		Jun	6993		Jun	7038		Jun	6981
	Jul	7069		Jul	7011		Jul	7060		Jul	7054
	Aug	7118		Aug	7045		Aug	7038		Aug	7121
	Sep	7060		Sep	7069		Sep	7078		Sep	7139
	Oct	7054		Oct	7051		Oct	7020		Oct	7139
	Nov	7026		Nov	7017		Nov	6993		Nov	7112
	Dec	7020		Dec	6974		Dec	6993		Dec	7078
1967	Jan	7026	1971	Jan	6959	1975	Jan	6993			
	Feb	6999		Feb	6947		Feb	6968			
	Mar	6944		Mar	6856		Mar	6962			
	Apr	6920		Apr	6929		Apr	6962			
	May	6956		May	7002		May	6941			
	Jun	7042		Jun	7029		Jun	6987			
	Jul	7081		Jul	7118		Jul	7038			
	Aug	7109		Aug	7106		Aug	7038			
	Sep	7102		Sep	7118		Sep	7081			
	Oct	7048		Oct	7081		Oct	7042			
	Nov	7087		Nov	7008		Nov	6993			
	Dec	7038		Dec	6990		Dec	6965			

Source: www.psmsl.org.

TABLE B.5 Temperature Readings from a Chemical Process ($^\circ$C)

94.84	99.32	103.04	95.55
94.95	94.99	102.31	91.45
98.75	97.98	100.83	92.56
101.38	100.8	102.26	94.9
101.79	102.6	102.44	100.48
99.17	103.39	· 99.29	104.02
95.94	99.45	101.48	104.7
93.92	97.99	104.15	103.95
95.72	96.78	105.54	103.55
96.15	95.61	103.14	98.96
95.6	99.03	102.7	100.16
99.01	99.68	103.78	101.92
99.07	99.46	102.03	101.9
94.37	97.7	103.19	104.63
90.76	99.14	106.47	104.13
92.13	101.72	107.71	100.33
97.87	99.85	105.75	97.06
99.75	98.26	105.26	95.48
102.13	101.02	102.31	94.95
98.08	101.78	99.14	94.72
98.31	101.65	98.30	98.30
100.34	101.85	101.64	97.63
96.55	99.54	104.42	102.32
91.94	99.72	101.75	99.25
95.06	96.89	102.37	95.72
96.97	93.29	103.78	96.09
95.72	92.13	100.62	95.97
96.70	96.39	97.95	97.28
100.81	97.01	95.31	99.29
101.4	96.88	95.18	102.44
100.77	98.30	100.34	103.75
101.19	98.34	103.12	104.33
96.30	95.78	98.93	102.43
91.75	98.83	99.02	102.77
95.57	100.86	99.38	104.21
98.6	104.26	95.41	102.8
101.16	102.48	95.04	104.15
105.62	96.81	98.42	104.47
106.50	92.96	101.87	104.85
103.78	95.85	105.71	100.19
102.52	96.55	106.19	97.82
103.30	93.12	103.83	100.11
101.42	91.78	102.15	99.95
100.33	92.88	98.34	99.10
98.39	95.03	94.41	103.40
98.62	99.11	94.17	109.51
100.70	100.00	95.52	108.45
102.93	97.41	94.67	106.95
101.83	98.32	94.24	105.77
101.93	101.97	96.95	102.34

200 Observations (read down).

TABLE B.6 Daily Average Exchange Rates between US Dollar and Euro

Day	Exchange Rate	Day	Exchange Rate	Day	Exchange Rate
1/1/2008	0.6803	2/21/2008	0.6802	4/12/2008	0.6332
1/2/2008	0.6855	2/22/2008	0.6784	4/13/2008	0.6328
1/3/2008	0.6824	2/23/2008	0.6748	4/14/2008	0.6331
1/4/2008	0.6791	2/24/2008	0.6747	4/15/2008	0.6342
1/5/2008	0.6783	2/25/2008	0.6747	4/16/2008	0.6321
1/6/2008	0.6785	2/26/2008	0.6748	4/17/2008	0.6302
1/7/2008	0.6785	2/27/2008	0.6733	4/18/2008	0.6278
1/8/2008	0.6797	2/28/2008	0.6650	4/19/2008	0.6308
1/9/2008	0.6800	2/29/2008	0.6611	4/20/2008	0.6324
1/10/2008	0.6803	3/1/2008	0.6583	4/21/2008	0.6324
1/11/2008	0.6805	3/2/2008	0.6590	4/22/2008	0.6307
1/12/2008	0.6760	3/3/2008	0.6590	4/23/2008	0.6277
1/13/2008	0.6770	3/4/2008	0.6581	4/24/2008	0.6269
1/14/2008	0.6770	3/5/2008	0.6578	4/25/2008	0.6335
1/15/2008	0.6736	3/6/2008	0.6574	4/26/2008	0.6392
1/16/2008	0.6726	3/7/2008	0.6532	4/27/2008	0.6400
1/17/2008	0.6764	3/8/2008	0.6502	4/28/2008	0.6401
1/18/2008	0.6821	3/9/2008	0.6514	4/29/2008	0.6396
1/19/2008	0.6831	3/10/2008	0.6514	4/30/2008	0.6407
1/20/2008	0.6842	3/11/2008	0.6507	5/1/2008	0.6423
1/21/2008	0.6842	3/12/2008	0.6510	5/2/2008	0.6429
1/22/2008	0.6887	3/13/2008	0.6489	5/3/2008	0.6472
1/23/2008	0.6903	3/14/2008	0.6424	5/4/2008	0.6485
1/24/2008	0.6848	3/15/2008	0.6406	5/5/2008	0.6486
1/25/2008	0.6824	3/16/2008	0.6382	5/6/2008	0.6467
1/26/2008	0.6788	3/17/2008	0.6382	5/7/2008	0.6444
1/27/2008	0.6814	3/18/2008	0.6341	5/8/2008	0.6467
1/28/2008	0.6814	3/19/2008	0.6344	5/9/2008	0.6509
1/29/2008	0.6797	3/20/2008	0.6378	5/10/2008	0.6478
1/30/2008	0.6769	3/21/2008	0.6439	5/11/2008	0.6461
1/31/2008	0.6765	3/22/2008	0.6478	5/12/2008	0.6461
2/1/2008	0.6733	3/23/2008	0.6481	5/13/2008	0.6468
2/2/2008	0.6729	3/24/2008	0.6481	5/14/2008	0.6449
2/3/2008	0.6758	3/25/2008	0.6494	5/15/2008	0.6470
2/4/2008	0.6758	3/26/2008	0.6438	5/16/2008	0.6461
2/5/2008	0.6750	3/27/2008	0.6377	5/17/2008	0.6452
2/6/2008	0.6780	3/28/2008	0.6329	5/18/2008	0.6422
2/7/2008	0.6833	3/29/2008	0.6336	5/19/2008	0.6422
2/8/2008	0.6856	3/30/2008	0.6333	5/20/2008	0.6425
2/9/2008	0.6903	3/31/2008	0.6333	5/21/2008	0.6414
2/10/2008	0.6896	4/1/2008	0.6330	5/22/2008	0.6366
2/11/2008	0.6896	4/2/2008	0.6371	5/23/2008	0.6346
2/12/2008	0.6882	4/3/2008	0.6403	5/24/2008	0.6350
2/13/2008	0.6879	4/4/2008	0.6396	5/25/2008	0.6346
2/14/2008	0.6862	4/5/2008	0.6364	5/26/2008	0.6346
2/15/2008	0.6852	4/6/2008	0.6356	5/27/2008	0.6343
2/16/2008	0.6823	4/7/2008	0.6356	5/28/2008	0.6346
2/17/2008	0.6813	4/8/2008	0.6368	5/29/2008	0.6379
2/18/2008	0.6813	4/9/2008	0.6357	5/30/2008	0.6416
2/19/2008	0.6822	4/10/2008	0.6354	5/31/2008	0.6442
2/20/2008	0.6802	4/11/2008	0.6320	6/1/2008	0.6431

Source: www.oanda.com.

TABLE B.7   Monthly US Unemployment Rates

	Jan	Feb	Mar	Apr	May	Jun	Jul	Aug	Sep	Oct	Nov	Dec
1980	6.3	6.3	6.3	6.9	7.5	7.6	7.8	7.7	7.5	7.5	7.5	7.2
1981	7.5	7.4	7.4	7.2	7.5	7.5	7.2	7.4	7.6	7.9	8.3	8.5
1982	8.6	8.9	9	9.3	9.4	9.6	9.8	9.8	10.1	10.4	10.8	10.8
1983	10.4	10.4	10.3	10.2	10.1	10.1	9.4	9.5	9.2	8.8	8.5	8.3
1984	8	7.8	7.8	7.7	7.4	7.2	7.5	7.5	7.3	7.4	7.2	7.3
1985	7.3	7.2	7.2	7.3	7.2	7.4	7.4	7.1	7.1	7.1	7	7
1986	6.7	7.2	7.2	7.1	7.2	7.2	7	6.9	7	7	6.9	6.6
1987	6.6	6.6	6.6	6.3	6.3	6.2	6.1	6	5.9	6	5.8	5.7
1988	5.7	5.7	5.7	5.4	5.6	5.4	5.4	5.6	5.4	5.4	5.3	5.3
1989	5.4	5.2	5	5.2	5.2	5.3	5.2	5.2	5.3	5.3	5.4	5.4
1990	5.4	5.3	5.2	5.4	5.4	5.2	5.5	5.7	5.9	5.9	6.2	6.3
1991	6.4	6.6	6.8	6.7	6.9	6.9	6.8	6.9	6.9	7	7	7.3
1992	7.3	7.4	7.4	7.4	7.6	7.8	7.7	7.6	7.6	7.3	7.4	7.4
1993	7.3	7.1	7	7.1	7.1	7	6.9	6.8	6.7	6.8	6.6	6.5
1994	6.6	6.6	6.5	6.4	6.1	6.1	6.1	6	5.9	5.8	5.6	5.5
1995	5.6	5.4	5.4	5.8	5.6	5.6	5.7	5.7	5.6	5.5	5.6	5.6
1996	5.6	5.5	5.5	5.6	5.6	5.3	5.5	5.1	5.2	5.2	5.4	5.4
1997	5.3	5.2	5.2	5.1	4.9	5	4.9	4.8	4.9	4.7	4.6	4.7
1998	4.6	4.6	4.7	4.3	4.4	4.5	4.5	4.5	4.6	4.5	4.4	4.4
1999	4.3	4.4	4.2	4.3	4.2	4.3	4.3	4.2	4.2	4.1	4.1	4
2000	4	4.1	4	3.8	4	4	4	4.1	3.9	3.9	3.9	3.9
2001	4.2	4.2	4.3	4.4	4.3	4.5	4.6	4.9	5	5.3	5.5	5.7
2002	5.7	5.7	5.7	5.9	5.8	5.8	5.8	5.7	5.7	5.7	5.9	6
2003	5.8	5.9	5.9	6	6.1	6.3	6.2	6.1	6.1	6	5.8	5.7
2004	5.7	5.6	5.8	5.6	5.6	5.6	5.5	5.4	5.4	5.5	5.4	5.4
2005	5.3	5.4	5.2	5.2	5.1	5	5	4.9	5	5	5	4.9
2006	4.7	4.8	4.7	4.7	4.6	4.6	4.7	4.7	4.5	4.4	4.5	4.4
2007	4.6	4.5	4.4	4.5	4.4	4.6	4.6	4.6	4.7	4.7	4.7	5
2008	5	4.8	5.1	5	5.4	5.5	5.8	6.1	6.2	6.6	6.9	7.4
2009	7.7	8.2	8.6	8.9	9.4	9.5	9.4	9.7	9.8	10.1	10	10
2010	9.7	9.7	9.7	9.9	9.7	9.5	9.5	9.6	9.6			

Source: data.bls.gov.

TABLE B.8   Monthly Residential Electricity Sales (MWh) and Average Residential Electricity Retail Price (c/kWh) in the United States

Year	Month	Residential Sales (M Wh)	Average Retail Price Residential (c/k Wh)	Year	Month	Residential Sales (M Wh)	Average Retail Price Residential (c/k Wh)
1990	1	95,420,231	7.17	1994	1	103,824,773	7.74
1990	2	74,498,370	7.48	1994	2	89,682,978	7.85
1990	3	71,901,767	7.57	1994	3	79,883,397	8.08
1990	4	65,190,618	7.69	1994	4	69,492,472	8.30
1990	5	62,881,008	7.96	1994	5	67,156,660	8.53
1990	6	73,899,811	8.10	1994	6	84,094,554	8.77
1990	7	90,935,492	8.18	1994	7	103,618,549	8.80
1990	8	88,563,829	8.24	1994	8	96,751,342	8.84
1990	9	86,239,965	8.15	1994	9	85,350,828	8.82
1990	10	69,595,203	8.04	1994	10	71,685,160	8.56
1990	11	66,428,238	7.80	1994	11	71,073,494	8.29
1990	12	78,464,162	7.61	1994	12	85,867,486	8.06
1991	1	93,854,576	7.41	1995	1	96,552,600	7.85
1991	2	79,444,016	7.61	1995	2	86,678,934	8.01
1991	3	73,862,548	7.79	1995	3	79,464,735	8.14
1991	4	65,887,836	7.98	1995	4	68,584,474	8.41
1991	5	67,266,873	8.14	1995	5	70,102,383	8.53
1991	6	80,947,065	8.32	1995	6	84,229,052	8.72
1991	7	94,569,114	8.38	1995	7	104,022,361	8.80
1991	8	92,966,539	8.41	1995	8	114,889,114	8.78
1991	9	84,536,461	8.37	1995	9	93,908,245	8.57
1991	10	69,256,478	8.32	1995	10	74,749,499	8.65
1991	11	70,906,273	7.95	1995	11	76,927,182	8.26
1991	12	81,919,575	7.81	1995	12	92,392,875	8.01
1992	1	91,472,367	7.70	1996	1	108,611,243	7.75
1992	2	82,166,086	7.78	1996	2	96,117,981	7.81
1992	3	73,753,733	8.01	1996	3	87,036,513	8.08
1992	4	68,450,124	8.03	1996	4	74,616,352	8.24
1992	5	64,782,539	8.38	1996	5	74,546,420	8.53
1992	6	70,906,637	8.61	1996	6	90,953,853	8.65
1992	7	88,765,464	8.54	1996	7	106,127,815	8.74
1992	8	88,505,744	8.56	1996	8	105,560,260	8.87
1992	9	79,592,966	8.59	1996	9	91,597,559	8.79
1992	10	69,962,112	8.44	1996	10	75,385,584	8.67
1992	11	70,061,583	8.14	1996	11	78,242,218	8.25
1992	12	87,519,433	7.86	1996	12	93,715,950	7.99
1993	1	93,792,370	7.73	1997	1	106,166,352	7.87
1993	2	83,424,707	7.79	1997	2	90,264,759	7.98
1993	3	83,053,786	7.80	1997	3	81,426,550	8.24
1993	4	69,702,639	8.13	1997	4	72,745,622	8.38
1993	5	63,882,663	8.56	1997	5	70,778,999	8.65
1993	6	76,580,103	8.74	1997	6	83,568,904	8.91
1993	7	101,040,539	8.73	1997	7	109,342,525	8.74
1993	8	102,191,754	8.73	1997	8	106,961,650	8.80
1993	9	88,905,699	8.79	1997	9	94,772,742	8.75
1993	10	71,786,478	8.76	1997	10	84,076,576	8.59
1993	11	72,725,759	8.21	1997	11	80,004,670	8.25
1993	12	87,694,328	7.90	1997	12	95,770,748	8.03

(continued overleaf)

TABLE B.8 (*Continued*)

Year	Month	Residential Sales (M Wh)	Average Retail Price Residential (c/k Wh)	Year	Month	Residential Sales (M Wh)	Average Retail Price Residential (c/k Wh)
1998	1	102,549,076	7.87	2002	1	116,891,801	8.07
1998	2	86,554,306	7.97	2002	2	96,592,933	8.19
1998	3	85,964,291	8.01	2002	3	95,318,667	8.17
1998	4	74,143,625	8.23	2002	4	85,407,607	8.37
1998	5	77,482,558	8.49	2002	5	87,318,539	8.64
1998	6	98,488,751	8.53	2002	6	107,169,995	8.73
1998	7	121,537,673	8.58	2002	7	133,694,801	8.82
1998	8	120,330,518	8.57	2002	8	134,332,468	8.72
1998	9	106,674,490	8.43	2002	9	115,833,033	8.59
1998	10	86,804,459	8.25	2002	10	94,531,343	8.47
1998	11	76,952,198	8.04	2002	11	88,822,303	8.31
1998	12	92,627,166	7.92	2002	12	109,266,379	8.08
1999	1	111,200,901	7.58	2003	1	124,272,788	8.00
1999	2	86,663,612	7.92	2003	2	110,937,370	8.02
1999	3	89,422,998	7.90	2003	3	99,561,312	8.35
1999	4	77,260,719	8.09	2003	4	83,448,244	8.82
1999	5	77,151,672	8.27	2003	5	87,804,372	8.99
1999	6	95,941,316	8.43	2003	6	101,285,792	9.25
1999	7	123,146,398	8.49	2003	7	130,673,857	9.21
1999	8	124,030,419	8.42	2003	8	134,467,799	9.22
1999	9	104,115,852	8.36	2003	9	113,848,983	8.92
1999	10	82,598,793	8.36	2003	10	89,780,860	8.85
1999	11	78,263,892	8.11	2003	11	86,658,681	8.72
1999	12	95,126,509	7.94	2003	12	113,083,851	8.30
2000	1	109,611,122	7.66	2004	1	127,121,383	8.24
2000	2	98,561,597	7.71	2004	2	112,464,055	8.33
2000	3	84,624,330	8.10	2004	3	98,947,207	8.62
2000	4	76,197,378	8.16	2004	4	85,376,865	8.93
2000	5	83,336,096	8.35	2004	5	90,598,212	9.07
2000	6	103,985,607	8.56	2004	6	112,335,133	9.29
2000	7	119,464,453	8.60	2004	7	129,305,347	9.36
2000	8	123,700,339	8.62	2004	8	126,423,340	9.50
2000	9	108,530,749	8.50	2004	9	112,337,694	9.39
2000	10	86,811,220	8.49	2004	10	93,466,169	9.05
2000	11	84,482,232	8.16	2004	11	89,649,741	8.96
2000	12	113,141,368	7.82	2004	12	113,956,432	8.58
2001	1	127,065,784	7.73	2005	1	125,287,859	8.52
2001	2	99,877,673	8.04	2005	2	106,666,913	8.76
2001	3	92,804,586	8.32	2005	3	104,065,203	8.87
2001	4	82,453,783	8.46	2005	4	86,749,160	9.22
2001	5	81,731,153	8.83	2005	5	87,384,112	9.56
2001	6	99,407,295	9.07	2005	6	116,627,438	9.79
2001	7	120,707,428	9.03	2005	7	144,476,336	9.77
2001	8	129,205,362	9.01	2005	8	146,904,780	9.93
2001	9	105,943,193	8.92	2005	9	126,515,635	9.94
2001	10	85,419,589	8.84	2005	10	102,685,879	9.76
2001	11	80,806,694	8.47	2005	11	91,686,571	9.76
2001	12	96,184,054	8.29	2005	12	120,177,222	9.27

TABLE B.8   *(Continued)*

Year	Month	Residential Sales (M Wh)	Average Retail Price Residential (c/k Wh)	Year	Month	Residential Sales (M Wh)	Average Retail Price Residential (c/k Wh)
2006	1	120,418,845	9.55	2008	7	143,269,372	12.05
2006	2	104,511,063	9.80	2008	8	138,764,503	12.06
2006	3	104,955,192	9.87	2008	9	117,588,918	11.90
2006	4	89,374,095	10.32	2008	10	96,092,696	11.81
2006	5	93,999,951	10.61	2008	11	95,665,266	11.43
2006	6	118,815,308	10.85	2008	12	125,002,668	10.90
2006	7	147,338,329	10.96	2009	1	135,904,280	10.99
2006	8	150,064,425	10.94	2009	2	115,431,542	11.18
2006	9	116,072,164	10.94	2009	3	106,467,461	11.33
2006	10	96,246,214	10.58	2009	4	91,395,346	11.55
2006	11	94,842,851	10.18	2009	5	94,083,823	11.80
2006	12	114,881,599	9.84	2009	6	114,177,785	11.84
2007	1	125,286,236	10.06	2009	7	137,467,468	11.90
2007	2	121,464,249	9.89	2009	8	138,290,140	12.00
2007	3	105,694,760	10.27	2009	9	115,216,561	12.00
2007	4	90,282,047	10.63	2009	10	98,398,564	11.70
2007	5	96,388,854	10.77	2009	11	92,613,854	11.33
2007	6	117,417,595	11.09	2009	12	123,422,556	10.93
2007	7	139,027,030	11.07	2010	1	147,848,708	10.54
2007	8	150,101,494	11.07	2010	2	123,329,790	10.93
2007	9	129,512,025	10.96	2010	3	112,057,413	11.20
2007	10	103,753,922	10.82	2010	4	88,111,138	11.75
2007	11	95,904,831	10.70	2010	5	94,776,950	11.96
2007	12	117,407,955	10.33	2010	6	126,974,815	11.92
2008	1	132,938,174	10.15	2010	7	155,325,187	12.01
2008	2	118,470,786	10.19				
2008	3	107,056,850	10.47				
2008	4	91,976,703	10.92				
2008	5	92,018,296	11.39				
2008	6	121,136,873	11.75				

Source: www.eia.doe.gov.

TABLE B.9 Monthly Outstanding Consumer Credits Provided by Commercial Banks in the United States (million USD)

Year	Month	Credit	Year	Month	Credit	Year	Month	Credit	Year	Month	Credit
1980	Jan	185,458	1984	Jan	213,461	1988	Jan	333,390	1992	Jan	366,160
	Feb	184,753		Feb	217,589		Feb	333,075		Feb	360,708
	Mar	183,415		Mar	219,840		Mar	333,855		Mar	357,166
	Apr	181,628		Apr	223,588		Apr	338,430		Apr	357,471
	May	179,552		May	229,961		May	340,195		May	356,312
	Jun	177,822		Jun	236,045		Jun	341,529		Jun	355,981
	Jul	177,658		Jul	239,880		Jul	343,669		Jul	357,018
	Aug	178,822		Aug	245,373		Aug	348,768		Aug	357,719
	Sep	179,789		Sep	248,051		Sep	351,421		Sep	357,300
	Oct	179,734		Oct	250,706		Oct	352,273		Oct	356,412
	Nov	179,216		Nov	252,654		Nov	354,935		Nov	357,412
	Dec	180,215		Dec	258,844		Dec	360,816		Dec	362,901
1981	Jan	178,158	1985	Jan	260,285	1989	Jan	353,344	1993	Jan	362,312
	Feb	176,335		Feb	262,680		Feb	353,025		Feb	359,813
	Mar	176,465		Mar	267,381		Mar	355,578		Mar	358,900
	Apr	176,975		Apr	271,745		Apr	359,803		Apr	362,295
	May	177,343		May	273,572		May	361,528		May	365,102
	Jun	179,065		Jun	276,513		Jun	363,704		Jun	366,632
	Jul	179,944		Jul	279,497		Jul	365,107		Jul	371,137
	Aug	181,201		Aug	282,757		Aug	368,832		Aug	375,059
	Sep	182,849		Sep	286,728		Sep	371,715		Sep	379,364
	Oct	182,116		Oct	287,981		Oct	373,483		Oct	381,660
	Nov	182,353		Nov	290,736		Nov	376,021		Nov	387,621
	Dec	184,193		Dec	297,227		Dec	383,309		Dec	395,704
1982	Jan	183,683	1986	Jan	297,500	1990	Jan	382,385	1994	Jan	394,842
	Feb	182,932		Feb	296,259		Feb	377,311		Feb	394,867
	Mar	182,922		Mar	296,270		Mar	372,789		Mar	397,802
	Apr	183,682		Apr	299,869		Apr	374,395		Apr	404,865
	May	183,897		May	301,951		May	373,770		May	407,966
	Jun	185,223		Jun	304,100		Jun	372,319		Jun	413,413
	Jul	185,450		Jul	306,303		Jul	373,236		Jul	419,979
	Aug	186,557		Aug	308,352		Aug	377,272		Aug	428,924
	Sep	188,072		Sep	311,443		Sep	379,467		Sep	434,560
	Oct	187,513		Oct	314,178		Oct	380,625		Oct	439,238
	Nov	187,241		Nov	314,969		Nov	380,586		Nov	445,114
	Dec	190,932		Dec	320,187		Dec	381,982		Dec	458,777
1983	Jan	188,756	1987	Jan	316,103	1991	Jan	376,400	1995	Jan	457,033
	Feb	186,259		Feb	313,062		Feb	374,255		Feb	455,749
	Mar	186,947		Mar	309,795		Mar	370,499		Mar	459,689
	Apr	187,581		Apr	313,583		Apr	371,625		Apr	466,365
	May	187,973		May	315,096		May	370,112		May	469,628
	Jun	190,915		Jun	316,822		Jun	368,385		Jun	471,591
	Jul	193,784		Jul	319,011		Jul	368,452		Jul	475,250
	Aug	197,482		Aug	322,671		Aug	368,426		Aug	484,446
	Sep	200,673		Sep	325,602		Sep	366,312		Sep	485,920
	Oct	203,203		Oct	326,917		Oct	364,599		Oct	485,565
	Nov	206,285		Nov	328,850		Nov	364,185		Nov	490,368
	Dec	213,663		Dec	334,090		Dec	370,176		Dec	502,328

TABLE B.9   (*Continued*)

Year	Month	Credit	Year	Month	Credit	Year	Month	Credit	Year	Month	Credit
1996	Jan	493,574	2000	Jan	506,793	2004	Jan	669,763	2008	Jan	808,706
	Feb	490,790		Feb	507,530		Feb	661,743		Feb	800,046
	Mar	490,807		Mar	505,673		Mar	658,036		Mar	796,317
	Apr	498,613		Apr	508,318		Apr	661,040		Apr	807,442
	May	498,795		May	510,676		May	667,048		May	807,859
	Jun	503,766		Jun	514,964		Jun	660,571		Jun	812,984
	Jul	507,884		Jul	515,102		Jul	664,144		Jul	820,315
	Aug	513,915		Aug	529,474		Aug	673,769		Aug	832,940
	Sep	513,906		Sep	531,032		Sep	676,319		Sep	844,087
	Oct	518,430		Oct	530,134		Oct	676,957		Oct	850,673
	Nov	520,593		Nov	537,425		Nov	674,508		Nov	863,194
	Dec	527,463		Dec	551,074		Dec	704,270		Dec	878,643
1997	Jan	522,154	2001	Jan	549,331	2005	Jan	698,237	2009	Jan	893,385
	Feb	513,753		Feb	544,584		Feb	691,785		Feb	879,077
	Mar	505,158		Mar	542,799		Mar	683,112		Mar	850,749
	Apr	511,280		Apr	550,755		Apr	690,490		Apr	843,841
	May	512,704		May	552,844		May	685,201		May	853,445
	Jun	512,017		Jun	550,374		Jun	683,951		Jun	837,783
	Jul	516,038		Jul	545,662		Jul	694,674		Jul	832,549
	Aug	518,003		Aug	548,030		Aug	705,421		Aug	840,395
	Sep	509,663		Sep	546,509		Sep	708,230		Sep	832,687
	Oct	508,541		Oct	550,854		Oct	701,905		Oct	837,286
	Nov	508,851		Nov	560,473		Nov	697,084		Nov	847,458
	Dec	515,064		Dec	568,437		Dec	707,039		Dec	855,311
1998	Jan	500,986	2002	Jan	567,381	2006	Jan	706,577	2010	Jan	849,478
	Feb	494,213		Feb	562,399		Feb	696,350		Feb	829,090
	Mar	493,846		Mar	561,548		Mar	697,480		Mar	1,161,466
	Apr	501,914		Apr	567,072		Apr	704,648		Apr	1,157,648
	May	499,216		May	568,930		May	713,087		May	1,152,223
	Jun	493,317		Jun	566,783		Jun	694,749		Jun	1,148,219
	Jul	493,306		Jul	569,687		Jul	701,261		Jul	1,147,999
	Aug	500,031		Aug	585,363		Aug	715,117		Aug	1,143,622
	Sep	499,624		Sep	589,192		Sep	710,502			
	Oct	504,274		Oct	591,441		Oct	712,209			
	Nov	501,555		Nov	594,839		Nov	725,239			
	Dec	512,007		Dec	602,570		Dec	741,200			
1999	Jan	512,402	2003	Jan	598,346	2007	Jan	742,420			
	Feb	504,946		Feb	598,672		Feb	725,923			
	Mar	499,258		Mar	593,231		Mar	723,278			
	Apr	500,394		Apr	595,530		Apr	729,133			
	May	499,066		May	601,382		May	735,749			
	Jun	484,471		Jun	603,684		Jun	737,869			
	Jul	485,049		Jul	603,382		Jul	748,568			
	Aug	483,988		Aug	610,810		Aug	763,852			
	Sep	480,406		Sep	614,124		Sep	771,518			
	Oct	481,922		Oct	608,947		Oct	771,267			
	Nov	488,642		Nov	639,768		Nov	788,005			
	Dec	507,763		Dec	669,386		Dec	804,096			

Source: US Federal Reserve.

TABLE B.10   100 Observations Simulated from an ARMA(1, 1) Process

1.371	4.311	0.080	6.418
1.182	4.664	−0.589	6.923
0.806	4.649	−1.670	6.354
0.911	5.360	−0.942	5.815
1.430	4.573	−1.057	4.072
2.058	3.430	−2.133	4.701
1.041	1.977	−0.398	7.042
0.314	1.998	0.178	7.417
−0.575	2.162	−1.938	4.933
−1.675	3.993	−3.022	3.441
−1.175	3.617	−2.616	3.718
−1.152	0.884	−2.103	2.410
−0.821	0.368	−0.578	3.700
−1.227	1.549	−0.214	5.727
−2.161	1.889	−0.269	6.981
−2.110	1.719	1.517	8.371
−1.371	0.650	2.927	8.614
−0.129	1.254	4.291	7.191
0.427	1.503	5.320	6.154
1.426	1.081	5.324	5.707
1.757	0.560	5.635	5.339
3.990	0.199	5.168	3.943
4.096	−0.330	5.881	2.349
1.895	−0.956	5.146	2.097
2.947	−0.669	4.968	2.046

TABLE B.11   Quarterly Rental Vacancy Rates in the United States

Year	Quarter	Rate	Year	Quarter	Rate	Year	Quarter	Rate	Year	Quarter	Rate
1956	1	6.2	1970	1	5.4	1984	1	5.6	1998	1	7.7
1956	2	5.9	1970	2	5.4	1984	2	5.5	1998	2	8.0
1956	3	6.3	1970	3	5.3	1984	3	6.0	1998	3	8.2
1956	4	5.8	1970	4	5.2	1984	4	6.3	1998	4	7.8
1957	1	5.3	1971	1	5.3	1985	1	6.3	1999	1	8.2
1957	2	5.3	1971	2	5.3	1985	2	6.2	1999	2	8.1
1957	3	5.7	1971	3	5.6	1985	3	6.8	1999	3	8.2
1957	4	5.8	1971	4	5.6	1985	4	6.7	1999	4	7.9
1958	1	6.3	1972	1	5.3	1986	1	6.9	2000	1	7.9
1958	2	6.5	1972	2	5.5	1986	2	7.3	2000	2	8.0
1958	3	6.3	1972	3	5.8	1986	3	7.5	2000	3	8.2
1958	4	6.6	1972	4	5.6	1986	4	7.7	2000	4	7.8
1959	1	6.8	1973	1	5.7	1987	1	7.4	2001	1	8.2
1959	2	7.3	1973	2	5.8	1987	2	7.5	2001	2	8.3
1959	3	7.2	1973	3	5.8	1987	3	8.1	2001	3	8.4
1959	4	7.0	1973	4	5.8	1987	4	7.8	2001	4	8.8
1960	1	8.0	1974	1	6.2	1988	1	8.0	2002	1	9.1
1960	2	8.0	1974	2	6.3	1988	2	7.7	2002	2	8.4
1960	3	8.3	1974	3	6.2	1988	3	7.8	2002	3	9.0
1960	4	8.4	1974	4	6.0	1988	4	7.3	2002	4	9.3
1961	1	8.9	1975	1	6.1	1989	1	7.5	2003	1	9.4
1961	2	8.8	1975	2	6.3	1989	2	7.4	2003	2	9.6
1961	3	8.6	1975	3	6.2	1989	3	7.6	2003	3	9.9
1961	4	8.5	1975	4	5.4	1989	4	7.1	2003	4	10.2
1962	1	8.5	1976	1	5.5	1990	1	7.5	2004	1	10.4
1962	2	8.1	1976	2	5.8	1990	2	7.0	2004	2	10.2
1962	3	8.0	1976	3	5.7	1990	3	7.2	2004	3	10.1
1962	4	8.1	1976	4	5.3	1990	4	7.2	2004	4	10.0
1963	1	8.0	1977	1	5.1	1991	1	7.5	2005	1	10.1
1963	2	8.2	1977	2	5.3	1991	2	7.3	2005	2	9.8
1963	3	8.3	1977	3	5.4	1991	3	7.6	2005	3	9.9
1963	4	8.3	1977	4	5.1	1991	4	7.3	2005	4	9.6
1964	1	8.1	1978	1	5.0	1992	1	7.4	2006	1	9.5
1964	2	8.1	1978	2	5.1	1992	2	7.7	2006	2	9.6
1964	3	8.4	1978	3	5.0	1992	3	7.3	2006	3	9.9
1964	4	8.3	1978	4	5.0	1992	4	7.1	2006	4	9.8
1965	1	8.5	1979	1	5.1	1993	1	7.8	2007	1	10.1
1965	2	8.2	1979	2	5.5	1993	2	7.6	2007	2	9.5
1965	3	7.8	1979	3	5.7	1993	3	7.0	2007	3	9.8
1965	4	8.5	1979	4	5.4	1993	4	6.9	2007	4	9.6
1966	1	8.3	1980	1	5.2	1994	1	7.5	2008	1	10.1
1966	2	7.4	1980	2	5.6	1994	2	7.4	2008	2	10.0
1966	3	7.4	1980	3	5.7	1994	3	7.2	2008	3	9.9
1966	4	7.7	1980	4	5.0	1994	4	7.4	2008	4	10.1
1967	1	7.3	1981	1	5.2	1995	1	7.4	2009	1	10.1
1967	2	6.9	1981	2	5.0	1995	2	7.7	2009	2	10.6
1967	3	7.0	1981	3	5.0	1995	3	7.7	2009	3	11.1
1967	4	6.2	1981	4	5.0	1995	4	7.7	2009	4	10.7
1968	1	6.1	1982	1	5.3	1996	1	7.9	2010	1	10.6
1968	2	6.2	1982	2	5.1	1996	2	7.8	2010	2	10.6
1968	3	5.9	1982	3	5.3	1996	3	8.0			
1968	4	5.4	1982	4	5.5	1996	4	7.7			
1969	1	5.6	1983	1	5.7	1997	1	7.5			
1969	2	5.7	1983	2	5.5	1997	2	7.9			
1969	3	5.5	1983	3	5.8	1997	3	7.9			
1969	4	5.1	1983	4	5.5	1997	4	7.7			

Source: US Census Bureau.

TABLE B.12 Wölfer Sunspot Numbers

Year	Number	Year	Number	Year	Number
1770	101	1803	43	1836	122
1771	82	1804	48	1837	138
1772	66	1805	42	1838	103
1773	35	1806	28	1839	86
1774	31	1807	10	1840	63
1775	7	1808	8	1841	37
1776	20	1809	2	1842	24
1777	92	1810	0	1843	11
1778	154	1811	1	1844	15
1779	125	1812	5	1845	40
1780	85	1813	12	1846	62
1781	68	1814	14	1847	98
1782	38	1815	35	1848	124
1783	23	1816	46	1849	96
1784	10	1817	41	1850	66
1785	24	1818	30	1851	64
1786	83	1819	24	1852	54
1787	132	1820	16	1853	39
1788	131	1821	7	1854	21
1789	118	1822	4	1855	7
1790	90	1823	2	1856	4
1791	67	1824	8	1857	23
1792	60	1825	17	1858	55
1793	47	1826	36	1859	94
1794	41	1827	50	1860	96
1795	21	1828	62	1861	77
1796	16	1829	67	1862	59
1797	6	1830	71	1863	44
1798	4	1831	48	1864	47
1799	7	1832	28	1865	30
1800	14	1833	8	1866	16
1801	34	1834	13	1867	7
1802	45	1835	57	1868	37
				1869	74

Source: BJR series E.

TABLE B.13   Viscosity Readings from a Chemical Process

8	8.8	9.3	9.1	9.0	10.0	9.6
8.0	8.6	9.9	9.5	9.0	9.8	8.6
7.4	8.6	9.7	9.4	9.4	9.8	8.0
8.0	8.4	9.1	9.5	9.0	9.7	8.0
8.0	8.3	9.3	9.6	9.0	9.6	8.0
8.0	8.4	9.5	10.2	9.4	9.4	8.0
8.0	8.3	9.4	9.8	9.4	9.2	8.4
8.8	8.3	9.0	9.6	9.6	9.0	8.8
8.4	8.1	9.0	9.6	9.4	9.4	8.4
8.4	8.2	8.8	9.4	9.6	9.6	8.4
8.0	8.3	9.0	9.4	9.6	9.6	9.0
8.2	8.5	8.8	9.4	9.6	9.6	9.0
8.2	8.1	8.6	9.4	10.0	9.6	9.4
8.2	8.1	8.6	9.6	10.0	9.6	10.0
8.4	7.9	8.0	9.6	9.6	9.6	10.0
8.4	8.3	8.0	9.4	9.2	9.0	10.0
8.4	8.1	8.0	9.4	9.2	9.4	10.2
8.6	8.1	8.0	9.0	9.2	9.4	10.0
8.8	8.1	8.6	9.4	9.0	9.4	10.0
8.6	8.4	8.0	9.4	9.0	9.6	9.6
8.6	8.7	8.0	9.6	9.6	9.4	9.0
8.6	9.0	8.0	9.4	9.8	9.6	9.0
8.6	9.3	7.6	9.2	10.2	9.6	8.6
8.6	9.3	8.6	8.8	10.0	9.8	9.0
8.8	9.5	9.6	8.8	10.0	9.8	9.6
8.9	9.3	9.6	9.2	10.0	9.8	9.6
9.1	9.5	10.0	9.2	9.4	9.6	9.0
9.5	9.5	9.4	9.6	9.2	9.2	9.0
8.5	9.5	9.3	9.6	9.6	9.6	8.9
8.4	9.5	9.2	9.8	9.7	9.2	8.8
8.3	9.5	9.5	9.8	9.7	9.2	8.7
8.2	9.5	9.5	10.0	9.8	9.6	8.6
8.1	9.9	9.5	10.0	9.8	9.6	8.3
8.3	9.5	9.9	9.4	9.8	9.6	7.9
8.4	9.7	9.9	9.8	10.0	9.6	8.5
8.7	9.1	9.5	8.8	10.0	9.6	8.7
8.8	9.1	9.3	8.8	8.6	9.6	8.9
8.8	8.9	9.5	8.8	9.0	10.0	9.1
9.2	9.3	9.5	8.8	9.4	10.0	9.1
9.6	9.1	9.1	9.6	9.4	10.4	9.1
9.0	9.1	9.3	9.6	9.4	10.4	
8.8	9.3	9.5	9.6	9.4	9.8	
8.6	9.5	9.3	9.2	9.4	9.0	
8.6	9.3	9.1	9.2	9.6	9.6	
8.8	9.3	9.3	9.0	10	9.8	

Source: BJR series D.

TABLE B.14   UK Midyear Population

Year	Population	Year	Population
1950	50,127,000	1981	56,382,597
1951	50,290,000	1982	56,330,706
1952	50,430,000	1983	56,364,615
1953	50,593,000	1984	56,435,179
1954	50,765,000	1985	56,584,077
1955	50,946,000	1986	56,750,920
1956	51,184,000	1987	56,927,038
1957	51,430,000	1988	57,095,731
1958	51,652,000	1989	57,251,270
1959	51,956,000	1990	57,410,719
1960	52,372,000	1991	57,573,614
1961	52,807,000	1992	57,735,940
1962	53,292,000	1993	57,854,243
1963	53,625,000	1994	58,005,687
1964	53,991,000	1995	58,186,804
1965	54,350,000	1996	58,344,567
1966	54,643,000	1997	58,494,073
1967	54,959,000	1998	58,680,368
1968	55,214,000	1999	58,909,682
1969	55,461,000	2000	59,139,969
1970	55,632,000	2001	59,374,727
1971	55,907,000	2002	59,601,520
1972	56,079,000	2003	59,823,619
1973	56,210,000	2004	60,126,953
1974	56,224,000	2005	60,487,347
1975	56,215,000	2006	60,846,809
1976	56,206,000	2007	61,249,260
1977	56,179,000	2008	61,642,600
1978	56,167,000	2009	61,996,848
1979	56,228,000	2010	62,348,447
1980	56,314,000		

Source: UK Office for National Statistics.

TABLE B.15   Unemployment and GDP data for the United Kingdom

Year	Quarter	Unemployment	GDP	Year	Quarter	Unemployment	GDP
1955	1	225	81.37	1963	1	535	96.16
	2	208	82.60		2	520	99.79
	3	201	82.30		3	489	101.14
	4	199	83.00		4	456	102.95
1956	1	207	82.87	1964	1	386	103.96
	2	215	83.60		2	368	105.28
	3	240	83.33		3	358	105.81
	4	245	83.53		4	330	107.14
1957	1	295	84.27	1965	1	306	108.07
	2	293	85.50		2	304	107.64
	3	279	84.33		3	321	108.87
	4	287	84.30		4	305	109.75
1958	1	331	85.07	1966	1	279	110.20
	2	396	83.60		2	282	110.20
	3	432	84.37		3	318	110.90
	4	462	84.50		4	414	110.40
1959	1	454	85.20	1967	1	463	111.00
	2	446	87.07		2	506	112.10
	3	426	88.40		3	538	112.50
	4	402	90.03		4	536	113.00
1960	1	363	92.30	1968	1	544	114.30
	2	342	92.13		2	541	115.10
	3	325	93.17		3	547	116.40
	4	312	93.50		4	532	117.80
1961	1	291	94.77	1969	1	532	116.80
	2	293	95.37		2	519	117.80
	3	304	95.03		3	547	119.00
	4	330	95.23		4	544	119.60
1962	1	357	95.07				
	2	401	96.40				
	3	447	96.97				
	4	483	96.50				

Source: BJR series P.

TABLE B.16 Monthly Crude Oil Production of OPEC Nations

Year	Algeria	Angola	Ecuador	Iran	Iraq	Kuwait	Libya	Nigeria	Qatar	Saudi Arabia	UAB	Venezuela
1990 January	1160	458	280	2700	2946	2003	1222	1731	390	5537	2052	1990
1990 February	1160	473	280	3000	2946	2003	1375	1731	401	5636	2027	2140
1990 March	1160	473	280	3000	2946	2184	1324	1731	422	5765	2052	2040
1990 April	1160	473	280	2900	2997	1958	1273	1830	422	5888	2097	2040
1990 May	1160	473	280	3200	3150	1958	1273	1731	385	5394	2107	2040
1990 June	1160	478	280	3100	3251	1762	1273	1731	385	5398	2047	2040
1990 July	1160	478	280	3050	3454	1858	1273	1731	390	5394	2047	2040
1990 August	1160	478	300	3300	1016	100	1426	1830	422	5789	1648	2090
1990 September	1190	488	290	3300	508	100	1426	1880	422	7660	2197	2290
1990 October	1210	473	290	3000	457	75	1579	1929	422	7729	2307	2275
1990 November	1210	478	290	3200	432	75	1528	1929	422	8224	2372	2320
1990 December	1210	473	290	3300	432	75	1528	1929	390	8481	2447	2340
1991 January	1230	493	298	3179	256	51	1500	1906	366	8075	2448	2396
1991 February	1230	503	303	3278	0	0	1500	1906	407	8134	2472	2396
1991 March	1230	503	303	3378	0	0	1450	1906	407	7936	2496	2396
1991 April	1230	503	303	3278	205	0	1450	1906	407	7341	2496	2346
1991 May	1230	503	282	3278	358	0	1450	1906	407	7341	2301	2346
1991 June	1230	503	282	3278	358	76	1450	1858	407	8085	2301	2346
1991 July	1230	493	303	3378	410	167	1450	1858	407	8407	2301	2346
1991 August	1230	493	303	3378	410	198	1450	1906	407	8397	2301	2346
1991 September	1230	493	303	3278	410	304	1500	1906	407	8332	2292	2346
1991 October	1230	503	303	3278	410	436	1500	1809	407	8382	2379	2396
1991 November	1230	503	303	3278	410	507	1550	1906	386	8372	2443	2396
1991 December	1230	503	303	3477	410	527	1550	1930	324	8571	2409	2446

1992 January	1230	505	297	3500	450	565	1550	1975	350	8790	2435	2390
1992 February	1230	505	297	3500	450	630	1550	1925	325	8640	2425	2340
1992 March	1230	505	318	3350	450	735	1450	1900	375	8260	2300	2190
1992 April	1230	530	318	3250	450	863	1500	1925	375	8213	2300	2190
1992 May	1210	510	318	3250	450	915	1450	1925	375	8265	2300	2290
1992 June	1210	544	318	3250	450	1015	1450	1925	375	8315	2275	2290
1992 July	1210	544	323	3300	450	1080	1450	1975	400	8350	2300	2290
1992 August	1210	544	333	3450	450	1130	1425	2000	425	8400	2330	2340
1992 September	1210	544	333	3450	450	1200	1475	2025	425	8450	2320	2390
1992 October	1210	544	333	3650	450	1280	1500	2050	440	8505	2310	2440
1992 November	1210	520	333	3650	450	1375	1500	2050	440	8500	2305	2440
1992 December	1210	520	328	3550	450	1550	1480	2100	440	8575	2305	2415
1993 January	1210	485	328	3650	500	1675	1425	2125	456	8500	2244	2484
1993 February	1210	505	328	3750	500	1865	1350	2105	436	8440	2254	2464
1993 March	1200	510	328	3700	500	1650	1350	2075	406	8300	2219	2412
1993 April	1200	544	328	3500	500	1645	1350	2025	406	8000	2219	2412
1993 May	1200	544	343	3650	500	1712	1350	2025	426	8000	2180	2412
1993 June	1200	483	348	3650	500	1775	1350	1995	406	8150	2180	2412
1993 July	1180	490	348	3800	500	1940	1350	1975	416	8240	2161	2464
1993 August	1180	490	348	3500	500	2045	1370	2025	416	8345	2161	2464
1993 September	1180	510	348	3650	530	2020	1370	2045	416	8270	2170	2453
1993 October	1180	510	358	3700	530	2045	1390	2005	416	8145	2170	2474
1993 November	1170	515	358	3550	540	2045	1370	2025	416	7995	2170	2474
1993 December	1170	521	358	3700	540	2050	1370	2175	416	8000	2170	2474

(continued overleaf)

TABLE B.16 (*Continued*)

Year	Algeria	Angola	Ecuador	Iran	Iraq	Kuwait	Libya	Nigeria	Qatar	Saudi Arabia	UAB	Venezuela
1994 January	1180	508	347	3618	545	1986	1370	2085	416	8068	2181	2564
1994 February	1180	523	347	3568	545	1988	1370	2085	402	8061	2205	2564
1994 March	1180	503	347	3668	545	1996	1370	2038	416	8068	2181	2564
1994 April	1180	523	352	3518	555	2011	1370	1962	416	8083	2181	2554
1994 May	1180	538	352	3568	555	2041	1370	1990	416	8063	2190	2574
1994 June	1180	538	362	3668	555	2041	1370	1981	425	8063	2210	2574
1994 July	1180	548	371	3568	555	2041	1380	1886	444	8073	2210	2595
1994 August	1180	548	371	3618	555	2041	1390	1545	407	8093	2210	2615
1994 September	1180	548	386	3668	555	2041	1370	1905	416	8153	2210	2615
1994 October	1180	553	381	3618	555	2036	1390	1971	360	8218	2171	2615
1994 November	1180	553	381	3717	555	2036	1390	1876	425	8218	2171	2615
1994 December	1180	553	381	3618	555	2041	1390	1862	435	8273	2200	2605
1995 January	1185	580	400	3585	560	2070	1390	1965	416	8120	2238	2600
1995 February	1185	580	400	3685	560	2070	1390	1945	434	8220	2238	2600
1995 March	1185	610	400	3485	560	2060	1390	1855	443	8110	2238	2600
1995 April	1185	630	400	3635	560	2070	1390	2015	443	8220	2238	2670
1995 May	1185	645	400	3835	560	2050	1390	2045	443	8400	2238	2790
1995 June	1185	660	390	3585	560	2050	1390	1925	443	8100	2238	2790
1995 July	1215	660	385	3535	560	2060	1390	1945	443	8410	2238	2790
1995 August	1215	660	375	3685	560	2075	1390	2000	443	8425	2238	2790
1995 September	1215	655	390	3635	560	2035	1390	2005	443	8315	2238	2790
1995 October	1215	690	390	3735	560	2065	1390	2025	443	8315	2238	2840
1995 November	1225	690	385	3635	560	2070	1390	2075	453	8020	2238	2840
1995 December	1225	690	390	3685	560	2015	1390	2110	453	8110	2175	2890

1996 January	1220	705	399	3735	550	2038	1400	1974	500	8118	2290	2829
1996 February	1220	705	399	3685	550	2057	1400	1992	500	8248	2265	2829
1996 March	1210	700	399	3715	550	2057	1400	2001	500	8248	2285	2877
1996 April	1230	715	399	3685	550	2067	1400	1974	505	8088	2250	2877
1996 May	1245	715	399	3635	550	2055	1400	2011	505	8135	2275	2877
1996 June	1250	715	399	3685	550	2065	1400	2011	505	8195	2270	2877
1996 July	1250	720	399	3685	550	2065	1400	1983	505	8295	2260	2926
1996 August	1250	720	383	3715	550	2040	1400	2001	505	8220	2260	2974
1996 September	1250	700	394	3735	550	2070	1400	1965	525	8200	2310	2974
1996 October	1260	700	394	3635	550	2075	1400	2024	525	8255	2310	3022
1996 November	1260	700	394	3685	550	2075	1400	2033	505	8255	2250	3070
1996 December	1260	710	394	3635	887	2077	1410	2038	545	8358	2305	3118
1997 January	1260	694	393	3685	1056	2009	1430	2098	496	8072	2300	3156
1997 February	1270	709	393	3685	1095	2002	1430	2126	496	8212	2330	3156
1997 March	1280	714	387	3685	1144	2029	1440	2062	496	8316	2360	3166
1997 April	1280	714	398	3685	1241	2031	1450	2126	496	8368	2360	3186
1997 May	1280	714	387	3635	1290	1953	1450	2089	513	8348	2210	3206
1997 June	1260	704	367	3735	589	1976	1450	2153	584	8341	2325	3226
1997 July	1280	709	367	3685	589	1995	1450	2144	580	8360	2325	3236
1997 August	1280	714	377	3685	1475	1995	1450	2162	580	8458	2325	3354
1997 September	1280	711	398	3485	1689	2000	1450	2117	580	8463	2325	3394
1997 October	1280	719	393	3635	1582	2000	1450	2208	580	8463	2325	3394
1997 November	1280	719	403	3685	1353	2000	1450	2171	597	8414	2305	3424
1997 December	1290	744	398	3685	760	2096	1450	2135	597	8521	2310	3453

(continued overleaf)

TABLE B.16 (*Continued*)

Year	Algeria	Angola	Ecuador	Iran	Iraq	Kuwait	Libya	Nigeria	Qatar	Saudi Arabia	UAB	Venezuela
1998 January	1290	735	388	3635	1261	2215	1463	2218	715	8765	2435	3440
1998 February	1290	730	378	3635	1703	2210	1463	2263	735	8760	2435	3410
1998 March	1290	730	378	3635	1825	2210	1463	2380	735	8460	2480	3410
1998 April	1270	710	378	3835	1985	2115	1412	2238	705	8585	2420	3240
1998 May	1250	710	378	3635	2245	2105	1372	2230	705	8625	2330	3240
1998 June	1240	710	378	3835	1920	2105	1372	2210	705	8325	2300	3210
1998 July	1230	710	329	3585	2355	2075	1372	2160	685	8275	2280	3070
1998 August	1220	730	378	3435	2555	2025	1352	2010	675	8225	2300	2990
1998 September	1220	735	383	3685	2555	1972	1347	2010	665	8173	2300	2940
1998 October	1220	765	388	3485	2555	1970	1347	1960	670	8220	2290	2990
1998 November	1220	770	378	3635	2505	2020	1362	2060	675	8170	2290	3040
1998 December	1220	785	368	3585	2305	2010	1362	2110	680	8110	2290	3040
1999 January	1230	758	370	3665	2515	1995	1360	2080	666	8065	2239	3019
1999 February	1240	758	380	3925	2655	2005	1360	2010	666	8165	2329	2999
1999 March	1250	768	350	3795	2430	2020	1360	2160	742	8220	2234	2960
1999 April	1210	749	350	3485	2655	1785	1320	2160	675	7665	2180	2800
1999 May	1190	734	370	3435	2705	1815	1300	2190	656	7665	2130	2780
1999 June	1180	734	367	3415	2355	1830	1290	2150	627	7610	2110	2760
1999 July	1180	724	365	3515	2805	1830	1290	2130	656	7610	2130	2760
1999 August	1190	729	380	3535	2855	1860	1290	2140	656	7710	2140	2760
1999 September	1190	739	380	3485	2855	1885	1300	2150	656	7735	2145	2760
1999 October	1190	749	390	3535	2670	1925	1310	2170	656	7845	2145	2760
1999 November	1190	749	380	3485	2205	1905	1320	2160	656	7865	2105	2780
1999 December	1190	749	390	3435	1405	1922	1330	2050	666	7863	2155	2780

2000 January	1205	760	390	3444	2215	1918	1330	2030	695	7863	2265	2985
2000 February	1205	760	390	3504	2595	1970	1380	2081	705	7865	2270	3049
2000 March	1205	760	390	3712	2215	1994	1390	2101	705	7865	2320	3049
2000 April	1245	750	375	3653	2655	2053	1400	2161	715	8100	2400	3103
2000 May	1255	760	370	3663	3055	2053	1400	2131	735	8200	2400	3135
2000 June	1265	750	373	3683	2565	2102	1420	2161	735	8250	2300	3156
2000 July	1265	750	401	3727	2525	2121	1425	2201	755	8390	2340	3178
2000 August	1275	760	410	3727	2995	2124	1420	2181	755	8823	2400	3189
2000 September	1265	740	425	3732	2875	2121	1430	2131	755	8975	2410	3189
2000 October	1285	723	420	3812	3005	2160	1440	2231	760	8800	2430	3264
2000 November	1280	723	415	3807	2815	2165	1440	2282	765	8900	2435	3264
2000 December	1295	721	380	3881	1355	2160	1445	2287	765	8800	2440	3296
2001 January	1337	746	420	3935	1705	2169	1450	2285	775	8700	2383	3240
2001 February	1305	749	408	3785	2157	2100	1400	2255	735	8320	2324	3167
2001 March	1305	731	425	3835	2805	2070	1390	2285	735	8300	2363	3135
2001 April	1289	739	425	3785	2879	1982	1380	2210	715	7950	2276	3052
2001 May	1305	733	420	3685	2854	1965	1360	2140	725	8000	2225	3020
2001 June	1326	728	424	3785	1086	2001	1370	2205	735	8050	2208	3031
2001 July	1337	713	406	3875	2108	1992	1380	2140	735	8250	2189	3020
2001 August	1337	701	408	3785	2825	2006	1380	2207	725	8070	2177	3010
2001 September	1305	710	404	3655	2626	1942	1350	2360	685	7800	2102	2842
2001 October	1284	754	401	3535	2860	1922	1320	2350	685	7670	2073	2874
2001 November	1295	785	402	3535	2756	1913	1310	2350	665	7670	2073	2863
2001 December	1295	820	403	3491	1990	1913	1310	2290	655	7600	2073	2874

(continued overleaf)

TABLE B.16 *(Continued)*

Year	Algeria	Angola	Ecuador	Iran	Iraq	Kuwait	Libya	Nigeria	Qatar	Saudi Arabia	UAB	Venezuela
2002 January	1221	910	390	3385	2315	1850	1260	2150	625	7300	2060	2630
2002 February	1215	950	383	3365	2545	1803	1280	2100	625	7210	2050	2600
2002 March	1235	940	391	3385	2515	1850	1290	2120	635	7310	2055	2620
2002 April	1245	940	389	3375	1215	1860	1300	2130	655	7455	2070	2530
2002 May	1275	930	388	3395	1865	1880	1310	2070	675	7450	2060	2730
2002 June	1285	895	390	3415	1525	1890	1320	2060	665	7500	2060	2735
2002 July	1305	870	402	3425	1835	1910	1330	2050	675	7700	2080	2735
2002 August	1315	910	396	3440	1505	1910	1330	2100	685	7730	2090	2765
2002 September	1345	885	407	3485	1825	1930	1350	2143	695	7880	2103	2955
2002 October	1395	895	393	3535	2425	1930	1350	2140	725	7900	2113	2980
2002 November	1383	815	392	3535	2395	1940	1350	2150	730	8100	2100	2972
2002 December	1445	820	392	3585	2325	1970	1350	2200	755	8050	2140	1020
2003 January	1490	820	385	3625	2549	1951	1375	2345	713	8499	2200	630
2003 February	1495	825	375	3699	2484	2011	1400	2395	737	8797	2250	1450
2003 March	1555	850	370	3724	1370	2256	1405	2060	737	9382	2450	2390
2003 April	1645	910	385	3719	53	2354	1430	1994	737	9521	2450	2555
2003 May	1645	920	330	3719	292	2241	1435	2082	737	9322	2400	2665
2003 June	1625	920	357	3719	452	2060	1430	2184	690	8628	2350	2640
2003 July	1645	920	375	3749	572	2060	1430	2219	690	8539	2350	2640
2003 August	1645	920	380	3749	1050	2060	1425	2295	690	8539	2340	2640
2003 September	1645	920	477	3749	1399	2060	1425	2395	690	8479	2300	2640
2003 October	1645	920	480	3749	1749	2158	1420	2395	690	8578	2330	2640
2003 November	1645	920	497	3798	1848	2158	1420	2446	737	8430	2350	2540
2003 December	1645	980	520	3912	1948	2256	1450	2497	737	8588	2400	2540

2004 January	1645	1030	503	3950	2103	2300	1450	2348	751	8700	2400	2540
2004 February	1645	1030	530	3950	2003	2300	1450	2348	761	8700	2420	2540
2004 March	1645	1030	510	3960	2203	2355	1450	2348	761	8400	2370	2540
2004 April	1645	1030	533	3970	2303	2350	1450	2348	761	8400	2220	2540
2004 May	1645	1030	540	3980	1903	2400	1450	2348	761	8500	2280	2540
2004 June	1665	1030	529	3990	1703	2400	1500	2395	799	9500	2510	2540
2004 July	1695	1030	532	4010	2003	2400	1550	2395	799	9500	2530	2540
2004 August	1695	1050	532	4030	1803	2400	1560	2302	799	9500	2600	2540
2004 September	1695	1070	532	4030	2303	2400	1560	2302	799	9500	2600	2540
2004 October	1695	1090	537	4035	2203	2400	1560	2302	799	9500	2602	2540
2004 November	1725	1100	534	4050	1703	2400	1600	2302	799	9500	2602	2640
2004 December	1725	1100	529	4060	1903	2400	1600	2210	799	9500	2602	2640
2005 January	1750	1110	532	4060	1903	2450	1600	2430	835	9500	2502	2640
2005 February	1755	1120	537	4080	1903	2500	1600	2480	835	9500	2502	2640
2005 March	1775	1140	528	4080	1903	2500	1620	2580	835	9500	2552	2640
2005 April	1775	1150	523	4090	1903	2500	1625	2640	835	9600	2602	2540
2005 May	1775	1170	526	4100	1903	2500	1630	2690	835	9600	2402	2540
2005 June	1805	1169	548	4210	1903	2500	1635	2695	835	9600	2402	2540
2005 July	1805	1211	548	4220	2003	2500	1635	2695	835	9600	2502	2540
2005 August	1825	1356	477	4230	1903	2500	1650	2590	835	9600	2552	2540
2005 September	1825	1400	542	4190	2053	2600	1650	2635	835	9600	2602	2540
2005 October	1825	1360	534	4150	1803	2600	1650	2695	835	9500	2602	2540
2005 November	1825	1400	535	4150	1703	2600	1650	2695	835	9500	2602	2540
2005 December	1825	1410	555	4100	1653	2600	1650	2695	835	9500	2602	2540

(continued overleaf)

TABLE B.16 (*Continued*)

Year	Algeria	Angola	Ecuador	Iran	Iraq	Kuwait	Libya	Nigeria	Qatar	Saudi Arabia	UAB	Venezuela
2006 January	1825	1420	553	4100	1603	2600	1650	2560	835	9400	2602	2540
2006 February	1825	1420	551	4050	1803	2550	1650	2410	835	9500	2602	2540
2006 March	1825	1420	528	4000	1903	2525	1680	2370	835	9350	2602	2540
2006 April	1825	1420	546	4000	1903	2525	1690	2370	835	9350	2602	2540
2006 May	1785	1320	547	3950	1903	2525	1700	2370	835	9200	2602	2540
2006 June	1795	1285	536	4030	2153	2550	1700	2465	835	9100	2602	2540
2006 July	1805	1460	543	4035	2203	2550	1700	2380	855	9300	2702	2440
2006 August	1805	1460	544	4035	2203	2550	1700	2430	885	9300	2702	2490
2006 September	1835	1438	533	4035	2153	2550	1700	2430	885	9000	2702	2490
2006 October	1835	1376	519	4060	2103	2550	1700	2530	885	8800	2702	2490
2006 November	1805	1452	511	4020	2003	2500	1650	2480	845	8800	2602	2490
2006 December	1805	1484	516	4020	2003	2450	1650	2480	835	8750	2602	2490
2007 January	1838	1584	517	4040	1753	2450	1680	2365	835	8750	2613	2380
2007 February	1833	1600	507	3900	2003	2420	1680	2390	825	8600	2573	2383
2007 March	1829	1640	482	3900	2053	2420	1680	2275	825	8600	2612	2445
2007 April	1825	1679	502	3900	2103	2420	1680	2400	825	8600	2611	2445
2007 May	1821	1695	512	3900	2103	2420	1680	2240	825	8600	2611	2444
2007 June	1828	1680	515	3900	2003	2420	1680	2230	835	8600	2610	2444
2007 July	1828	1710	510	3900	2053	2445	1700	2380	865	8600	2610	2444
2007 August	1824	1730	508	3900	1903	2500	1700	2380	865	8600	2659	2444
2007 September	1831	1791	517	3900	2203	2500	1720	2380	865	8800	2709	2440
2007 October	1842	1889	514	3900	2303	2500	1740	2330	869	8800	2711	2440
2007 November	1852	1940	518	3900	2253	2520	1740	2400	883	9000	2242	2440
2007 December	1852	1986	532	3900	2303	2550	1740	2430	888	9100	2659	2440

2008 January	1826	1992	520	4000	2203	2550	1790	2230	892	9200	2709	2440
2008 February	1826	1997	519	4000	2353	2600	1790	2100	916	9200	2709	2440
2008 March	1825	2003	508	4000	2353	2600	1790	2330	920	9200	2710	2430
2008 April	1825	2009	510	4000	2353	2600	1769	2130	934	9100	2710	2420
2008 May	1825	2015	499	4000	2453	2600	1745	2060	938	9400	2710	2410
2008 June	1824	2013	495	4000	2453	2607	1745	2140	942	9450	2710	2400
2008 July	1824	2009	498	4100	2505	2614	1720	2120	947	9700	2710	2390
2008 August	1824	1937	503	4100	2456	2622	1645	2216	951	9600	2711	2380
2008 September	1824	1871	498	4100	2328	2629	1745	2210	955	9400	2711	2370
2008 October	1824	1990	497	4100	2328	2629	1745	2185	925	9400	2661	2360
2008 November	1824	1990	502	4100	2359	2486	1700	2180	885	8959	2561	2350
2008 December	1824	1940	508	4100	2360	2493	1650	2080	885	8518	2561	2340
2009 January	1758	1915	504	4007	2212	2350	1650	2192	860	8113	2411	2340
2009 February	1757	1840	498	3963	2313	2350	1650	2162	935	8068	2412	2340
2009 March	1757	1840	497	3970	2365	2350	1650	2060	910	8072	2412	2340
2009 April	1757	1840	495	4030	2366	2350	1650	2217	910	8077	2412	2240
2009 May	1757	1840	486	4044	2418	2350	1650	2212	910	8081	2412	2240
2009 June	1756	1840	491	4050	2419	2350	1650	2059	910	8335	2412	2240
2009 July	1806	1890	483	4053	2470	2350	1650	2051	910	8540	2413	2240
2009 August	1806	1950	477	4056	2472	2350	1650	2193	945	8440	2413	2240
2009 September	1806	1950	475	4060	2473	2350	1650	2240	945	8340	2413	2240
2009 October	1806	1990	475	4063	2425	2350	1650	2290	951	8340	2413	2240
2009 November	1806	1990	477	4067	2375	2350	1650	2370	962	8340	2413	2140
2009 December	1806	1990	470	4076	2375	2350	1650	2450	974	8240	2414	2040

(continued overleaf)

TABLE B.16 (*Continued*)

Year	Algeria	Angola	Ecuador	Iran	Iraq	Kuwait	Libya	Nigeria	Qatar	Saudi Arabia	UAB	Venezuela
2010 January	1810	2040	463	4088	2475	2350	1650	2480	969	8240	2414	2090
2010 February	1809	2060	469	4100	2475	2350	1650	2420	1036	8240	2414	2140
2010 March	1809	2070	479	4112	2375	2350	1650	2430	1055	8240	2414	2090
2010 April	1809	2070	477	4120	2375	2350	1650	2360	1072	8240	2414	2110
2010 May	1809	2030	478	4120	2375	2350	1650	2310	1091	8340	2415	2140
2010 June	1808	1980	494	4127	2425	2350	1650	2410	1113	8440	2415	2140
2010 July	1808	1970	488	4133	2325	2350	1650	2410	1136	8540	2415	2140

Source: U.S. Energy Information Administration.

TABLE B.17   Quarterly Dollar Sales of Marshall Field & Company ($1000)

Year	Quarter	Sales	Year	Quarter	Sales
1960	Q1	50,147	1968	Q1	73,567
1960	Q2	49,325	1968	Q2	79,360
1960	Q3	57,948	1968	Q3	97,324
1960	Q4	76,781	1968	Q4	136,036
1961	Q1	48,617	1969	Q1	78,071
1961	Q2	50,898	1969	Q2	84,382
1961	Q3	58,517	1969	Q3	100,422
1961	Q4	77,691	1969	Q4	139,632
1962	Q1	50,862	1970	Q1	79,823
1962	Q2	53,028	1970	Q2	84,750
1962	Q3	58,849	1970	Q3	96,724
1962	Q4	79,660	1970	Q4	140,586
1963	Q1	51,640	1971	Q1	97,015
1963	Q2	54,119	1971	Q2	103,137
1963	Q3	65,681	1971	Q3	114,402
1963	Q4	85,175	1971	Q4	152,135
1964	Q1	56,405	1972	Q1	101,984
1964	Q2	60,031	1972	Q2	108,562
1964	Q3	71,486	1972	Q3	121,749
1964	Q4	92,183	1972	Q4	160,499
1965	Q1	60,800	1973	Q1	108,152
1965	Q2	64,900	1973	Q2	114,499
1965	Q3	76,997	1973	Q3	127,292
1965	Q4	103,337	1973	Q4	171,965
1966	Q1	67,279	1974	Q1	115,673
1966	Q2	69,869	1974	Q2	125,995
1966	Q3	81,029	1974	Q3	136,255
1966	Q4	107,192	1974	Q4	169,677
1967	Q1	69,110	1975	Q1	111,905
1967	Q2	76,084	1975	Q2	127,872
1967	Q3	87,160	1975	Q3	145,367
1967	Q4	113,829	1975	Q4	188,635

Source: Foster (1978).

BIBLIOGRAPHY

Abraham, B. and J. Ledolter (1983). *Statistical Methods for Forecasting*, New York: John Wiley and Sons.

Akaike, H. (1974). "A new look at the statistical model identification," *IEEE Transactions Automatic Control*, AC-19, pp. 716–723.

Anscombe, F. J. (1973). "Graphs in statistical analysis," *The American Statistician*, 27, pp. 17–21.

Bisgaard, S. and M. Kulahci (2005a). "Checking process stability with the variogram," *Quality Engineering*, 17(2), pp. 323–327.

Bisgaard, S. and M. Kulahci (2005b). "The effect of autocorrelation on statistical process control procedures," *Quality Engineering*, 17(3), pp. 481–489.

Bisgaard, S. and M. Kulahci (2005c). "Interpretation of time series models," *Quality Engineering*, 17(4), pp. 653–658.

Bisgaard, S. and M. Kulahci (2006a). "Process regime changes," *Quality Engineering*, 19(1), pp. 83–87.

Bisgaard, S. and M. Kulahci (2006b). "Studying input output relationships, I," *Quality Engineering*, 18(2), pp. 273–281.

Bisgaard, S. and M. Kulahci (2006c). "Studying input output relationships, II," *Quality Engineering*, 18(3), pp. 405–410.

Bisgaard, S. and M. Kulahci (2007a). "Beware of the effect of autocorrelation in regression," *Quality Engineering*, 19(2), pp. 143–148.

Bisgaard, S. and M. Kulahci (2007b). "Practical time series modeling I," *Quality Engineering*, 19(3), pp. 253–262.

Bisgaard, S. and M. Kulahci (2007c). "Practical time series modeling II," *Quality Engineering*, 19(4), pp. 394–400.

Bisgaard, S. and M. Kulahci (2008a). "Using a time series model for process adjustment and control," *Quality Engineering*, 20(1), pp. 134–141.

Bisgaard, S. and M. Kulahci (2008b). "Forecasting with seasonal time series models," *Quality Engineering*, 20(2), pp. 250–260.

Bisgaard, S. and M. Kulahci (2008c). "Box–Cox transformations and time series modeling—part I," *Quality Engineering*, 20(3), pp. 376–388.

Bisgaard, S. and M. Kulahci (2008d) "Box–Cox transformations and time series modeling—part II," *Quality Engineering*, 20(4), pp. 516–523.

Bisgaard, S. and M. Kulahci (2009). "Time series model selection and parsimony," *Quality Engineering*, 21(3), 341–353.

Box, G. E. P. (1991). "Understanding exponential smoothing: a simple way to forecast sales and inventory," *Quality Engineering*, 3(4), pp. 561–566.

Box, G. E. P. and D. R. Cox (1964). "An analysis of transformations," *Journal of the Royal Statistical Society, Series B*, 26, pp. 211–243.

Box, G. E. P., J. S. Hunter and W. G. Hunter (2005). *Statistics for Experiments: Design, Innovation and Discovery*, 2nd Edition, New York: John Wiley and Sons.

Box, G. E. P. and G. M. Jenkins (1970). *Time Series Analysis: Forecasting and Control*, San Francisco, CA: Holden-Day.

Box, G. E. P. and G. M. Jenkins (1973). "Some comments on a paper by Chatfield and Prothero and on a review by Kendall," *Journal of the Royal Statistical Society, Series A*, 136, pp. 337–352.

Box, G. E. P., G. M. Jenkins and G. C. Reinsel (2008). *Time Series Analysis: Forecasting and Control*, 4th Edition, Englewood Cliffs, NJ: Prentice Hall.

Box, G. E. P. and A. Luceno (1997). *Statistical Control by Monitoring and Feedback Adjustment*, New York: John Wiley and Sons.

Box, G. E. P. and P. Newbold (1971). "Some comments on a paper of Coen, Gomme and Kendall," *Journal of the Royal Statistical Society, Series A (General)*, 134(2), pp. 229–240.

Box, G. E. P. and G. C. Tiao (1965). "A change in level of a non-stationary time series," *Biometrika*, 52, pp. 181–192.

Box, G. E. P and G. C. Tiao (1977). "Canonical analysis of multiple time series," *Biometrika*, 64(2), pp. 355–365.

Brockwell, P. J. and R. A. Davis (1991). *Time Series: Theory and Methods*, 2nd Edition, New York: Springer.

Brockwell, P. J. and R. A. Davis (2002). *Introduction to Time Series and Forecasting*, 2nd Edition, New York: Springer.

Brown, R. G. (1962). *Smoothing, Forecasting and Prediction of Discrete Time Series*, Englewood Cliffs, NJ: Prentice-Hall.

Chatfield, C and D. L. Prothero (1973). "Box–Jenkins seasonal forecasting: problems in a case study," *Journal of the Royal Statistical Society, Series A*, 136, pp. 295–336.

Cleveland, W. S. (1993). *Visualizing Data*, NJ: Hobart Press.

Cleveland, W. S. (1994). *The Elements of Graphing Data*, 2nd Edition, NJ: Hobart Press.

Cleveland, W. S. and R. McGill (1987). "Graphical perception: the visual decoding of quantitative information on statistical graphs (with discussion)," *Journal of the Royal Statistical Society, Series A*, 150, pp. 192–229.

Coen, P. J., E. D. Gomme and M. G. Kendall (1969). "Lagged relationships in economic forecasting," *Journal of the Royal Statistical Society, Series A (General)*, 132(1), pp. 133–152.

Daniel, C. (1976). *Applications of Statistics to Industrial Experimentation*, New York: Wiley.

Dickey, D. A. and W. A. Fuller (1979). "Distribution of the estimates for autoregressive time series with a unit root," *Journal of American Statistical Association*, 74, pp. 427–431.

Draper, N. R. and H. Smith (1998). *Applied Regression Analysis*, 3rd Edition, New York: Wiley.

Engle, R. F. and C. W. J. Granger (1991). *Long-Run Economic Relationships*, Oxford: Oxford University Press.

Foster, G. (1978). *Financial Statement Analysis*, Englewood Cliffs, NJ: Prentice Hall.

Goldberg, S. (2010). *Introduction to Difference Equations*, New York: Dover Publications.

Granger, C. W. J. and P. Newbold (1974). "Spurious regression in econometrics," *Journal of Econometrics*, 2, pp. 111–120.

Granger, C. W. J. and P. Newbold (1986). *Forecasting Economic Time Series*, 2nd Edition, New York: Academic Press.

Hamilton, J. D. (1994). *Time Series Analysis*, Princeton, NJ: Princeton University Press.

Haslett, J. (1997). "On the sample variogram and the sample autocovariance for non-stationary time series," *The Statistician*, 46, pp. 475–485.

Holt, C. C. (1957). "Forecasting Trends and Seasonals by Exponentially Weighted Moving Averages," *O. N. R. Memorandum 52*, Pittsburgh, PA: Carnegie Institute of Technology.

Hurvich, C. M. and C. Tsai (1989). "Regression and time series model selection in small samples," *Biometrika*, 76(2), pp. 297–307.

Jenkins, G. M. (1979). *Practical Experiences with Modelling and Forecasting Time Series*, Jersey, Channel Islands: Gwilym Jenkins & Partners Ltd.

Johnson, R. A. and D. W. Wichern (2002). *Applied Multivariate Statistical Analysis*, 4th Edition, Upper Saddle River, NJ: Prentice Hall.

Jones, R. H. (1985). "Time Series Analysis with Unequally Spaced Data," *Handbook of Statistics*, Vol. 5, pp. 157–178, Amsterdam: North Holland.

Ledolter, J. and B. Abraham (1981). "Parsimony and its importance in time series forecasting," *Technometrics*, 23, pp. 411–414.

Makridakis, S. G., S. C. Wheelwright and R. J. Hyndman (2003). *Forecasting: Method and Applications*, 3rd Edition, New York: Wiley.

Matheron, G. (1963). "Principles of geostatistics," *Economic Geology*, 58(8), 1246–1266.

Montgomery, D. C., C. L. Jennings and M. Kulahci (2008). *Introduction to Time Series Analysis and Forecasting*, New York: Wiley.

Montgomery, D. C., E. A. Peck and G. G. Vining (2006). *Introduction to Linear Regression Analysis*, 4th Edition, New York, Wiley.

Muth, J. F. (1960). "Optimal properties of exponentially weighted forecasts," *Journal of American Statistical Association*, 55, pp. 299–306.

Nelson, C. R. and C. I. Plosser (1982). "Trends and random walks in macroeconomic time series," *Journal of Monetary Economics*, 10(2), pp. 139–162.

Peña, D., G. C. Tiao and R. S. Tsay,, Eds (2001). *A Course in Time Series Analysis*, New York: John Wiley and Sons.

Playfair, W. (1786). *Commercial and Political Atlas*. London: Debrett, Robinson and Sewell.

Quenouille, M. H. (1957). *Analysis of Multiple Time Series*, New York, Hafner.

Reinsel, G. (1997). *Elements of Multivariate Time Series Analysis*, 2nd Edition, New York, Springer.

Santner, T. J., B. J. Williams and W. I. Notz (2003). *The Design and Analysis of Computer Experiments*, New York: Springer.

Schwartz, G. (1978). "Estimating the dimension of a model," *Annals of Statistics*, 6, 461–464.

'Student' (1914). "The elimination of spurious correlation due to position in time or space," *Biometrika*, 10(1), pp. 179–180.

Tiao, G. C. and G. E. P. Box (1981). "Modeling multiple time series with applications," *Journal of American Statistical Association*, 76, pp. 802–816.

Tintner, G. (1940). *The Variate Difference Method*, Bloomington, IN: Principia Press.

Tufte, E. R. (1990). *Envisioning Information*, 4th Edition, Connecticut: Graphics Press.

Tufte, E. R. (1997). *Visual Explanations: Images and Quantities, Evidence and Narrative*, Connecticut: Graphics Press.

Wichern, D. W. (1973). "The behavior of sample autocorrelation function for an integrated moving average process," *Biometrika*, 60, pp. 235–239.

Wichern, D. W. and R. H. Jones (1977). "Assessing the impact of market disturbances using intervention analysis," *Management Science*, 24(3), pp. 329–337.

Wilson, G. T. (1973). "Contribution to discussion of "Box–Jenkins seasonal forecasting: problems in a case study," by C. Chatfield and D. L. Prothero," *Journal of the Royal Statistical Society, Series A*, 136, pp. 315–319.

Winters, P. R. (1960). "Forecasting sales by exponentially weighted moving averages," *Management Science*, 6, pp. 235–239.

Wold, H. O. (1938). *A Study in the Analysis of Stationary Time Series*, 2nd Edition, 1954, Uppsala, Sweden: Almquist & Wiksell.

Yule, G. U. (1927). "On a method of investigating periodicities in disturbed series, with reference to Wolfer's Sunspot Numbers," *Philosophical Transactions of the Royal Society of London, Series A*, 226, pp. 267–298.

INDEX

Time Series Analysis and Forecasting by Example, First Edition. Søren Bisgaard and Murat Kulahci.
© 2011 John Wiley & Sons, Inc. Published 2011 by John Wiley & Sons, Inc.

WILEY SERIES IN PROBABILITY AND STATISTICS
ESTABLISHED BY WALTER A. SHEWHART AND SAMUEL S. WILKS

Editors: *David J. Balding, Noel A. C. Cressie, Garrett M. Fitzmaurice, Harvey Goldstein, Iain M. Johnstone, Geert Molenberghs, David W. Scott, Adrian F. M. Smith, Ruey S. Tsay, Sanford Weisberg*
Editors Emeriti: *Vic Barnett, J. Stuart Hunter, Joseph B. Kadane, Jozef L. Teugels*

The *Wiley Series in Probability and Statistics* is well established and authoritative. It covers many topics of current research interest in both pure and applied statistics and probability theory. Written by leading statisticians and institutions, the titles span both state-of-the-art developments in the field and classical methods.

Reflecting the wide range of current research in statistics, the series encompasses applied, methodological and theoretical statistics, ranging from applications and new techniques made possible by advances in computerized practice to rigorous treatment of theoretical approaches.

This series provides essential and invaluable reading for all statisticians, whether in academia, industry, government, or research.

† ABRAHAM and LEDOLTER · Statistical Methods for Forecasting
AGRESTI · Analysis of Ordinal Categorical Data, *Second Edition*
AGRESTI · An Introduction to Categorical Data Analysis, *Second Edition*
AGRESTI · Categorical Data Analysis, *Second Edition*
ALTMAN, GILL, and McDONALD · Numerical Issues in Statistical Computing for the Social Scientist
AMARATUNGA and CABRERA · Exploration and Analysis of DNA Microarray and Protein Array Data
ANDĚL · Mathematics of Chance
ANDERSON · An Introduction to Multivariate Statistical Analysis, *Third Edition*
* ANDERSON · The Statistical Analysis of Time Series
ANDERSON, AUQUIER, HAUCK, OAKES, VANDAELE, and WEISBERG · Statistical Methods for Comparative Studies
ANDERSON and LOYNES · The Teaching of Practical Statistics
ARMITAGE and DAVID (editors) · Advances in Biometry
ARNOLD, BALAKRISHNAN, and NAGARAJA · Records
* ARTHANARI and DODGE · Mathematical Programming in Statistics
* BAILEY · The Elements of Stochastic Processes with Applications to the Natural Sciences
BALAKRISHNAN and KOUTRAS · Runs and Scans with Applications
BALAKRISHNAN and NG · Precedence-Type Tests and Applications
BARNETT · Comparative Statistical Inference, *Third Edition*
BARNETT · Environmental Statistics
BARNETT and LEWIS · Outliers in Statistical Data, *Third Edition*
BARTOSZYNSKI and NIEWIADOMSKA-BUGAJ · Probability and Statistical Inference
BASILEVSKY · Statistical Factor Analysis and Related Methods: Theory and Applications
BASU and RIGDON · Statistical Methods for the Reliability of Repairable Systems
BATES and WATTS · Nonlinear Regression Analysis and Its Applications

*Now available in a lower priced paperback edition in the Wiley Classics Library.
†Now available in a lower priced paperback edition in the Wiley–Interscience Paperback Series.

BECHHOFER, SANTNER, and GOLDSMAN · Design and Analysis of Experiments for Statistical Selection, Screening, and Multiple Comparisons

BELSLEY · Conditioning Diagnostics: Collinearity and Weak Data in Regression

† BELSLEY, KUH, and WELSCH · Regression Diagnostics: Identifying Influential Data and Sources of Collinearity

BENDAT and PIERSOL · Random Data: Analysis and Measurement Procedures, *Fourth Edition*

BERRY, CHALONER, and GEWEKE · Bayesian Analysis in Statistics and Econometrics: Essays in Honor of Arnold Zellner

BERNARDO and SMITH · Bayesian Theory

BHAT and MILLER · Elements of Applied Stochastic Processes, *Third Edition*

BHATTACHARYA and WAYMIRE · Stochastic Processes with Applications

BILLINGSLEY · Convergence of Probability Measures, *Second Edition*

BILLINGSLEY · Probability and Measure, *Third Edition*

BIRKES and DODGE · Alternative Methods of Regression

BISGAARD and KULAHCI · Time Series Analysis and Forecasting by Example

BISWAS, DATTA, FINE, and SEGAL · Statistical Advances in the Biomedical Sciences: Clinical Trials, Epidemiology, Survival Analysis, and Bioinformatics

BLISCHKE AND MURTHY (editors) · Case Studies in Reliability and Maintenance

BLISCHKE AND MURTHY · Reliability: Modeling, Prediction, and Optimization

BLOOMFIELD · Fourier Analysis of Time Series: An Introduction, *Second Edition*

BOLLEN · Structural Equations with Latent Variables

BOLLEN and CURRAN · Latent Curve Models: A Structural Equation Perspective

BOROVKOV · Ergodicity and Stability of Stochastic Processes

BOULEAU · Numerical Methods for Stochastic Processes

BOX · Bayesian Inference in Statistical Analysis

BOX · R. A. Fisher, the Life of a Scientist

BOX and DRAPER · Response Surfaces, Mixtures, and Ridge Analyses, *Second Edition*

* BOX and DRAPER · Evolutionary Operation: A Statistical Method for Process Improvement

BOX and FRIENDS · Improving Almost Anything, *Revised Edition*

BOX, HUNTER, and HUNTER · Statistics for Experimenters: Design, Innovation, and Discovery, *Second Editon*

BOX, JENKINS, and REINSEL · Time Series Analysis: Forcasting and Control, *Fourth Edition*

BOX, LUCEÑO, and PANIAGUA-QUIÑONES · Statistical Control by Monitoring and Adjustment, *Second Edition*

BRANDIMARTE · Numerical Methods in Finance: A MATLAB-Based Introduction

† BROWN and HOLLANDER · Statistics: A Biomedical Introduction

BRUNNER, DOMHOF, and LANGER · Nonparametric Analysis of Longitudinal Data in Factorial Experiments

BUCKLEW · Large Deviation Techniques in Decision, Simulation, and Estimation

CAIROLI and DALANG · Sequential Stochastic Optimization

CASTILLO, HADI, BALAKRISHNAN, and SARABIA · Extreme Value and Related Models with Applications in Engineering and Science

CHAN · Time Series: Applications to Finance with R and S-Plus®, *Second Edition*

CHARALAMBIDES · Combinatorial Methods in Discrete Distributions

CHATTERJEE and HADI · Regression Analysis by Example, *Fourth Edition*

CHATTERJEE and HADI · Sensitivity Analysis in Linear Regression

CHERNICK · Bootstrap Methods: A Guide for Practitioners and Researchers, *Second Edition*

CHERNICK and FRIIS · Introductory Biostatistics for the Health Sciences

CHILÈS and DELFINER · Geostatistics: Modeling Spatial Uncertainty

---

*Now available in a lower priced paperback edition in the Wiley Classics Library.

†Now available in a lower priced paperback edition in the Wiley–Interscience Paperback Series.

---

*Now available in a lower priced paperback edition in the Wiley Classics Library.
†Now available in a lower priced paperback edition in the Wiley–Interscience Paperback Series.

*Now available in a lower priced paperback edition in the Wiley Classics Library.
†Now available in a lower priced paperback edition in the Wiley–Interscience Paperback Series.

*Now available in a lower priced paperback edition in the Wiley Classics Library.

†Now available in a lower priced paperback edition in the Wiley–Interscience Paperback Series.

---

*Now available in a lower priced paperback edition in the Wiley Classics Library.

†Now available in a lower priced paperback edition in the Wiley–Interscience Paperback Series.

---

*Now available in a lower priced paperback edition in the Wiley Classics Library.
†Now available in a lower priced paperback edition in the Wiley–Interscience Paperback Series.

CPSIA information can be obtained
at www.ICGtesting.com
Printed in the USA
BVHW04*0843170818
524187BV00003B/1/P